AF779203

Improvement Technology on Building Seismic Toughness

Improvement Technology on Building Seismic Toughness

Editors

Bo Wang
Bo Fu
Xinxin Wei

MDPI

Basel • Beijing • Wuhan • Barcelona • Belgrade • Novi Sad • Cluj • Manchester

Editors

Bo Wang
Chang'an University
Xi'an
China

Bo Fu
Chang'an University
Xi'an
China

Xinxin Wei
Ruhr-Universität Bochum
Bochum
Germany

Editorial Office
MDPI
St. Alban-Anlage 66
4052 Basel, Switzerland

This is a reprint of articles from the Special Issue published online in the open access journal *Buildings* (ISSN 2075-5309) (available at: https://www.mdpi.com/journal/buildings/special_issues/build_seismic_tough).

For citation purposes, cite each article independently as indicated on the article page online and as indicated below:

Lastname, A.A.; Lastname, B.B. Article Title. *Journal Name* **Year**, *Volume Number*, Page Range.

ISBN 978-3-7258-0559-4 (Hbk)
ISBN 978-3-7258-0560-0 (PDF)
doi.org/10.3390/books978-3-7258-0560-0

© 2024 by the authors. Articles in this book are Open Access and distributed under the Creative Commons Attribution (CC BY) license. The book as a whole is distributed by MDPI under the terms and conditions of the Creative Commons Attribution-NonCommercial-NoDerivs (CC BY-NC-ND) license.

Contents

buildings

MDPI

Article

Investigation on the Seismic Performance of RC Frame Buildings with Rubber Isolation Bearing Installed in Staircases

Liang Bai [1,*], Xingwen Liang [2], Li Xin [3], Mingxiao Liu [1], Ziming Yu [1] and Yage Chu [1]

1 School of Civil Engineering, Chang'an University, Xi'an 710061, China
2 School of Civil Engineering, Xi'an University of Architecture and Technology, Xi'an 710055, China
3 China Northwest Architecture Design and Research Institute Co., Ltd., Xi'an 710018, China
* Correspondence: bailiang@chd.edu.cn

Abstract: A rubber isolation bearing installed in a staircase (RBS) was proposed to mitigate seismic damages to the component of staircases and improve their seismic performance. A series of quasi-static tests performed with respect to the horizontal and vertical mechanical properties of RBS as well as the numerical modelling were verified with experimental results. Next, the finite element models of typical types of reinforced concrete frames (without stairs, with cast-in-place reinforced concrete stairs, and with RBS stairs) were established. The modal dynamic and non-linear elasto-plastic analyses of system-level building models were herein conducted, respectively. The effect of the RBS on the dynamic response of the whole structure and the mechanical performance of stair components was revealed. The analysis results indicated that the RBS provided the desired floor-to-floor relative drift and effectively isolate the seismic action from the stair flight. Therefore, the presence of RBS can effectively mitigate the influence of a staircase on the entire building's structure, weaken the diagonal strut effect of flights, and reduce the inner forces of staircase components.

Keywords: rubber isolation bearing; staircase; isolation properties; dynamic response; numerical simulation

Citation: Bai, L.; Liang, X.; Xin, L.; Liu, M.; Yu, Z.; Chu, Y. Investigation on the Seismic Performance of RC Frame Buildings with Rubber Isolation Bearing Installed in Staircases. *Buildings* **2023**, *13*, 616. https://doi.org/10.3390/buildings13030616

Academic Editor: Asimina Athanatopoulou-Kyriakou

Received: 19 January 2023
Revised: 14 February 2023
Accepted: 24 February 2023
Published: 26 February 2023

Copyright: © 2023 by the authors. Licensee MDPI, Basel, Switzerland. This article is an open access article distributed under the terms and conditions of the Creative Commons Attribution (CC BY) license (https://creativecommons.org/licenses/by/4.0/).

1. Introduction

Stairways, the primary vertical emergency exit routes in most buildings, are critically significant for building accessibility and survivability in emergency events. They should remain operable following even a severe earthquake and the ensuing post-earthquake disaster in order to support the occupant evacuation and recovery response [1]. However, the seismic behavior of stairways is complicated, and staircases have performed poorly in prior earthquakes [2–5]. In some cases, stairways were damaged more severely than the main structures, such as in the observations made following the 2003 Algeria earthquake, where the presence of short columns near the staircase was one of the main causes of severe stair damage [6]. In the 2008 Wenchuan earthquake [2], the damages to buildings, especially stairwells with reinforced concrete frames, were severe. The majority of deaths and injuries were caused by collapsed stairways. In the 2011 Christchurch earthquake [3,4], many stairways exhibited significant damage in buildings where inter-story movements of the stairs were restrained. Seismically damaged stairways, in short, caused disruptions to the building functionality, delayed rescue operations, and threatened safety, as shown in Figures 1 and 2.

Despite such proof of the vulnerability of stairs as well as the implications related to stair damage, studies on the seismic behavior of stair systems have been only occasionally conducted. In the work of Li et al. [2], the seismic behavior of reinforced-concrete stairs during the 2008 Wenchuan earthquake was studied. This research indicated that various types of seismic damage to the stair components or the major structures occurred because of the interaction between the primary structural systems and stairways. Cosenza et al. [7] investigated the seismic behavior of stairs in existing RC buildings and suggested that the stairs increase the stiffness and strength of the structure, which results in a reduction in the natural vibration period and the attraction of seismic forces. In addition, Tegos et al. [8] attempted to quantify the influence of a stair system on the structure response with the numerical simulation method. Thus far, limited experimental data are available to validate the findings of these and other numerical studies. For complementing these studies, tests on the seismic performance of stair systems are required. Simmons and Bull [9] conducted quasi-static cyclic tests of three full-height reinforced concrete straight-run stair flights conforming to the New Zealand design code. Higgins [10] performed tests of two full-sized prefabricated steel scissor stairs constructed in accordance with the US design standards, to assess the seismic inter-story drift response combined with the factored gravity loads. These experimental studies promoted the understanding regarding the seismic performance of stair systems. However, they were both conducted in an isolated configuration, using cyclic loading to replicate earthquake floor drifts.

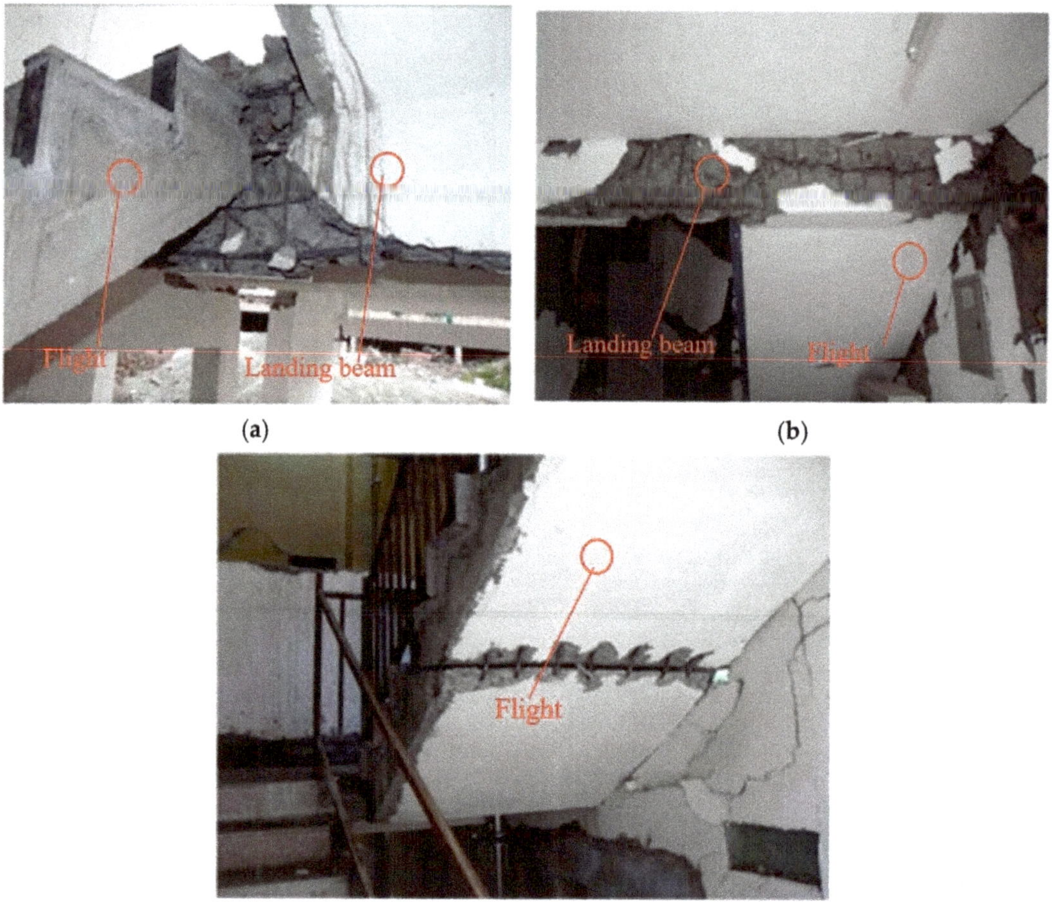

(a) **(b)** **(c)**

Figure 1. Staircase damage in Wenchuan earthquake. (**a**) Damage along the whole landing beam length. (**b**) Shear damage and deformation of landing beam. (**c**) Damage due to cyclic tension and compression [2].

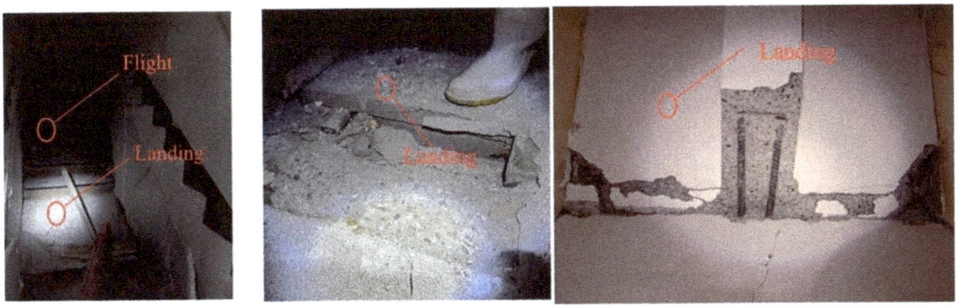

(**a**) (**b**) (**c**)

Figure 2. Staircase damage in Christchurch earthquake. (**a**) Collapse. (**b**) Top landing damage. (**c**) Bottom landing damage [4].

In the design practice, stairways are typically attached to their major structure and span as a floor-to-floor system. Nevertheless, the seismic design forces for stairways can be easily determined from code provisions [11], detailing why it is difficult to keep these floor-to-floor systems damage-free with relative deformation between their lower and upper floors. The current condition of practice often imposes specific provisions to one end of the stairways (e.g., slotted or ductile connections and seismic gaps) to provide the desired floor-to-floor relative drift. However, more commonly, the designers simply fix both the bottom and top of the stair system to the lower and upper floors of the supporting structure. This practice has demonstrated catastrophic results, as the stairways lack protection from the large floor-to-floor relative displacements [11,12].

Feng et al. [13] and Zhao et al. [14] proposed a design to separate the staircase from the overall structure and suggested cutting off the transfer path of horizontal load in the staircase. In addition, Jiang et al. [15] put forward detailed requirements to separate the staircase from the overall structure, and the sliding support installed in the staircase was also proposed. Subsequently, shaking table tests of a reinforced concrete stair with sliding bearing were conducted by Zhao et al. [16]. The test results showed that the presence of sliding bearing can improve the seismic performance of staircases, however, several vertical detachment and horizontal sways occurred between the lower end of the flight and the sliding bearing, and there existed a risk of falling off the flight during a strong earthquake.

Seismic isolation is the separation of the superstructure from its base to negotiate the destructive movement of the ground [17,18]. By comparison with the conventional seismic design approach, which is based on an increased strengthening of the structures, the seismic isolation concept is aimed at a significant reduction in dynamic response induced by the earthquake. To date, extensive isolation devices, including rubber bearings (RB), lead plug rubber bearings (LRB), frictional-sliding bearings, and roller bearings, have been developed [19–23]. Though those isolation bearings have been applied to the area of base isolation, the application of the seismic isolation devices into the staircases is still rare. Compared to LRB and rolling bearings, the RB has the advantages of simpler construction and lower cost. Additionally, as shown in the following Figure 3a, the isolation bearings are set between the flight and the landing beam, and the space for installation is limited. The application of RB could solve this problem and is more convenient to install in staircases. To mitigate seismic damages to the component of staircases and improve their seismic performance, this work introduces a rubber isolation bearing installed in a staircase (RBS), as shown in Figure 3a, which is composed of a rubber layer, connecting steel plate, cover steel plate, and anchor bolts. The rubber layer was vulcanized to the cover steel plate. The connecting steel plate and cover steel plate were connected by sunken bolts. Figure 3b,c depict the specific layout and detailed view of the RBS, respectively. As seen in Figure 3b, the RBS is set between the flight and the landing beam, and the flight and landing beam are connected by anchor bolts with the RBS. When under a horizontal seismic action, the RBS will generate relative deformation, which could accommodate the desired floor-to-floor relative drift, effectively weaken the diagonal strut effect of the flight, and mitigate the effect of the staircase on the whole structure, consequently protecting the staircase from damage.

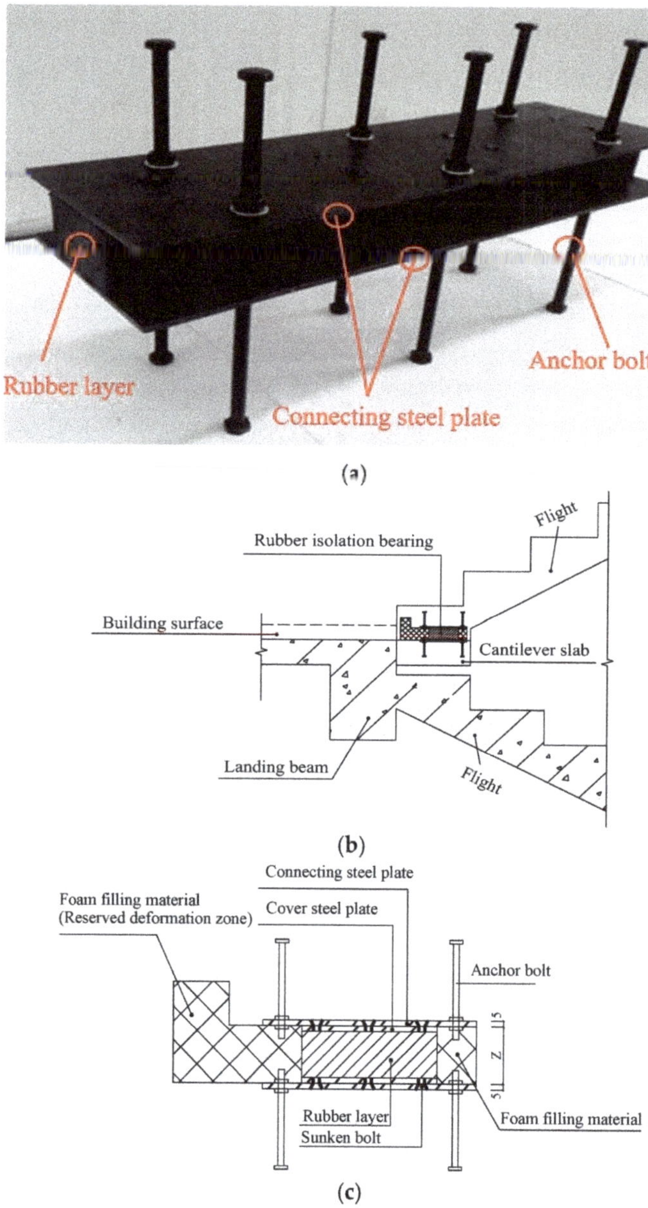

Figure 3. Illustration of the RBS. (**a**) Photograph of the RBS. (**b**) Specific layout of the RBS. (**c**) Detailed view of the RBS. Z is the height of RBS, including the thickness of rubber layer (50 mm thick) and cover steel plate (5 mm thick).

In this study, the horizontal and vertical mechanical properties of RBS were determined through experimental study, and numerical modeling was validated with the experimental result. Subsequently, finite element models of three types of reinforced concrete frames (without stairs, with cast-in-place reinforced concrete stairs, and with RBS stairs) were established in order to assess the seismic performance of the proposed RBS. Response spectrum and non-linear time–history analyses were conducted to analyze the effect of the RBS on the dynamic characteristics of the whole structure. The dynamic responses (i.e., natural vibration period, base shear, story shear, story drift angle, and vertex acceleration) were obtained for each model, and the inner forces of stair components were deeply analyzed.

2. Mechanical Performance Test of RBS

2.1. Specimen Description

The details of the specimens are presented in Figure 4 and the elevation view is shown in Figure 5. The specimens are composed of a rubber layer, cover steel plates, and connecting steel plate. The density of natural rubber is 930 kg/m^3 and the hardness of the rubber was determined to be 58 using the rubber hardness tester. Additionally, the elastic modulus of rubber is 2.63 MPa and the shear modulus is 0.81 MPa. The rubber layer is bonded to the cover steel plates both above and below the layer. The connecting steel plates and cover steel plates are connected by sunken bolts. The main function of the connecting steel plate is to connect the RBS to the loading device. In fact, the thickness of the rubber layer of the RBS products and test specimens is the same and the value is 50 mm; however, the thickness of the cover steel plate and connecting steel plate is increased to a certain extent in the test to ensure the reliability of connection between the test specimens and the loading device. In engineering practice, the production of the RBS adopts a modular design to cope with the demands of building standardization. The RBS is designed to be a series of modular units. Different modular units can be assembled according to the width of the staircase in order to meet the practical application requirements.

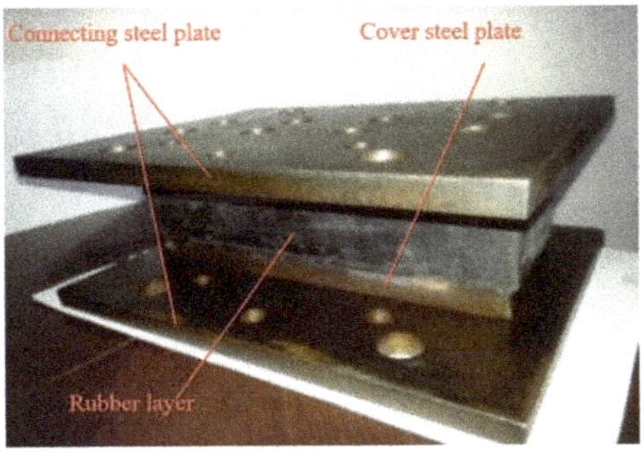

Figure 4. Test specimen.

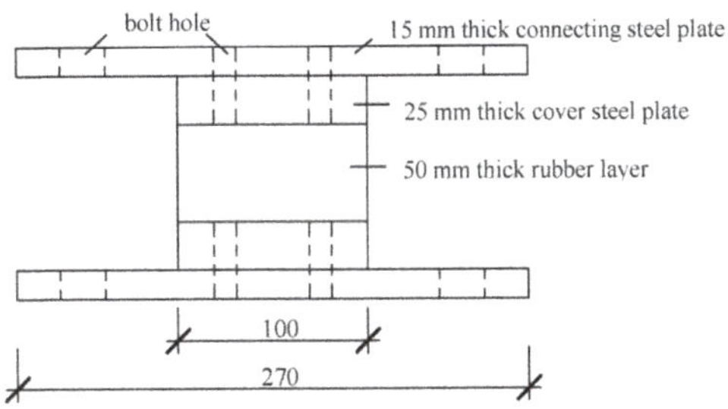

(**a**)

Figure 5. *Cont.*

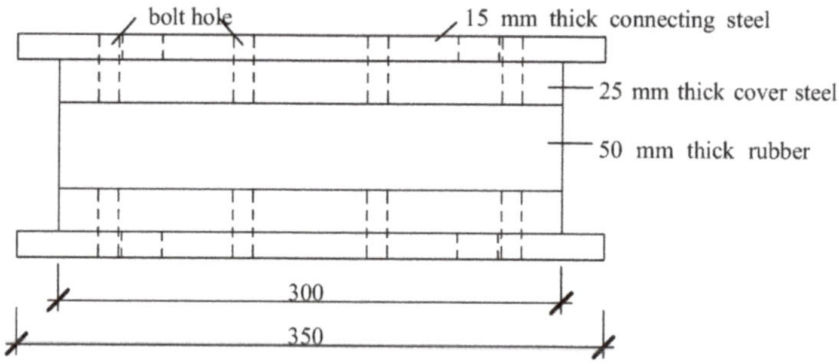

(b)

Figure 5. Elevation view of specimens. (**a**) Front view. (**b**) Side view (Unit: mm).

2.2. Test Setup and Methodology

All the tests of horizontal mechanical behavior are carried out using an electrohydraulic servo compression–shear loading system, which is designed to work on specimens, including RBS-1and RBS-2. In the horizontal direction, a sinusoidal wave loading is used and the maximum sampling frequency is 10 Hz. The loading device is shown in Figure 6. The tests of vertical mechanical behavior are also conducted using this loading system, by providing loading force in the vertical direction only.

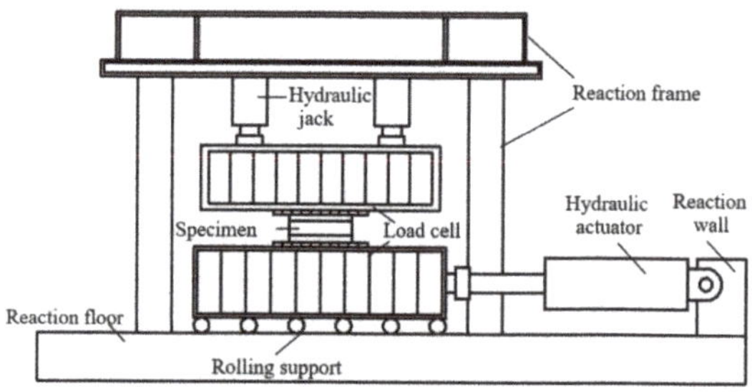

(a)

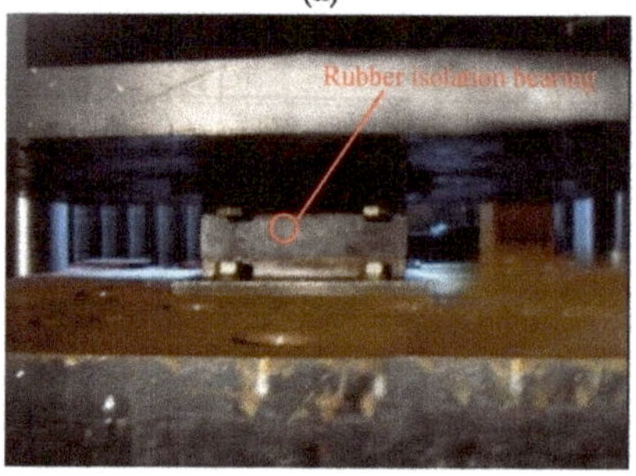

(b)

Figure 6. Test loading device. (**a**) Schematic of test loading device. (**b**) Photograph of loading device during testing.

The test contents include shear behavior, correlation test, and vertical performance, which are used to evaluate the RBS mechanical performance and provide the mechanical parameters for the following numerical analysis.

2.3. Horizontal Shearing Performance

Specimen RBS-1 was used to conduct the test of horizontal shearing performance. The specific test loading process was performed by using the standard test method for rubber bearings [24]. The vertical design compression stress of RBS is 0.2 MPa. While under constant axial load, the sinusoidal wave lateral loading was used to load four cycles. The target lateral displacement of this paper took 50 mm, corresponding to a shear strain of 100%, and the loading frequency was 0.15 Hz. The horizontal load–displacement curve is shown in Figure 7a. The horizontal stiffness was obtained by using the data of the third cycle of the four cyclic loadings, and the value is 0.471 kN/mm.

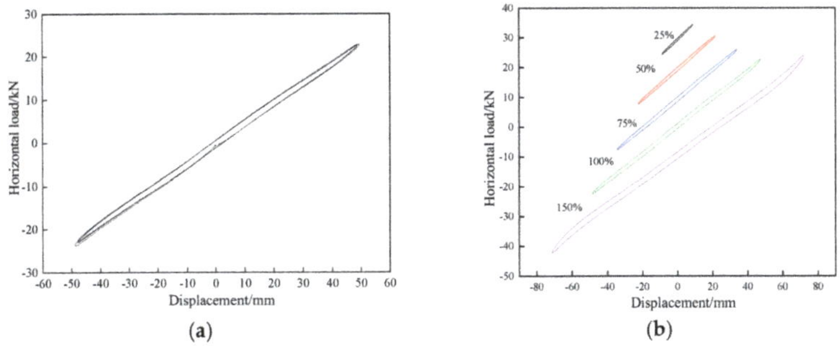

Figure 7. Horizontal test result. (**a**) Horizontal load–displacement curve at 100% shear strain. (**b**) Horizontal load–displacement curves under different shear strains.

Specimen RBS-2 was used to study the RBS compression–shear behaviors under varying shear strain. The influence of different shear strains on the RBS shear behavior was tested. According to the standard test method of rubber bearings [24], the target lateral displacements for each of the cycles were 12.5 mm, 25 mm, 37.5 mm, 50 mm, and 75 mm, corresponding to shear strains of 25%, 50%, 75%, 100%, and 150%, respectively. When the target displacement reached 75 mm, the RBS is still in good condition. The horizontal load–displacement curve under different shear strains was obtained, as shown in Figure 7b. The force-displacement results were used to calculate the horizontal stiffness K_h and equivalent damping ratio h_{eq}. The horizontal stiffness corresponding to 100% shear strain is 0.472 kN/mm, and the equivalent damping ratio is 0.021. As shown in Figure 8, the horizontal stiffness and equivalent damping of RBS decreases gradually with the increase in shear strain.

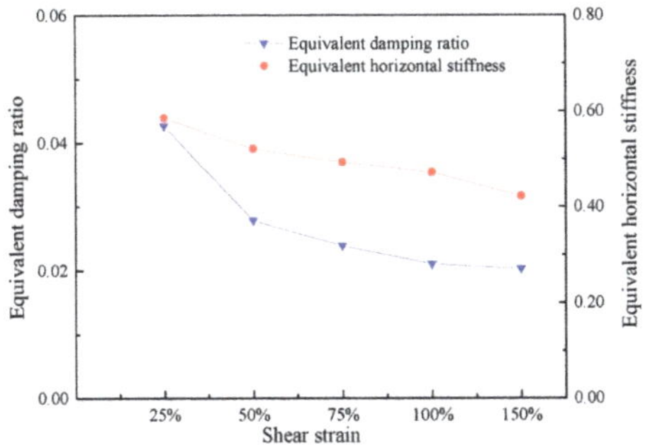

Figure 8. RBS compression–shear behaviors under varying shear strain.

2.4. Vertical Mechanical Properties

Specimen RBS-3 was used to study the vertical mechanical properties of the RBS. The test loading process was performed by using the standard test method of rubber bearings [24]. The vertical stress is ±30% σ_0, where σ_0 is the design compressive stress. The permanent and live loads to which the RBS was subjected were calculated in accordance with the load code for the design of building structures [25]. The design compressive stress was obtained, which is 0.2 MPa. After four cyclic loadings, the vertical load–displacement relationship curve was obtained. Using the data of the third cycle of the four cyclic loadings, the vertical stiffness can be obtained, and the value is 5.2 kN/mm. Under the vertical design force, the vertical deformation of RBS is 1.15 mm.

3. Numerical Modeling and Verification of RBS Bearings

The general-purpose software package for finite element analysis, ABAQUS/CAE6.10-1 [26], was employed to simulate the shear behavior of the RBS bearings tested in this study, and the experimental results are used to calibrate the model.

The existing studies [27,28] demonstrated that the strength degradation of LRB under cyclic loading should be considered when modeling the behavior of LRB. In comparison, the rubber materials were characterized by their isotropic property and nonlinear elastic and incompressible behavior; therefore, the strength degradation of rubber materials under cyclic loading was not involved. The Mooney–Rivlin model [29] was used to simulate the mechanical properties of rubber materials in this study which considers the aging effect of the rubber materials.

The rubber components of the bearings were simulated with eight-node continuum elements (C3D8H) that use a first order hybrid formulation in which the displacement and pressure fields are integrated independently to prevent the volumetric locking caused by the nearly incompressible behavior of rubber. The Mooney-Rivlin [29] model was adopted for the rubber material because of the direct relationship between the engineering properties of the rubber and the model parameters. The Mooney-Rivlin model assumes that rubber is isotropic, nonlinear elastic, and incompressible or nearly incompressible and is stated by the strain energy potential W, that is

$$W = C_{10}(I_1 - 3) + C_{01}(I_2 - 3) \tag{1}$$

In which I_1 = the first stress invariant, I_2 = the second stress invariant, and C_{10} and C_{01} = material parameters. The material parameters which were input into the ABAQUS program were calculated as C_{10} = 0.32 MPa and C_{01} = 0.08 MPa, based on the following Equations (2)–(5).

According to the incompressibility of the rubber material, the Poisson's ratio of rubber is obtained, where ν = 0.5; hence, the relation between C_{10}, C_{01}, and elastic modulus E_0 can be obtained as follows:

$$E_0 = 6(C_{10} + C_{01}) \tag{2}$$

An empirical equation between C_{01} and C_{10}: $C_{01} = 0.25C_{10}$ is proposed in the literature of Gent [30]. The following Equation (3) can be obtained.

$$E_0 = 6 \times 1.25C_{10} \tag{3}$$

The relation between hardness (H_r) and the elastic moduli (E_0) of the rubber material is proposed according to the experimental data [31].

$$\lg E_0 = 0.0238H_r - 0.9905 \tag{4}$$

Substituting Equation (3) into Equation (4), the relation between H_r and C_{10} can be obtained as follows:

$$\lg(7.5C_{10}) = 0.0238H_r - 0.9905 \tag{5}$$

where H_r was determined to be 58 in the above experiment. Substituting H_r into Equation (5), C_{10} can be obtained; furthermore, C_{01} is also obtained by the equation $C_{01} = 0.25C_{10}$.

Figure 9a showed three–dimensional views of the mesh for the RBS bearing, which was selected as the mesh with the fewest number of elements that provided a converged solution from a mesh sensitivity analysis.

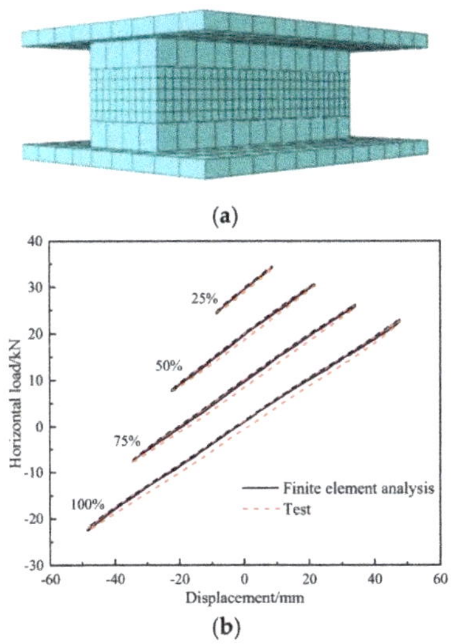

Figure 9. Finite element analysis of RBS. (**a**) Three–dimensional view of the mesh for the RBS bearing. (**b**) Comparison of horizontal load–displacement curves between finite element analysis and test.

The finite element model of the RBS was used to simulate the horizontal shearing performance of the bearing. The comparisons of horizontal load-displacement curves between finite element analysis and test were shown in Figure 9b. The results indicated the comparison between the numerical performance parameters of RBS under varying shear strains and the experimental results. Comparing the finite element analysis results with the experimental results, the error rate is within 0.2%, which indicates that the finite element analysis is in good agreement with the experimental results.

4. Analysis of Isolation Properties and Discussion

In order to investigate the effect of RBS on the dynamic characteristic of typical structures and the mechanical behavior of stair components, the response spectrum and non-linear time–history analyses were carried out using the ETABS and ABAQUS 6.10 software, respectively. The finite element models of three types of reinforced concrete frames, without stairs (RC-woS), with cast-in-place reinforced concrete stairs (RC-wS), and with RBS stairs (RC-wRBS), were established. The response spectrum analysis was performed to investigate the structural response in the elastic stage, whereas the non-linear time–history analysis was conducted to study the elastoplastic response of the structures.

4.1. Prototype RC Frame Building

A typical building was designed with the aim of performing the numerical simulations and capturing the influence of the RBS on the seismic behavior of the whole structure and stair members. The typical plan view of the building was shown in Figure 10. The building has a total height of 27.3 m and 7 stories, with the story height being 3.9 m. The dimensions of the building in the plan are 54 × 18 m. The column dimension is 0.65 × 0.65 m, the beam cross section is 0.30 × 0.60 m, and the thickness of the slab is 0.12 m. The cross-sectional size of the staircase components is presented in Table 1, and the total building mass is 4885 t. The bottom frame column of the concrete type used for this building structure is C35 type, and the rest are C30 type.

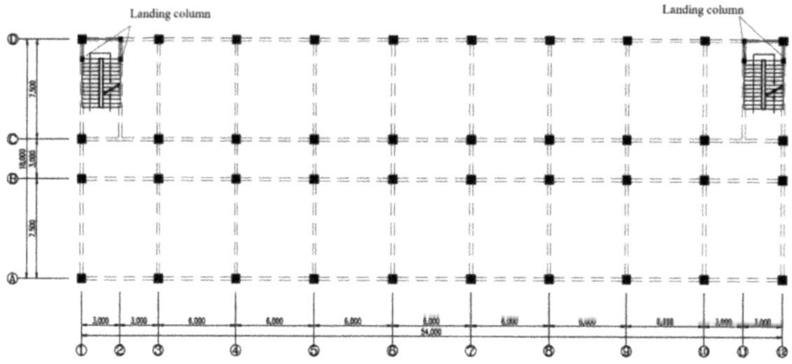

Figure 10. Plan view of the typical building.

Table 1. Dimensions of staircase components.

Cross-Sectional Dimension			
Landing Column	**Landing Beam**	**Thickness of the Flight**	**Thickness of the Landing Slabs**
300 mm × 300 mm	250 mm × 400 mm	120 mm	120 mm

The typical buildings considered in this study are reinforced concrete frame structures, which are designed in accordance with the Code for Seismic Design of Buildings [32]. The seismic fortification intensity varies in different regions, and the corresponding relationship between seismic fortification intensity and designed basic seismic acceleration value of ground motion can be obtained from ligature [32]. The typical building in this study was assumed to be located in a region with an eight-degree seismic design intensity. The corresponding peak ground acceleration (PGA) value of the design earthquake (i.e., probability of exceedance of 10% in 50 years) is around 0.20 g. According to the code [32], the site category is divided into site-class I_0, I_1, II, III, and IV. The typical building in this study was assumed to be located in an area with site-class II, where the shear wave velocity of soil is between 250 m/s and 500 m/s, and the thickness of site covering layer is more than 5 m.

4.2. Response Spectrum Analysis

4.2.1. Analytical Model

The finite element models of RC-woS, RC-wS, and RC-wRBS were established with the ETABS software [33]. The columns and beams were modeled using the beam element, the floor slab was modeled by the membrane element, and the flight was modeled by the shell element. In order to describe the mechanical behavior of the RBS isolator, a nonlinear finite element NL link "Rubber Isolator" was adopted. Based on the previous experimental results, the equivalent horizontal stiffness of the RBS obtained from the test was input in finite element models as the element attribute of the Rubber Isolator. The response spectrum analysis of RC-woS, RC-wS, and RC-wRBS was carried out under frequent earthquakes, and the finite element model of RC-wRBS was shown in Figure 11.

4.2.2. Numerical Results

(1) Natural vibration period of structure

A modal dynamic analysis was conducted on the models of RC-woS, RC-wS, and RC-wRBS to understand the influence of the stairs and RBS on the dynamical behavior of the whole structure. In Table 2, the modal analysis results are summarized and, in particular, the first three modal periods are reported. The presence of the stairs in model RC-wS increases the transversal stiffness, and the period drastically decreases with respect to the result of model RC-woS without stairs. This confirms the stiffness increase caused by the presence of the stairs, which is consistent with the conclusion of Kumbhar et al. [34]. However, the comparison between models RC-wS and RC-wRBS shows that the presence of the RBS elongates the whole building period: as the introduction of the RBS weakens the diagonal strut effect of the flight, it also decreases the transversal stiffness of the whole structure.

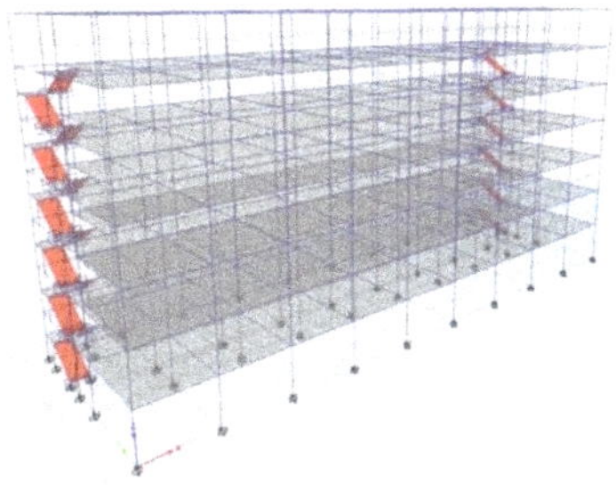

Figure 11. Finite element model of RC-wRBS in ETABS.

Table 2. Natural vibration period of each model.

Period (s)	RC-woS	RC-wS	RC-wRBS
First mode	1.504	1.311	1.441
Second mode	1.476	1.223	1.434
Third mode	1.376	1.094	1.340

(2) Story stiffness

The contribution ratio of stairs to the stiffness of the whole structure is 13~30% in the transversal direction Y (parallel-to-stair-run direction); by comparison, the ratio in the longitudinal direction X (perpendicular-to-stair-run direction) is lower, as shown in Figure 12. In addition, Figure 13 shows the story stiffness distribution of each model, including both the transversal direction and longitudinal direction. It is revealed that the transversal stiffness of the whole structure increased significantly due to the presentation of staircase. However, the transversal stiffness of the second floor of model RC-wRBS is reduced by 28.4% with respect to that of model RC-wS. This is because the introduction of the RBS almost eliminates the influence of the staircase on the transversal stiffness of the whole structure.

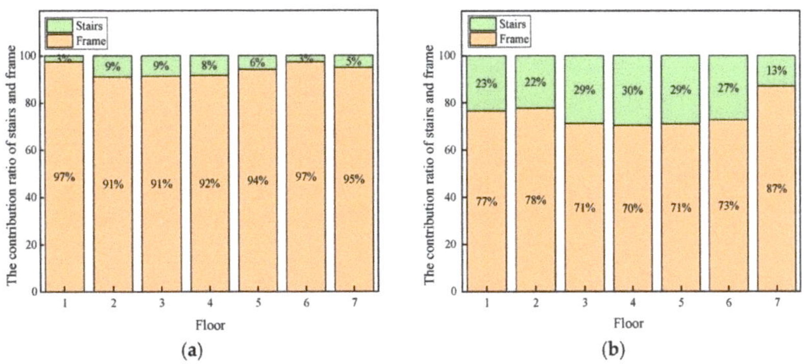

Figure 12. Contribution ratio of stairs and frame to stiffness of whole structure. (**a**) Longitudinal direction X. (**b**) Transversal direction Y.

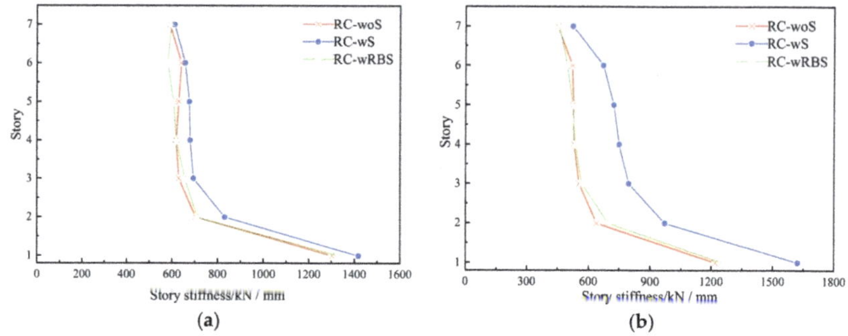

Figure 13. Distribution curves of story stiffness. (**a**) Longitudinal direction X. (**b**) Transversal direction Y.

(3) Story shear

The story shear distribution of each model is shown in Figure 14. It can be seen that the staircase has little effect on the longitudinal story shear of the whole structure. Taking the transversal story shear of the second floor as an example, the story shear of model RC-wS increases by 16.9% compared with that of model RC-woS. It indicates that the presence of the stairs increases the transversal story shear of the whole structure. Compared with the longitudinal direction, the reduction in the story shear for the transversal direction is distinct. The story shear of model RC-wRBS is reduced by 13.5% with respect to that of model RC-wS because the RBS can considerably reduce the one-way slanting bracing effect of the flight and decrease the transversal stiffness of the whole structure.

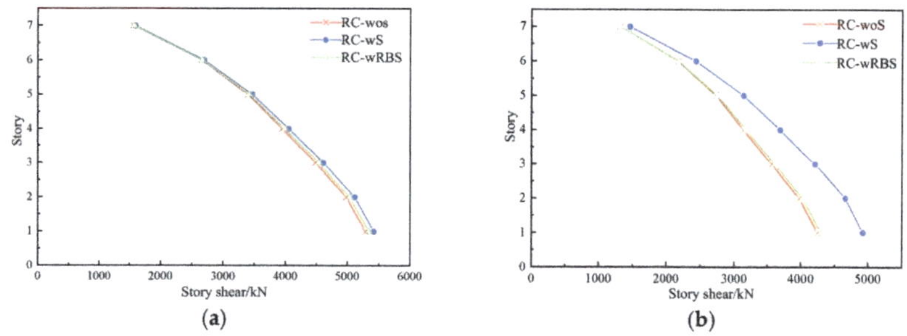

Figure 14. Distribution curves of story shear. (**a**) Longitudinal direction X. (**b**) Transversal direction Y.

(4) Story drift

The distribution curves of the story drift of each model are shown in Figure 15. It can be seen from the figure that the staircase has little influence on the story drift in the longitudinal direction. In the transversal direction, the maximum story drift angles of models RC-woS, RC-wS, and RC-wRBS are 1/606, 1/718, and 1/568, respectively. From the comparison of the results, we can find that the maximum story drift angle of model RC-wS decreases by 15.6% with respect to that of model RC-woS. The main reason is that the presence of the stairs in model RC-wS increases the transversal stiffness of the whole structure, and thus, it decreases the inter-story deformation in the transversal direction. However, the maximum story drift of model RC-wRBS is close to that of model RC-woS, which indicates that the presence of the RBS can reduce the influence of the staircase on the transversal deformation of the whole structure.

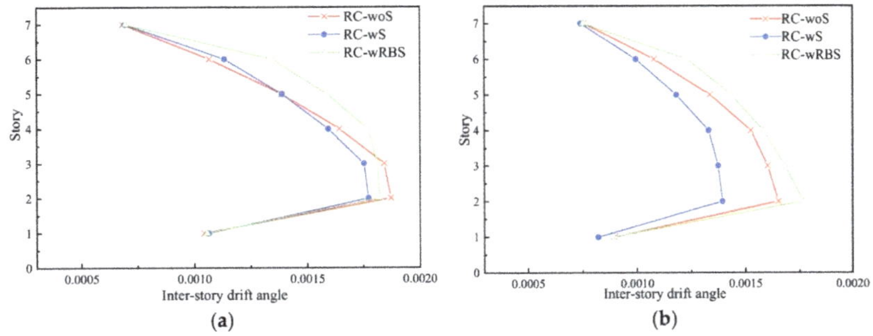

Figure 15. Distribution curves of story drift angle. (**a**) Longitudinal direction X. (**b**) Transversal direction Y.

(5) Pushover analysis

The Pushover analysis of the models RC-ws and RC-wRBS was performed by ETABS software. The development of the plastic hinge of the model is shown in Figure 16.

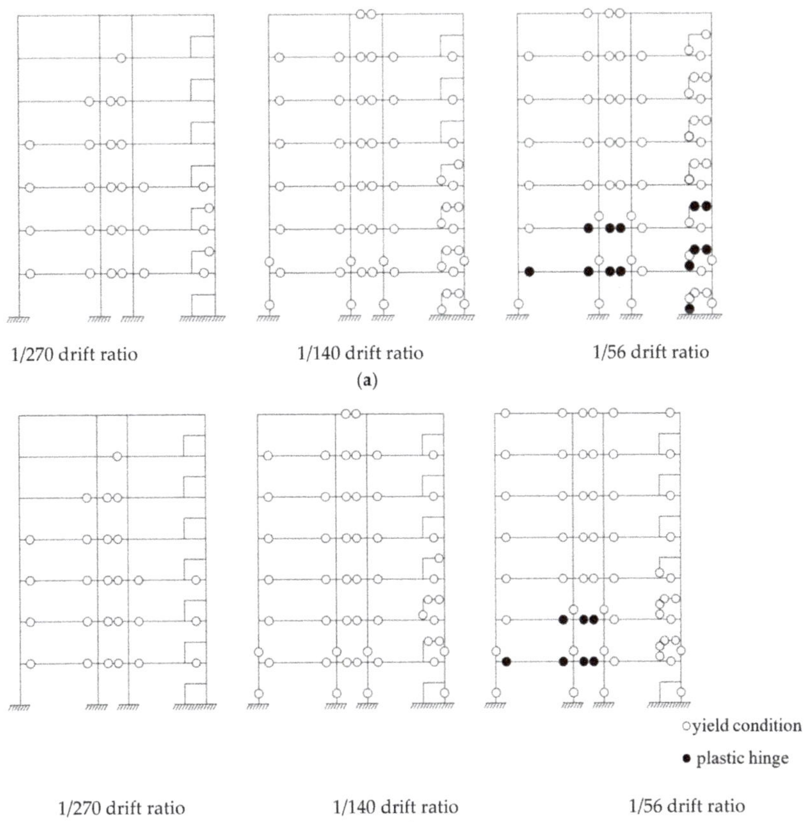

Figure 16. Distribution of plastic hinges. (**a**) Model RC-ws. (**b**) Model RC-wRBS.

When the drift ratio was 1/270, several frame beams in both models reached yield condition first. When the drift ratio reached 1/140, most of the frame beams and columns reached the yield condition. In contrast, the number of landing columns reaching yield condition in model RC-wRBS was less than that in model RC-ws. When the drift ratio was 1/56, the plastic hinges formed at the ends of the landing beams and columns in the model RC-ws. For model RC-wRBS, the landing beams and columns were still under yielding condition, and there was no plastic hinge formed. The above results indicated that the presence of the RBS could prevent the collapse of the staircases and protect the staircases from damage by severe earthquake.

4.2.3. Parametric Analysis

This section is focused on the parametric analysis of the relative contribution of the frame and the stairs to the overall horizontal stiffness, respectively. The different relative contribution to the overall horizontal stiffness may be presented by the variation in stair quantity, and then an identical building was designed except for the number of stairs, i.e., the building with a single stair named RCA and the other with more stairs named RCB, as shown in Figure 17. Since the change in location of staircases causes a difference in structural rigidity distribution, the staircases are symmetrically arranged. The finite element models were established likewise, with cast-in-place reinforced concrete stairs (RCA-wS, RCB-wS) and with RBS stairs (RCA-wRBS, RCB-wRBS).

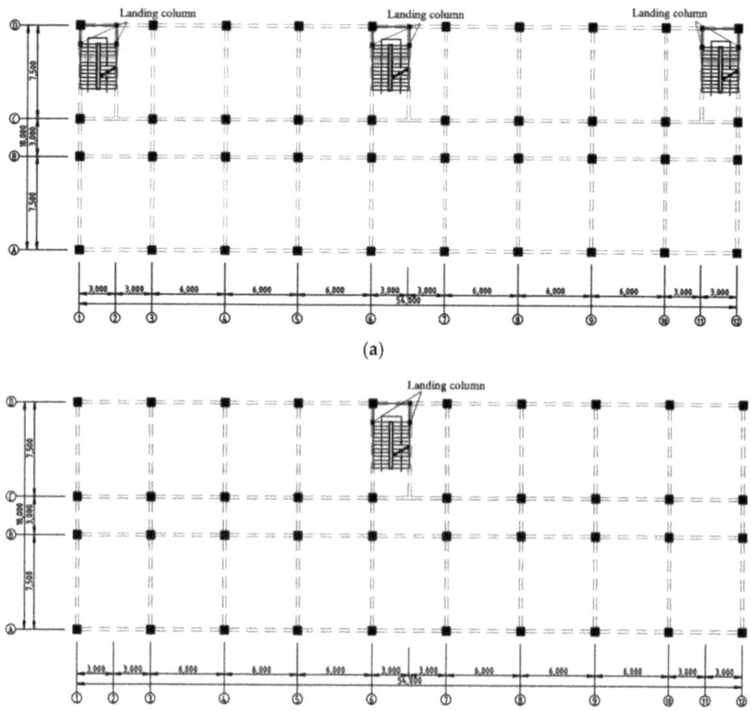

Figure 17. Plan view of the RC frame with different number of stairs. (**a**) RCB. (**b**) RCA.

Since the influence of the stairs on the whole structure is mainly concentrated on the transversal direction Y, the parametric analysis in the transversal direction was mainly conducted. The relative contribution of the stairs to the overall horizontal stiffness increased with the increasing staircase quantity, as shown in Figure 18. The transversal story stiffness distribution and story shear of each model was obtained as shown in Figures 19 and 20. For the RCA model with a single stair, the transversal stiffness of the second floor of model

RCA-wRBS is reduced by 16.4% with respect to that of model RCA-wS. For the RCB model with more stairs, the transversal stiffness of the second floor of model RCB-wRBS is reduced by 35.8% with respect to that of model RCB-wS. The reduction rate of transversal stiffness increases as the relative contribution of stairs increase. Similarly, the transversal story shear of the second floor of model RCA-wRBS is only 8% lower than that of model RCA-wS; however, the value of RCB-wRBS is 18% lower than that of model RCB-wS. It is stated that the reduction degree of transversal story shear caused by the RBS is more obvious with the increasing relative contribution of stairs to the overall horizontal stiffness.

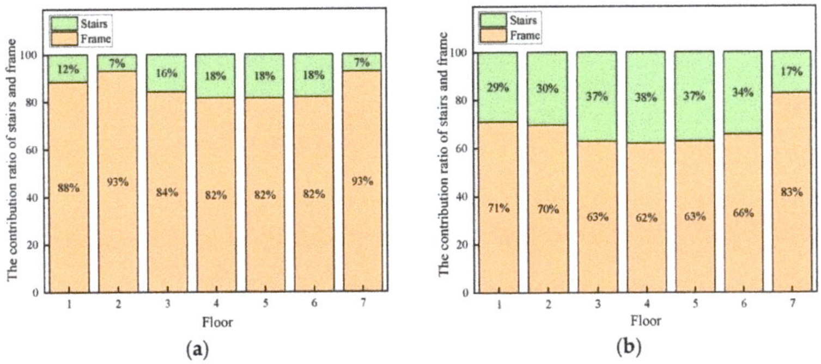

Figure 18. Contribution ratio to stiffness of the whole structure along transversal direction Y. (**a**) RCA. (**b**) RCB.

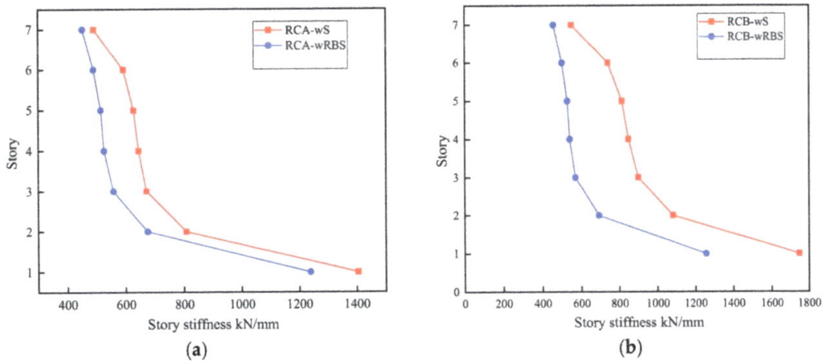

Figure 19. Distribution curves of story stiffness along transversal direction Y. (**a**) RCA. (**b**) RCB.

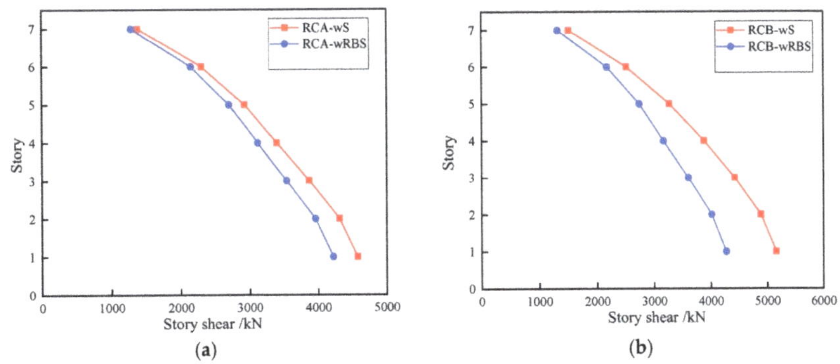

Figure 20. Distribution curves of story shear along transversal direction Y. (**a**) RCA. (**b**) RCB.

4.3. Dynamic Elasto-Plastic Analysis

4.3.1. Analysis Model

Due to the article length limit, the finite element models of RC-woS, RC-wS, and RC-wRBS were only established using the finite element software ABAQUS to compare the dynamic response of these three models under elasto-plastic condition. Using the fiber element model PQ-fiber, based on the work of Lu et al. [35], the beam, column, and steel rebar were modeled by setting the parameters of the UConcrete02 and USteel02 materials, respectively. The shell element was used to simulate the floor and flight, and the RBS bearings were simulated based on the Section 3, where detailed model parameters were introduced. The finite element model of RC-wRBS and the local model of the stairs are shown in Figure 21. The non-linear time-history analyses of RC-woS, RC-wS, and RC-wRBS were conducted.

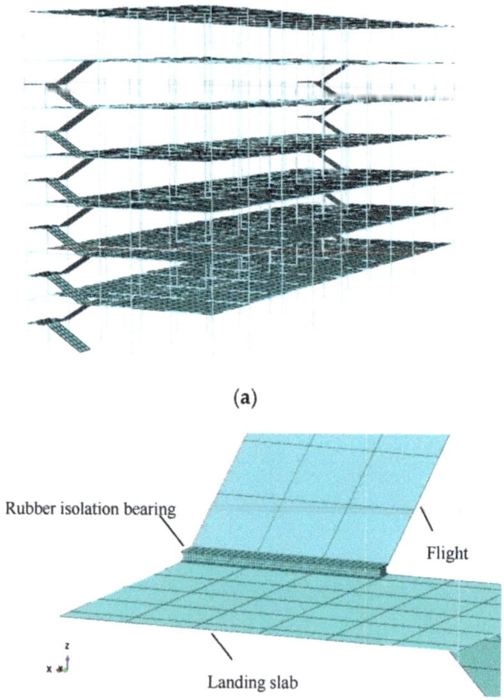

(a)

(b)

Figure 21. Finite element model of RC-wRBS in ABAQUS. (**a**) Whole structural model. (**b**) Local model of stairs with RBS.

4.3.2. Dynamic Response

Selection of appropriate ground motion records for nonlinear dynamic analysis of structures plays a crucial role to estimate structural responses reasonably. Various criteria are used in the selection of ground motion records for nonlinear time history analyses; spectral matching is the primary consideration, and maximum displacement demands are also considered [36,37]. According to the Code for Seismic Design of Buildings (GB50011-2010) [32], at least two actual strong earthquake records and one artificially simulated acceleration time–history curve should be selected for time-history analysis. The El-Centro wave, Tangshan wave, and an artificial wave (with components in the north–south (N–S) and west–east (W–E) directions) were used to conduct a time–history analysis.

In this study, the details about selected waves were as follows. The EL-Centro wave was recorded in El-Centro, southern California in 1940, whose magnitude was 7.1, and the acceleration interval was 0.02 s. The peak acceleration of the N–S component was 341.7 cm/s^2, and the E–W component was 210.1 cm/s^2. The Tangshan wave was recorded at the Beijing Hotel observation site in Tangshan City, and the magnitude was 7.8. The peak acceleration of the N–S component was 55.49 cm/s^2, and the E–W component was 65.9 cm/s^2. The acceleration interval was 0.01 s. An artificial wave was generated according to the Code for Seismic Design of Buildings (GB50011-2010) [32]; the design's basic acceleration of ground motion was 0.2 g, the site class was II, and the intensity was 8-degree. For a rare earthquake, the maximum acceleration value for the seismic acceleration of the ground was adjusted to 400 cm/s^2 when conducting a time–history analysis. The comparison between the response spectra of the selected seismic waves and the standard response spectra is shown in Figure 22, indicating that the response spectrum of selected waves was compatible with the code-based target spectrum.

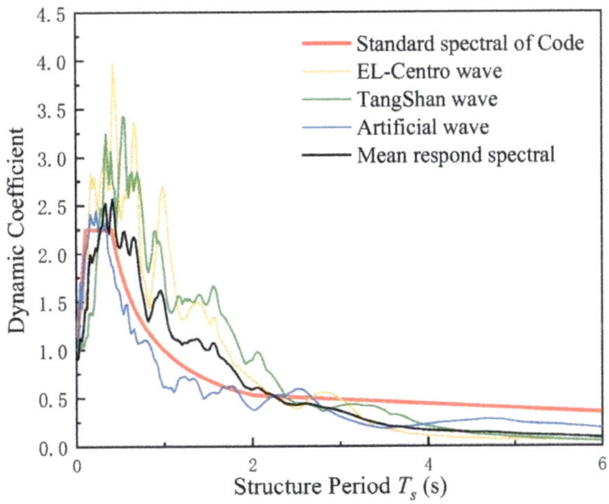

Figure 22. Comparison between the response spectra of the selected seismic waves and the standard response spectra in GB50011-2010.

For the rare earthquake of 8-degree intensity, the maximum value for the seismic acceleration of the ground was adjusted to 400 cm/s^2 when conducting time–history analysis, as the code proposed. The duration of each ground motion was 20 s.

The results of the response spectrum analysis show that the influence of the stairs on the whole structure is mainly concentrated on the transversal direction Y in Section 4.2.2. Therefore, in the nonlinear time–history analysis, the dynamic response of the whole structure in the transversal direction was mainly studied.

(1) Story drift angle

Figure 23 showed the envelope values of the story drift angle (transversal direction Y) of models RC-woS, RC-wS, and RC-wRBS under a rare earthquake. As can be seen from the figure, the maximum story drift angles of the three models appear on the second story. Among the three models, model RC-wS has the smallest story drift angle, which shows that the presence of the stairs increases the transversal stiffness of the structure. However, the story drift angle of model RC-wRBS is close to that of model RC-woS, which indicates that the presence of the RBS can eliminate the influence of the stairs on the lateral displacement of the whole structure.

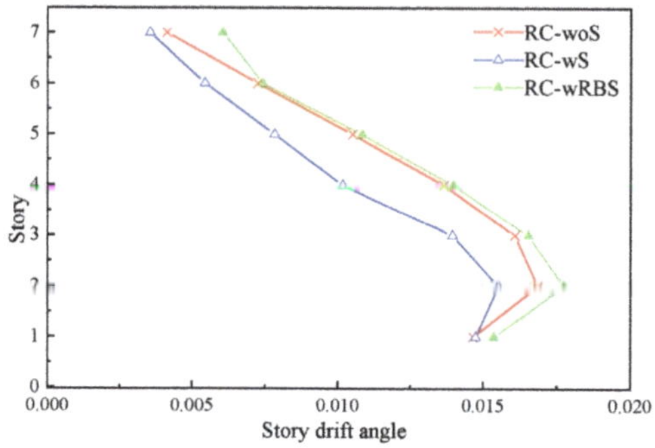

Figure 23. Envelope value of the story drift angle under a rare earthquake (transversal direction).

(2) Base shear

The maximum transversal base shear of each model is listed in Table 3. It can be seen from the table that the transversal base shear of model RC-wRBS is reduced by 19.5% at most, compared with that of model RC-wS, which indicates that the presence of the RBS reduces the transversal base shear. Furthermore, it improves the seismic performance of the whole structure.

Table 3. Maximum transversal base shear of the whole structure (kN).

Model	El-Centro Wave	Tangshan Wave	Artificial Wave
RC-woS	9764.8	10,468.7	8775.39
RC-wS	11,374.4	13,437.6	10,976.5
RC-wRBS	10,374.5	10,952.1	9214.77

(3) Vertex acceleration

The maximum vertex acceleration of the transversal direction and the amplification coefficients of each model are listed in Table 4. The amplification amplitude refers to the ratio between output maximum acceleration on the top floor and input ground acceleration. It can be seen from the table that the vertex acceleration of model RC-wRBS is very close to that of model RC-wS, which indicates that the presence of the RBS is not effective in reducing the vertex acceleration response.

Table 4. Maximum vertex acceleration of the structure along the transversal direction Y (cm/s^2).

Seismic Wave		El-Centro Wave	Tangshan Wave	Artificial Wave
RC-woS	Vertex acceleration	606.9	581.5	610.9
	Amplification factor	1.52	1.45	1.53
RC-wS	Vertex acceleration	708.8	615.9	651.6
	Amplification factor	1.77	1.54	1.63
RC-wRBS	Vertex acceleration	693.8	600.5	635.2
	Amplification factor	1.73	1.50	1.59

4.3.3. Force Analysis of Stair Component

It can be seen that the maximum story drift angle occurred on the second floor; therefore, the RBS and structural components located at the second floor were selected to analyze.

(1) Deformation analysis of RBS

The RBS located at the axis Ⓓ/①-② of the second floor was analyzed, as shown in Figure 10. The time–history curve of relative horizontal displacement between the upper and lower ends of the RBS is shown in Figure 24. It can be seen that the maximum relative horizontal displacement of the upper and lower ends of the RBS is 38 mm. This indicates that the RBS provides the desired floor-to-floor relative drift and can effectively isolate the seismic action from the stair flight. According to the horizontal test result of RBS, it is still on the good condition at the target displacement of 75 mm, corresponding to the 150% shear strain. Therefore, the drift demand from a global analysis can be achieved. The time-history curve of vertical deformation of the RBS was also analyzed, as shown in Figure 25. The vertical displacement of RBS under earthquake action is lower, and the maximum value is less than 5 mm.

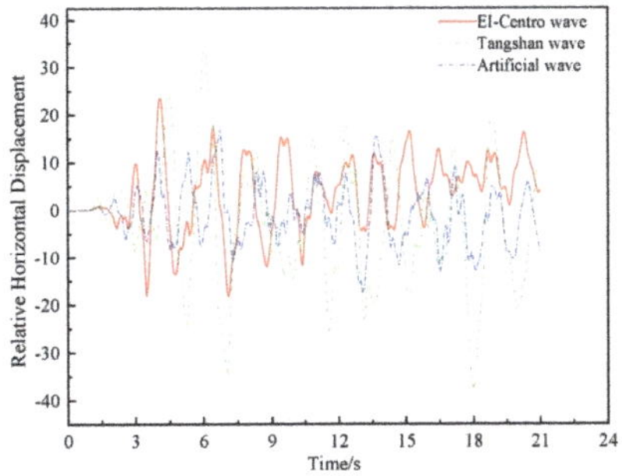

Figure 24. Time−history curve of relative horizontal displacement between upper and lower ends of the RBS.

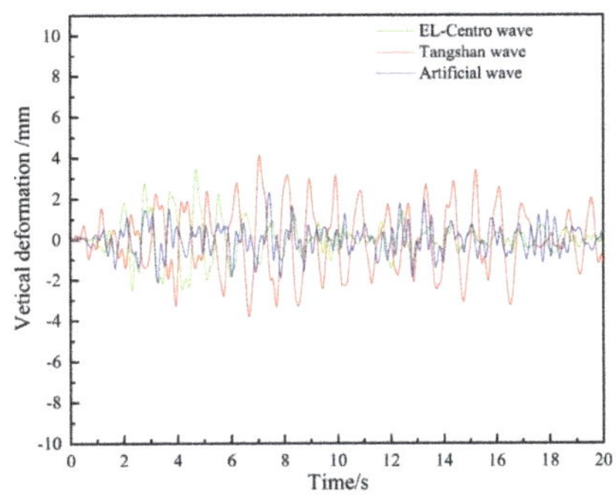

Figure 25. Time−history curve of vertical deformation of the RBS.

(2) Landing column

In Chinese engineering practice, the landing columns are commonly used in staircases to support the landing beam. The landing column located at the axis Ⓕ/② of the second floor was analyzed. The envelope values of axial force, shear force, and bending moment of the landing column under an earthquake wave are listed in Table 5. Compared with model RC-wS, the axial force of model RC-wRBS is reduced by 44.5% and the transversal shear and bending moment of the landing column decrease by 56.5% and 40.9%, respectively. This indicated that the presence of the RBS significantly reduces the forces of the landing column.

Table 5. Envelope values of axial force, shear force, and bending moment of staircase column.

Model	Axial Force (kN)	Shear (kN)	Bending Moment (kN·m)
RC-wS	306.9	132.3	68.9
RC-wRBS	170.4	57.6	40.7

(3) Landing beam

The landing beam located at the axis Ⓓ/①-② of the second floor was analyzed. Under the earthquake wave, the envelope values of the shear force and bending moment of the landing beam are listed in Table 6. It can be seen from the table that the mid-span shear and bending moment of the landing beam of model RC-wRBS are reduced by 93% and 76.2%, respectively, and the shear and bending moment at the end of the beam are reduced by 65.7% and 63.8%, respectively, compared with those of model RC-wS. This indicates that the forces of the landing beam are significantly reduced when the RBS is installed in the staircase.

Table 6. Envelope values of shear force and bending moment of landing beam.

Model	Midspan		End of the Landing Beam	
	Shear (kN)	Bending Moment (kN·m)	Shear (kN)	Bending Moment (kN·m)
RC-wS	315	34.9	201	32
RC-wRBS	21.9	8.3	69	11.6

(4) Stair flight

The flight located at the axis ①–② of the second floor was analyzed. Under an earthquake wave, the time–history curve of axial force of the flight is shown in Figure 26, and the axial pressure on the flight of model RC-wRBS decreased by 74.9% and the axial tension decreased by 80.6% compared with model RC-wS.

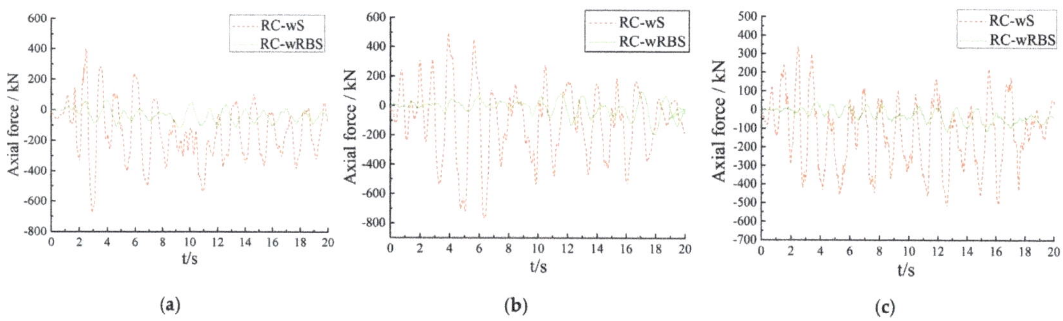

Figure 26. Time—history curves of axial force of the flight. (**a**) EI-Centro wave. (**b**) Tangshan wave. (**c**) Artificial wave.

With regard to model RC-wS, the flights and the landing beams are fixed together. The damage to the stair flights was mostly caused by the interaction of bending moment–axial force (M–N) due to the diagonal truss action of the stairway. Following the Wenchuan earthquake, field reconnaissance [2] was carried out for direct evaluation of building performance, and it was found that several types of damage were due to the tension and compression generated by the complex diagonal truss action of the stairway, such as cracking or spalling of concrete and buckling of the reinforcing bars in the middle of the flights or at the stair flight–landing junction. For model RC-wRBS, the boundary conditions of the stair flights are changed because of the addition of RBS. The time–history curve of relative displacement between the flights and the landing beams was analyzed, as shown in Figure 27. The relative deformation along transversal direction Y and longitudinal direction X is obvious, and the deformation along vertical direction Z is slight. However, there is no relative rotor angle between the flights and the landing beams around the longitudinal X axis based on output results. It shows that the translational degree of freedom is released and the rotation degree of freedom is restricted. This is the main reason why the application of the RBS can reduce the repeated action of tension and compression effectively, which can avoid the failure of the flights.

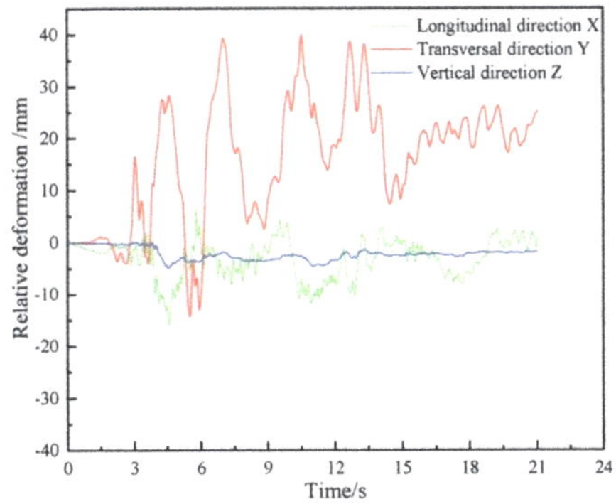

Figure 27. Time−history curve of relative displacement between the flights and the landing beams under EI-Centro wave.

(5) Frame column located in staircase

The frame column in the staircase, located at the axis Ⓕ/① of the second floor, was analyzed. The envelope values of shear force and bending moment of the frame column are listed in Table 7. It can be concluded from the table that the envelope values of shear force and moment of the frame column in the staircase of model RC-wRBS were reduced by 28.5% and 35.9%, respectively, compared with those of model RC-wS. This indicates that the presence of the RBS can reduce the shear force and bending moment of the frame column in the staircase and effectively avoid the destruction of this frame column under the earthquake action.

Table 7. Envelope values of shear force and bending moment of the frame column in staircase.

Model	Shear (kN)	Bending Moment (kN·m)
RC-wS	475.7	487.9
RC-wRBS	340.3	312.8

5. Discussion

In design practice, the stairways are typically attached to their major structure and span as a floor-to-floor system. The bottom and top of the stair system are fixed to the lower and upper layers of the supporting structures, which results in the diagonal strut effect of the staircase. The existing studies [13,14] showed that the diagonal strut effect can significantly change the natural vibration period and the story drift angle of the whole structure. Moreover, the internal forces of the frame members joined with the staircases greatly increase due to the diagonal strut effect of the staircases and the short columns formed with the frame columns, becoming the weakness of the structures. Therefore, reasonable measures should be taken to make the stairs not participate in the overall force of the structure. This study indicated that the presence of RBS can effectively mitigate the negative influence of a staircase on the whole building structure, weaken the diagonal strut effect of flights, and reduce the inner forces of staircase components. Although the cost of the RBS is about RMB 5000, including raw materials, fabrication and construction, etc., this study demonstrated that the introduction of RBS can reduce the inner forces of components under an earthquake action, and consequently, the section size and re-bars area of structural members has decreased. Additionally, the RBS can effectively prevent severe damage to the staircase components during an earthquake. Therefore, the RBS can provide reliable and economical solutions for staircase designs in seismic fortification areas.

6. Conclusions

This study proposed a rubber isolation bearing installed in a staircase (RBS). A series of quasi-static tests were conducted to investigate the mechanical properties of the RBS. Then, a finite element analysis was used to simulate the shear behavior of the RBS, and the experimental results were used to verify the validity of the finite element model. Subsequently, the effect of the RBS on the whole structure and on the stair components was further investigated by numerical simulations of system-level building models. The following conclusions are drawn on the basis of the results of these investigations:

(1) The equivalent horizontal stiffness and equivalent damping ratio of the RBS decreases with an increase in shear deformation; when the target displacement reaches to 150% of shear strain, the RBS is still on the good condition. The maximum relative deformation of RBS from a global analysis is less than the target values; in addition, the displacement demands can be satisfied.

(2) The RBS can considerably reduce the one-way slanting bracing effect of the flights and decrease the transversal stiffness (parallel-to-stair-run direction) of the whole structure. It is the main reason for mitigating the story shear of transversal direction. The Pushover analysis indicated that the introduction of RBS could prevent the collapse of the staircases and protect the staircases from damage by severe earthquake. Base on the parametric analysis result, the reduction degree of transversal story shear caused by the RBS is more obvious with increasing relative contribution of the stairs to the overall horizontal stiffness.

(3) The RBS provides the desired floor-to-floor relative drift and effectively isolates the seismic action from the stair flight. Therefore, the presence of the RBS can reduce the inner forces of the staircase components under an earthquake action, including the landing column, landing beam, flight, and frame column of the staircase.

This study presented a new method for improving the seismic behavior of staircases and mainly investigated the role of the proposed bearings in the seismic response of the whole structure. This work has been primarily focused on numerical simulations of system-level building models with and without an incorporated RBS. To date, very little test data are available to validate the findings of these and other numerical studies. Therefore, the corresponding tests of the whole structure involving RBS should be performed to provide sufficient test data for the numerical studies.

Author Contributions: Conceptualization, X.L.; Methodology, L.B. and L.X.; Formal analysis, M.L., Z.Y. and Y.C.; Investigation, M.L., Z.Y. and Y.C.; Resources, Z.Y.; Data curation, M.L. and Y.C.; Writing—original draft, M.L.; Writing—review & editing, L.B.; Visualization, L.B., X.L. and L.X.; Supervision, L.B.; Funding acquisition, L.X. All authors have read and agreed to the published version of the manuscript.

Funding: The study in this paper was supported by the National Natural Science Foundation of China (51208058, 51708035) and the Natural Science Foundation of Shaanxi Province (2022JM-242), and was funded by the Youth Innovation Team of Shanxi Provincial Education Department (21JP006) and the Research and Development Project of China State Construction Engineering Corporation (CSCEC-2015-Z-41).

Data Availability Statement: The written data used to support the findings of this study are included within the article. The digital data used to support the findings of this study are available from the corresponding author upon request.

Acknowledgments: The study in this paper was supported by the National Natural Science Foundation of China (51208058, 51708035) and the Natural Science Foundation of Shaanxi Province (2022JM-242), and was funded by the Youth Innovation Team of Shanxi Provincial Education Department (21JP006) and the Research and Development Project of China State Construction Engineering Corporation (CSCEC-2015-Z-41). This financial support is gratefully acknowledged.

Conflicts of Interest: The authors declare that there are no conflict of interest regarding the publication of this paper.

References

1. ICC. *International Building Code*; International Code Council: Washington, DC, USA, 2012.
2. Bi, X.L.; Khalid, M.M. Seismic performance of reinforced-concrete stairways during the 2008 Wenchuan earthquake. *J. Perform. Constr. Facil.* **2013**, *27*, 721–730.
3. Bull, D. *Stair and Access Ramps between Floors in Multi-Storey Buildings, a Report of the Canterbury Earthquakes Royal Commission*; Royal Commission: Christchurch, New Zealand, 2011.
4. Kam, W.; Pampanin, S. The seismic performance of RC buildings in the 22 February 2011 Christchurch earthquake. *Struct. Concr.* **2011**, *12*, 223–233. [CrossRef]
5. Roha, C.; Axley, J.W.; Bertero, V.V. *The Performance of Stairways in Earthquakes*; Report No.UCB/EERC-82/15; Earthquake Engineering Research Center; University of California: Berkeley, CA, USA, 1982.
6. Bechtoula, H.; Ousalem, H. The 21 May 2003 Zemmouri (Algeria) earthquake: Damages and disaster responses. *J. Adv. Concr. Technol.* **2005**, *3*, 161–174. [CrossRef]
7. Cosenza, E.; Verderame, G.M.; Zambrano, A. Seismic performance of stairs in the existing reinforced concrete building. In Proceedings of the 14th World Conference on Earthquake Engineering, Beijing, China, 12–17 October 2008.
8. Tegos, I.A.; Panoskaltsis, V.P.; Tegou, S.D. Analysis and design of staircases against seismic loadings. In Proceedings of the 4th Eccomas Thematic Conference on Computational Methods in Structural Dynamics and Earthquake Engineering, Kos Island, Greece, 12–14 June 2013.
9. Simmons, P.W.; Bull, D.K. *The Safety of Single Storey Straight Stair Flights with Mid-Height Landings under Simulated Seismic Displacements*; Research Report 2000-09; Department of Civil Engineering, University of Cantebury: Christchurch, New Zealand, 2000.
10. Higgins, C. Prefabricated steel stair performance under combined seismic and gravity loads. *J. Struct. Eng.* **2009**, *135*, 122–129. [CrossRef]
11. *ASCE 7-10*; Minimum Design Loads for Buildings and Other Structures. American Society of Civil Engineers: Reston, VA, USA, 2010.
12. Zhang, Y.; Tan, P.; Ma, H.; Donà, M. Improving the Seismic Performance of Staircases in Building Structures with a Novel Isolator. *CMES-Comp. Model. Eng.* **2020**, *124*, 415–431. [CrossRef]
13. Feng, Y.; Wu, X.B.; Xionget, Y.Q. Seismic performance analysis and design suggestion for frame buildings with cast-in-place staircases. *Earthq. Eng. Eng. Vib.* **2013**, *12*, 209–219. [CrossRef]
14. Zhao, Y.X.; Yang, Q.; Li, L.Y. Seismic Design by Semi-Slip-Supported Stairs for Reasonable Structural System of the Frame Building. *Ind. Constr.* **2012**, *42*, 94–97.
15. Jiang, H.J.; Wang, B.; Lv, X.L. Seismic damage analysis and design suggestions for staircases in RC frame structures. *J. Vib. Shock* **2013**, *32*, 22–28.
16. Zhao, J.; Hou, P.C.; Liu, M.; Jun, T.; Zhiqiang, G.; Lingyun, P. Shaking table test on structural model of concrete frame-staircase with sliding supports. *J. Build. Struct.* **2014**, *35*, 53–59.
17. Kandemir, E.C.; Mortazavi, A. Optimization of Seismic Base Isolation System Using a Fuzzy Reinforced Swarm Intelligence. *Adv. Eng. Softw.* **2022**, *174*, 103323. [CrossRef]
18. Warn, G.P.; Ryan, K.L. A Review of Seismic Isolation for Buildings: Historical Development and Research Needs. *Buildings* **2012**, *2*, 300–325. [CrossRef]
19. Chen, C.; Lei, J.; Liu, Z. A Ternary Seismic Metamaterial for Low Frequency Vibration Attenuation. *Material* **2022**, *15*, 1246. [CrossRef]
20. Buckle, I.G.; Eeri, M.; Mayes, R.L. Seismic isolation: History, application, and performance—A world view. *Earthq. Spectra* **1990**, *6*, 161–201. [CrossRef]
21. Alfarah, B. Suitability of Seismic Isolation for Buildings Founded on Soft Soil. Case Study of a RC Building in Shanghai. *Buildings* **2020**, *10*, 241.
22. Zhou, Y.; Zhang, Z.D. Experimental and analytical investigations on compressive behavior of thick rubber bearings for mitigating subway-induced vibration. *Eng. Struct.* **2022**, *270*, 114879. [CrossRef]
23. Wu, D.; Lin, J.; Xiong, Y. Analysis of Seismic Isolation Performance of X-Shaped Rubber Bearings (XRBs). *Buildings* **2022**, *12*, 1102. [CrossRef]
24. *GB/T20688.1-2007*; Rubber Bearings-Part 1: Seismic-Protection Isolators Test Methods. China Standard Press: Beijing, China, 2007.
25. *GB50009-2012*; Ministry of Housing and Urban-Rural Development of the People's Republic of China; Load Code for the Design of Building Structures. China Architecture and Building Press: Beijing, China, 2012.
26. DSSC. *Abaqus/CAE6.10-1*, 6.10-1 ed.; Dassault Systèmes Simulia Corporation: Mayfield Heights, OH, USA, 2010.
27. Zheng, W.; Wang, H.; Tan, P.; Li, J.; Liu, Y. Numerical modeling and experimental validation of Sliding-LRBs considering hysteretic strength degradation. *Eng. Struct.* **2022**, *262*, 114374. [CrossRef]
28. Liu, Y.; Zhong, W.; Mercan, O.; Tan, P.; Zhou, F. A new nonlinear model to describe the degradation law of the mechanical properties of lead-rubber bearings under high-speed horizontal loading. *Struct. Control. Health Monit.* **2021**, *28*, e2836. [CrossRef]
29. Keerthiwansa, R.; Javorik, J.; Kledrowetz, J. Elastomer testing: The risk of using only uniaxial data for fitting the Mooney-Rivlin hyper elastic-material model. *Mater. Technol.* **2018**, *52*, 3–8.
30. Gent, A.N. *Engineering with Rubber: How to Design Rubber Components*; Hanser Gardner Publications: Munich, Germany, 2001.

31. Ma, Y.H.; Zhao, G.F.; Luo, J.R.; Cui, J.; Zhou, F.L. Influence of uneven aging on mechanical property of natural rubber bearing for offshore bridge and simplified aging prediction method. *J. Cent. South Univ.* **2016**, *47*, 3498–3506.
32. *GB50011-2010*; Ministry of Housing and Urban-Rural Development of the People's Republic of China. Code for Seismic Design of Buildings; China Architecture and Building Press: Beijing, China, 2010.
33. Computers and Structures Inc. *ETABS Structural Analysis Programs—Theoretical and User's Manual*; Release No. 13.20; Computers and Structures Inc.: Berkeley, CA, USA, 2013.
34. Kumbhar, O.G.; Kumar, R.; Adhikary, S. Effect of staircase on seismic performance of RC frame building. *Earthq. Struc.* **2015**, *9*, 375–390. [CrossRef]
35. Lu, X.L.; Jiang, Q.; Liao, Z.W. *Elasto-Plastic Analysis of Buildings against Earthquake*; China Architecture and Building Press: Beijing, China, 2015.
36. Kayhan, A.H.; Demir, A.; Palanci, M. Multi-functional solution model for spectrum compatible ground motion record selection using stochastic harmony search algorithm. *Bullet. Earthq. Eng.* **2022**, *20*, 6407–6440. [CrossRef]
37. Kayhan, A.H.; Demir, A.; Palanci, M. Statistical evaluation of maximum displacement demands of SDOF systems by code-compatible nonlinear time history analysis. *Soil Dyn. Earthq. Eng.* **2018**, *115*, 513–530. [CrossRef]

Disclaimer/Publisher's Note: The statements, opinions and data contained in all publications are solely those of the individual author(s) and contributor(s) and not of MDPI and/or the editor(s). MDPI and/or the editor(s) disclaim responsibility for any injury to people or property resulting from any ideas, methods, instructions or products referred to in the content.

buildings

MDPI

Article

Shaking Table Test of a Base-Isolated Frame Structure under Near-Fault Ground Motions

Qing Lyu [1], Baijie Zhu [2,3], Wensheng Lu [4], Bo Fu [5,*], Liangkun Liu [6], Wei Qian [1], Ming Zhou [1] and Zhenya Zhang [1]

1 School of Civil and Transportation Engineering, Ningbo University of Technology, Ningbo 315211, China
2 Key Laboratory of Earthquake Engineering and Engineering Vibration, Institute of Engineering Mechanics, China Earthquake Administration, Harbin 150080, China
3 Key Laboratory of Earthquake Disaster Mitigation, Ministry of Emergency Management, Harbin 150080, China
4 State Key Laboratory of Disaster Reduction in Civil Engineering, Tongji University, Shanghai 200092, China
5 School of Civil Engineering, Chang'an University, Xi'an 710061, China
6 School of Environment and Civil Engineering, Dongguan University of Technology, Dongguan 523808, China
* Correspondence: 90_bofu@chd.edu.cn

Abstract: A five-story moment frame structural model with a base isolation system was tested on a shaking table. The isolation system comprised both linear natural rubber bearing (LNR) and nonlinear viscous dampers (NLVDs). Seven ground motions were employed: including three far-fault (FF) and four near-fault (NF) earthquake ground motions. The performance of the isolation system was evaluated by measuring the displacement and base shear of the isolation bearings. Furthermore, the axial force and displacement of the NLVDs were measured. The evolution of the fundamental dynamic frequency of the frame during the test was also determined. During strong earthquakes, NF ground motions caused larger story drifts and floor accelerations of the superstructure than FF ground motions. The displacement and base shear of the isolation base was very large when the isolated structure was subjected to Kobe_TAK000 and ChiChi_TCU102/278 pulse-like NF ground motions. Furthermore, the LNR s experienced tension and uplift when the PGA of input earthquake ground motions was larger than 0.80 g. Although the NLVDs performed very well in combination with the LNRs, the severe responses of the isolation bearings were caused by NF ground motion with a pulse period T_p neighboring the fundamental period of the isolated structure.

Keywords: base isolation; near-fault (NF) ground motions; linear natural rubber bearing (LNR); nonlinear viscous damper (NLVD); shaking table test; seismic responses

Citation: Lyu, Q.; Zhu, B.; Lu, W.; Fu, B.; Liu, L.; Qian, W.; Zhou, M.; Zhang, Z. Shaking Table Test of a Base-Isolated Frame Structure under Near-Fault Ground Motions. *Buildings* **2022**, *12*, 2258. https://doi.org/10.3390/buildings12122258

Academic Editor: Weixin Ren

Received: 7 November 2022
Accepted: 15 December 2022
Published: 18 December 2022

Copyright: © 2022 by the authors. Licensee MDPI, Basel, Switzerland. This article is an open access article distributed under the terms and conditions of the Creative Commons Attribution (CC BY) license (https://creativecommons.org/licenses/by/4.0/).

1. Introduction

Base isolation is an effective passive control technique employed to protect buildings from earthquake-related damage. However, near-fault (NF) earthquakes with high-energy pulses cause more serious damage to isolated structures than far-fault (FF) earthquakes. NF earthquakes usually refer to an earthquake that is no more than 20 km away from the fault. Large displacement demand of isolation bearings will be caused by NF earthquakes with long-period pulses because a large amount of energy was put into the base isolation structures. The base isolation structure yield fewer cycles of high inelastic deformations to dissipate the energy, which may lead to severe damage or failure of base isolation systems [1,2]. Thus, numerous analytical studies have been performed on the seismic performance of base-isolated structures subjected to NF ground motions. Elastomeric bearings can effectively reduce the seismic response of the superstructure under FF ground motions containing medium-high frequency by affording large energy content. However, the seismic response of the base isolation structure will be amplified under NF ground motions in comparison with that of the corresponding fixed-base structure [3]. Jangid and Kelly [4] analyzed the damping effect on the seismic performance of isolation systems under NF ground motions, and they determined that there exists optimum damping of the isolation system to minimize acceleration responses of superstructures. Jangid [5] investigated seismic responses of buildings isolated by lead-rubber bearings (LRB) under NF ground motions. Parameters including isolation period, bearing yield strength, and superstructure flexibility were employed to analyze the seismic performance of the superstructure and isolation system. The results indicate that for the bearing with low yield strength, significant displacement will be caused by NF ground motions. Providakis [6] conducted a nonlinear statistical analysis and showed that base isolated buildings could experience extreme nonlinear behavior under NF ground motions. Mazza and Vulcano [7] determined that supplemental viscous damping at the base effectively reduces the isolator displacement, but seismic responses of the superstructure could not be guaranteed in all cases, especially for relatively short pulse periods. The seismic performance of base-isolated buildings under bi-directional NF and FF ground motions was also investigated by Bhagat and Wijeyewickrema [8], and the results showed that NF ground motions afford larger superstructure and isolation system seismic demands than FF ground motions. Fling-step and forward-directivity characteristics of NF ground motions have an important effect on the seismic response of the base-isolated building [9]. NF ground motions with fling-step characteristics induce larger seismic demands on the base-isolated superstructure than other ground motions.

Buckling or rupture of isolation bearings could be induced under NF ground motions containing long-period pulses [10]. Moreover, the frequency component of the ground motion transmitted to the superstructure can become critical when the pulse intensity induces plastic deformations of the superstructures [11]. Additionally, the structural response could be amplified due to the long duration of the pulse [12]. NF ground motions may afford one or more displacement pulses that will cause large isolator displacement. Large isolator displacements can be resolved using large isolators, but it is not economical. Viscous dampers or large lead plugs are usually used for deducing the displacement demands on isolators. However, these damping mechanisms are nonlinear and become less effective at large displacements.

Although supplemental damping can reduce displacements of isolation bearings, it increases the seismic response of the higher vibration modes of the superstructure. Too large a supplemental damping ratio could significantly increase the inter-story drifts and floor accelerations of the superstructures, consequently reducing the benefits of the base isolation system [4,13]. Moreover, if large supplemental damping is employed to control the displacement of isolation bearings at a rare earthquake level, the superstructure could not benefit from the base isolation system at a low-level earthquake because the isolation layer becomes heavily damped. Thus, the isolation system will not be effective when the building is subject to a moderate earthquake if a high damping value is employed for controlling large bearing displacement under rare earthquakes.

Numerous analytical studies have been recently conducted into the effect of the isolation parameters on the seismic of steel frames and the optimal design of isolation devices for steel frames [14–16]. The effect of the distribution of a viscous damper for steel frame structures was also investigated [17]. Some analytical studies investigated the use of a base-isolation system with passive control devices [13,18–20] to protect framed buildings under strong NF ground motions. The influence of characteristics of ground motion and parameters of the base isolation systems on the structural seismic responses was investigated [13,18–20]. To protect frame buildings from NF ground motions with large displacement, the challenge is the selection of mechanical properties that will improve the response of structures subjected to high-frequency spikes and low-frequency pulses. Base-isolation systems with passive control devices have afforded satisfactory performances for framed buildings under strong NF ground motions [18].

Previous studies paid special attention to the sensitivity of the structural response to the characteristics of both ground motions and base-isolation systems via numerical calculation. However, few studies have experimentally investigated the performance of isolation systems and superstructures subjected to NF ground motions. Herein, the seismic performance of a five-story frame structure isolated using LNRs under FF and NF ground motions is investigated via the shaking table test. Seven ground motions are simulated, including three FF ground motions and four NF ground motions, which were selected from the Imperial Valley, Kobe, and ChiChi earthquakes. Additionally, the base shear and vertical force of the isolation bearings, as well as the maximum deformation and residual displacement of the viscous dampers, are evaluated under a rare earthquake. The floor acceleration and inter-story drift ratio are also investigated.

2. Shaking Table Test Model

2.1. Structural Model

The shaking table test was conducted on scaled structural models that are similar to the prototype structure and subjected to the same ground motions. The model structures were constructed to have a length scale factor of S_l = 1/4, considering the dimensions and payload capacity of the shaking table. Since the model structures comprised the same steel materials as the prototype structure, their elastic modulus scale factor was S_E = 1.0. The acceleration scale was chosen to have a value of unity S_a = 1. Additional masses were used to obtain the total mass of the models $S_E/S_a \cdot S_l$ times lower than that of the full-scale prototype while scaling times by $S_l^{0.5} \cdot S_a^{-0.5}$; thus, the model experiences the same accelerations as the prototype structure and S_l times lower displacements. Table 1 presents the detailed scale factors used in the testing.

Table 1. Typical similitude factors of the model structure.

Quantity	Scaling Law	Similitude Factors	
Length	S_l	1/4	Controlling size
Displacement	S_l	1/4	
Elastic modulus	S_E	1.0	Controlling material
Stress	S_σ	1.0	
Strain	$S_\varepsilon = S_\sigma / S_E$	1.0	
Acceleration	S_a	1.0	Controlling acceleration
Mass density	$S_\rho = S_E/(S_a \cdot S_l)$	4.0	
Mass	$S_m = S_E \cdot S_l^2 / S_a$	1/16	
Force	$S_F = S_E \cdot S_l^2$	1/16	
Damping	$S_C = S_E \cdot S_l^{1.5} \cdot S_a^{-0.5}$	1/8	
Time	$S_T = S_l^{0.5} \cdot S_a^{-0.5}$	0.5	
Frequency	$S_f = S_l^{-0.5} \cdot S_a^{0.5}$	2.0	
Velocity	$S_v = (S_l \cdot S_a)^{0.5}$	1/2	

According to the similarity theory, the required scaling factor for the material density is $S_\rho = 1/S_l$ = 4. However, the provided material density is the same as the prototype or S_ρ = 1 because the same material as the prototype was used to manufacture the model structure. Three concrete blocks with a weight of 8 kN were added on each floor to ensure the scale factor of the mass density satisfying similarity laws.

Figure 1 displays the scaled structural model. The model structure has five stories and a total height of 4.5 m from the base. The model structure has three bays in the longitudinal direction and one bay in the transverse direction; each bay is 1.2 m wide with total dimensions of 1.2 × 3.6 m^2. Concrete blocks were used to add mass to satisfy similitude laws, bringing the total weight, including the base, to 180 kN.

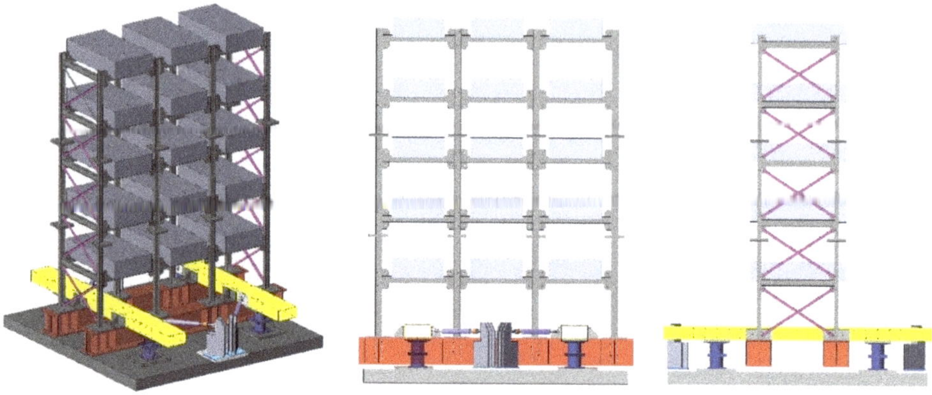

Figure 1. Model structure: (**a**) Steel structure model on the shaking table; (**b**) front and (**c**) side elevations of the steel structure model on the shaking table.

All columns and beams have a constant cross-section (I100 × 68 × 4.5 × 7.6). All bracings in the transverse direction are L40 × 4. All the beam–column joints are connected by bolts; thus, the model can be disassembled and transported using a truck. All beams and columns are rigidly connected by bolts at the flange and welded, ensuring the effective transformation of the bending moment and shear force. The superstructure is bolted to a basement with a grid of two I400 × 400 × 12 × 20 and two 400 × 200 × 10 tubes.

2.2. Base-Isolation Systems

Four low-damping rubber bearings (Figure 2a) were installed under the basement tubes to support the substructure. Four additional nonlinear viscous dampers (Figure 2c) were installed between the basement and the damper reaction support. Tables 2 and 3 list the mechanical properties of the rubber bearings and nonlinear viscous dampers, respectively.

Table 2. Mechanical properties of the LNRs.

Bearing Properties	Characterized Value in Model Domain
Characteristic Strength	7.53 kN
Post-yield Stiffness	182 kN/m
Initial Stiffness	2586 kN/m
Yield Displacement	2.0 mm
Effective Stiffness	200 kN/m
Designed displacement	80 mm
Maximum displacement	120 mm

Table 3. Mechanical properties of the NLVDs.

Damper Properties	Characterized Value in Model Domain
Damping coefficient C	30 kN s/m
α	0.5
Designed displacement	75 mm
Maximum displacement	112.5 mm
Maximum Force	30 kN
Maximum Velocity	1 m/s

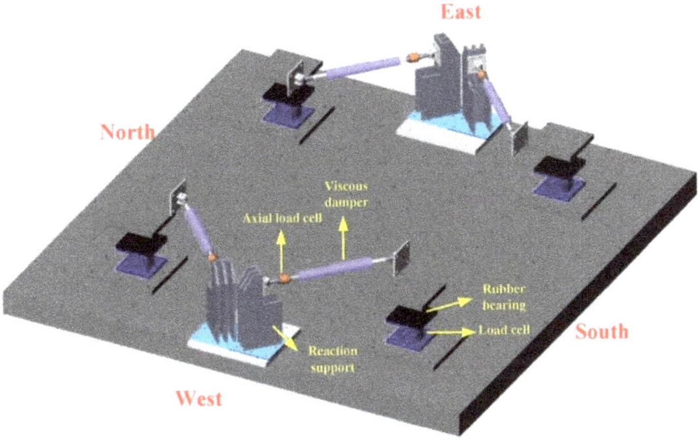

(a)

(b)

(c)

Figure 2. Base isolation system: (**a**) Layout of base isolation system; (**b**) LNR; (**c**) NLVD.

3. Test Program and Sensor Set-Up

3.1. Input Ground Motions

Table 4 shows the seven employed records for the shaking table test. The seven ground motions from the Kobe (1995), Chi-Chi (1999), and Imperial Valley (1979) earthquake events were chosen because they are all well-known in the field of earthquake engineering. The ground motion selected from Imperial Valley Earthquake at Delta station is a FF ground motion, and at Bonds Corner station is a NF ground motion. These two ground motions selected from the Imperial Valley Earthquake are referred to as Imp_F and Imp _N ground motions. The ground motion selected from the ChiChi Earthquake at TCU067 and TCU102 station is a NF ground motion, and at TCU045 station is a FF ground motion. The ground motion at TCU102 station had a pulse period of 9.632 s. These three ground motions selected from the ChiChi Earthquake are referred to as ChiChi_F, ChiChi_N, and ChiChi_NP ground motions, respectively. The ground motion selected from Kobe Earthquake at Shin Osaka is a FF ground motion, and at Takatori station is a pulse-like (T_p = 1.554 s) NF ground motion. These two ground motions selected from Kobe Earthquake are referred to as Kobe_F and Kobe_NP ground motions. Plots of the acceleration time histories for each motion scaled with a PGA of 1 m/s^2 are provided in Figure 3. The response spectrums of the seven ground motions are plotted in Figure 4.

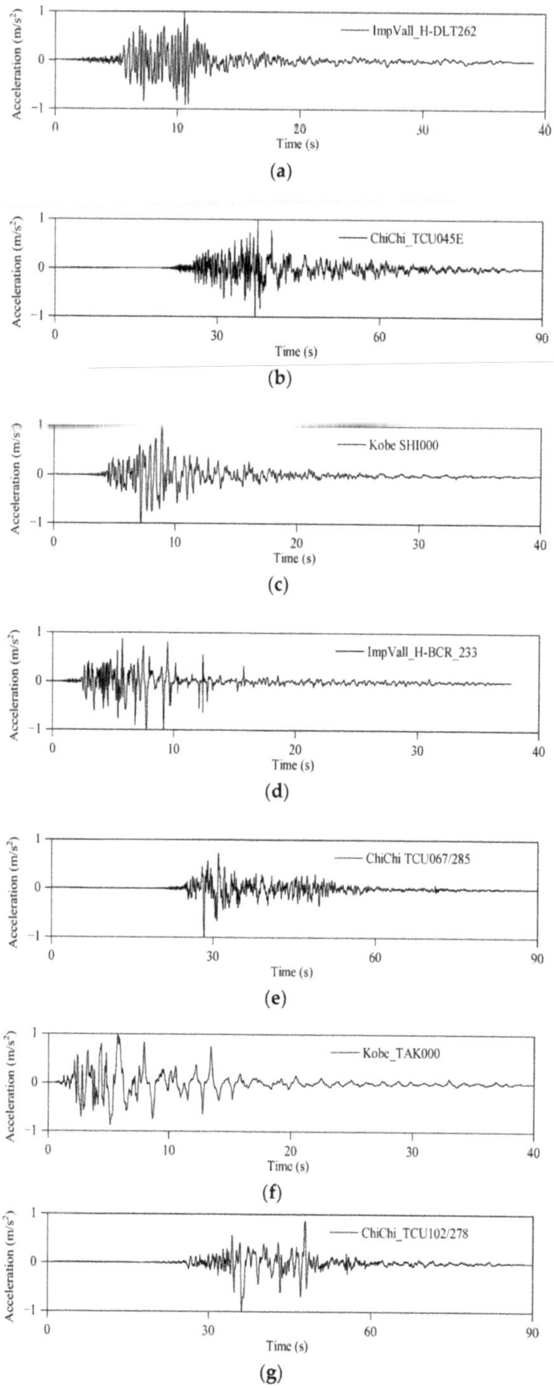

Figure 3. History of acceleration of the earthquake ground motions: (**a**) Imp_F; (**b**) Chihi_F; (**c**) Kobe_F; (**d**) Imp_N; (**e**) ChiChi_N; (**f**) Kobe_NP; (**g**) ChiChi_NP.

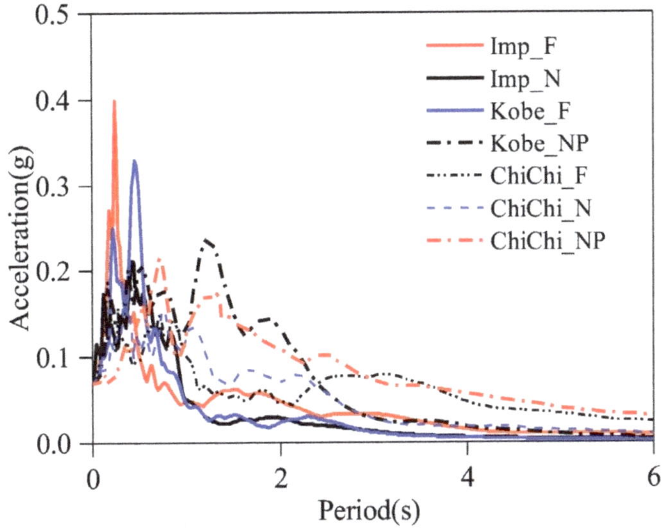

Figure 4. Acceleration response spectrum.

Table 4. Ground motions.

Earthquake	Station	Component Name	Component Name Abbreviation	Magnit-ude	PGA (g)	PGV (cm/s)	T_p (s)	R_{rup} (km)
Imperial Valley, 1979	Delta	ImpVall_H-DLT262	Imp_F	6.5	0.35	33	-	22.03
ChiChi, 1999	TCU045	ChiChi_TCU045E	ChiChi_F	7.6	0.51	39	-	26
Kobe, 1995	Shin Osaka	Kobe SHI000	Kobe_F	6.9	0.24	38	-	19.15
Imperial Valley, 1979	Bonds Corner	ImpVall_H-BCR233	Imp_N	6.5	0.76	44.3	-	2.66
ChiChi, 1999	TCU067	ChiChi_TCU067/285	ChiChi_N	7.6	0.56	91.8	-	0.62
Kobe, 1995	Takatori	Kobe_TAK000	Kobe_NP	6.9	0.28	120.67	1.554	1.47
ChiChi, 1999	TCU102	ChiChi_TCU102/278	ChiChi_NP	7.6	0.29	106.6	9.632	1.49

3.2. Test Program

The test program is listed in Table 5. White noise was used to identify the frequency and dynamic mode of the model structure. The PGA value of the white noise was 0.10 g. The PGAs of the earthquake ground motions were gradually increased from 0.1 g to 0.8 g.

3.3. Sensor Set-Up

Instruments were installed to record the actual shaking table motion, the isolation system response (isolation bearing horizontal deformations, horizontal and vertical forces, displacements, and axial forces of the viscous damper), and the superstructure response (absolute accelerations and displacements of each floor). The responses in the longitudinal direction were mainly measured.

Figure 5 displays the sensor arrangement. In the longitudinal direction, two accelerometers and two displacement sensors were installed at each of the five-floor levels and the basement. Two accelerometers and displacement transducers were installed at the northeast and southeast corners (Figure 5b) to record the torsional responses of the isolation system. In the transverse direction, accelerometers and displacement transducers were placed in the basement and each of the five stories. Six component (three force and three moment readings) load cells (Figure 5c) were installed under the isolation bearings to measure the shear force and axial force of the isolation bearings. Four dampers were arranged between the model structure basement and the damper reaction support. To measure the axial force of the damper, a uniaxial load cell was connected in series with the viscous damper.

Additionally, a laser displacement transducer was fixed on the outside of the damper barrel to measure the axial deformation of the viscous damper (Figure 5d).

Table 5. Test program.

Case Number	Case Name	Input	Input PGA
01	WN_01	White noise	0.10 g
02	Imp_F_0.10 g	Imp_F	0.10 g
03	ChiChi_F_0.10 g	ChiChi_F	0.10 g
04	Kobe_F_0.10 g	Kobe_F	0.10 g
05	Imp_N_0.10 g	Imp_N	0.10 g
06	ChiChi_N_0.10 g	ChiChi_N	0.10 g
07	Kobe_NP_0.10 g	Kobe_NP	0.10 g
08	ChiChi _NP_0.10 g	ChiChi _NP	0.10 g
09	WN 02	White noise	0.10 g
10	Imp_F_0.20 g	Imp_F	0.20 g
11	ChiChi_F_0.20 g	ChiChi_F	0.20 g
12	Kobe_F_0.20 g	Kobe_F	0.20 g
13	Imp_N_0.20 g	Imp_N	0.20 g
14	ChiChi_N_0.20 g	ChiChi_N	0.20 g
15	Kobe_NP_0.20 g	Kobe_NP	0.20 g
16	ChiChi _NP_0.20 g	ChiChi _NP	0.20 g
17	WN 03	White noise	0.10 g
18	Imp_F_0.30 g	Imp_F	0.30 g
19	ChiChi_F_0.30 g	ChiChi_F	0.30 g
20	Kobe_F_0.30 g	Kobe_F	0.30 g
21	Imp_N_0.30 g	Imp_N	0.30 g
22	ChiChi_N_0.30 g	ChiChi_N	0.30 g
23	Kobe_NP_0.30 g	Kobe_NP	0.30 g
24	ChiChi _NP_0.30 g	ChiChi _NP	0.30 g
25	WN 04	White noise	0.10 g
26	Imp_F_0.40 g	Imp_F	0.40 g
27	ChiChi_F_0.40 g	ChiChi_F	0.40 g
28	Kobe_F_0.40 g	Kobe_F	0.40 g
29	Imp_N_0.40 g	Imp_N	0.40 g
30	ChiChi_N_0.40 g	ChiChi_N	0.40 g
31	Kobe_NP_0.40 g	Kobe_NP	0.40 g
32	ChiChi _NP_0.40 g	ChiChi _NP	0.40 g
33	WN 05	White noise	0.10 g
34	Imp_F_0.60 g	Imp_F	0.60 g
35	ChiChi_F_0.60 g	ChiChi_F	0.60 g
36	Kobe_F_0.60 g	Kobe_F	0.60 g
37	Imp_N_0.60 g	Imp_N	0.60 g
38	ChiChi_N_0.60 g	ChiChi_N	0.60 g
39	Kobe_NP_0.60 g	Kobe_NP	0.60 g
40	ChiChi _NP_0.60 g	ChiChi _NP	0.60 g
41	WN 06	White noise	0.10 g
42	Imp_F_0.80 g	Imp_F	0.80 g
43	ChiChi_F_0.80 g	ChiChi_F	0.80 g
44	Kobe_F_0.80 g	Kobe_F	0.80 g
45	Imp_N_0.80 g	Imp_N	0.80 g
46	ChiChi_N_0.80 g	ChiChi_N	0.80 g
47	Kobe_NP_0.80 g	Kobe_NP	0.80 g
48	ChiChi _NP_0.80 g	ChiChi _NP	0.80 g
49	WN 07	White noise	0.10 g

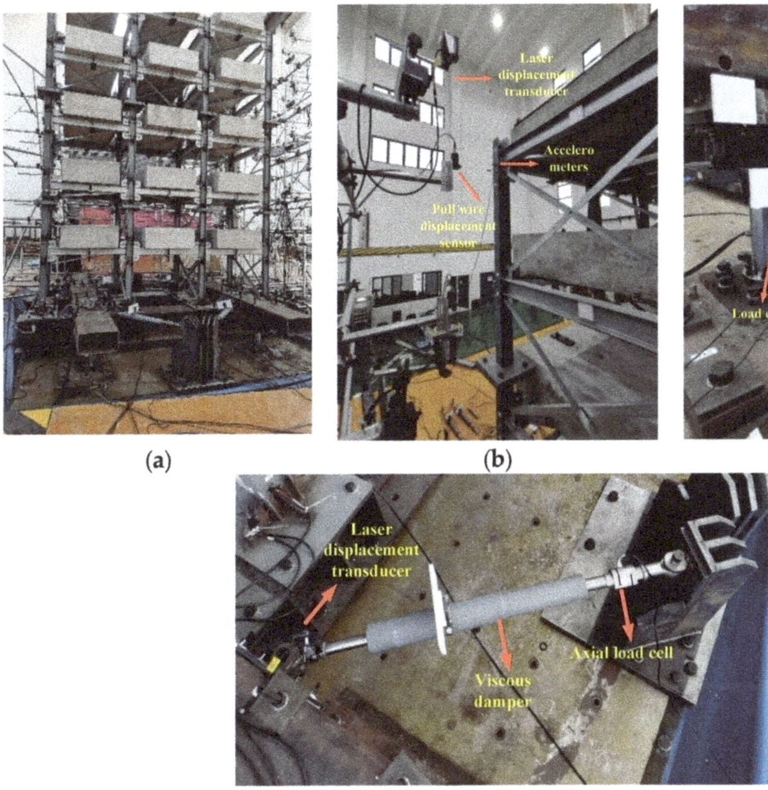

Figure 5. Sensors of viscous dampers: (**a**) Test model structure; (**b**) Sensors of the superstructure; (**c**) Load cell—six components; (**d**) Sensors of NLVDs.

4. Shaking Table Test Results

4.1. Dynamic Modes

The white noise test results were used to identify the dynamic properties of the isolated structure in the longitudinal direction. The structure was subjected to a white noise excitation with frequencies ranging from 0 to 50 Hz and a PGA of 0.10 g. The transfer functions shown in Figure 6 were obtained as the ratio of the Fourier transform of the horizontal acceleration of each floor to the Fourier transform of the horizontal acceleration of the shaking table.

Based on the transfer function of the test acceleration results, the least square method was used to fit the transfer function curve. The frequency was obtained by searching the peak values of the real part of the transfer function curve, and the damping ratio was determined using the half-power broadband method. The modal frequencies corresponded to the local maxima of the amplitude of the transfer function. By measuring the amplitude of the transfer function at each floor, the mode shapes were determined from the ratios of the amplitudes for the frequency corresponding to one vibration mode. Additionally, the phase angles of the transfer function were determined to obtain the sign of the vibration mode shape. The damping ratios in each mode were calculated using the half-power bandwidth method around the transfer function peaks. The first five modes are shown in Figure 7, and the corresponding frequencies and damping ratios are listed in Table 6. It can be seen from the data in the table that the arrangement of NLVDs enhanced the damping ratio of the first three modes of the isolated structure.

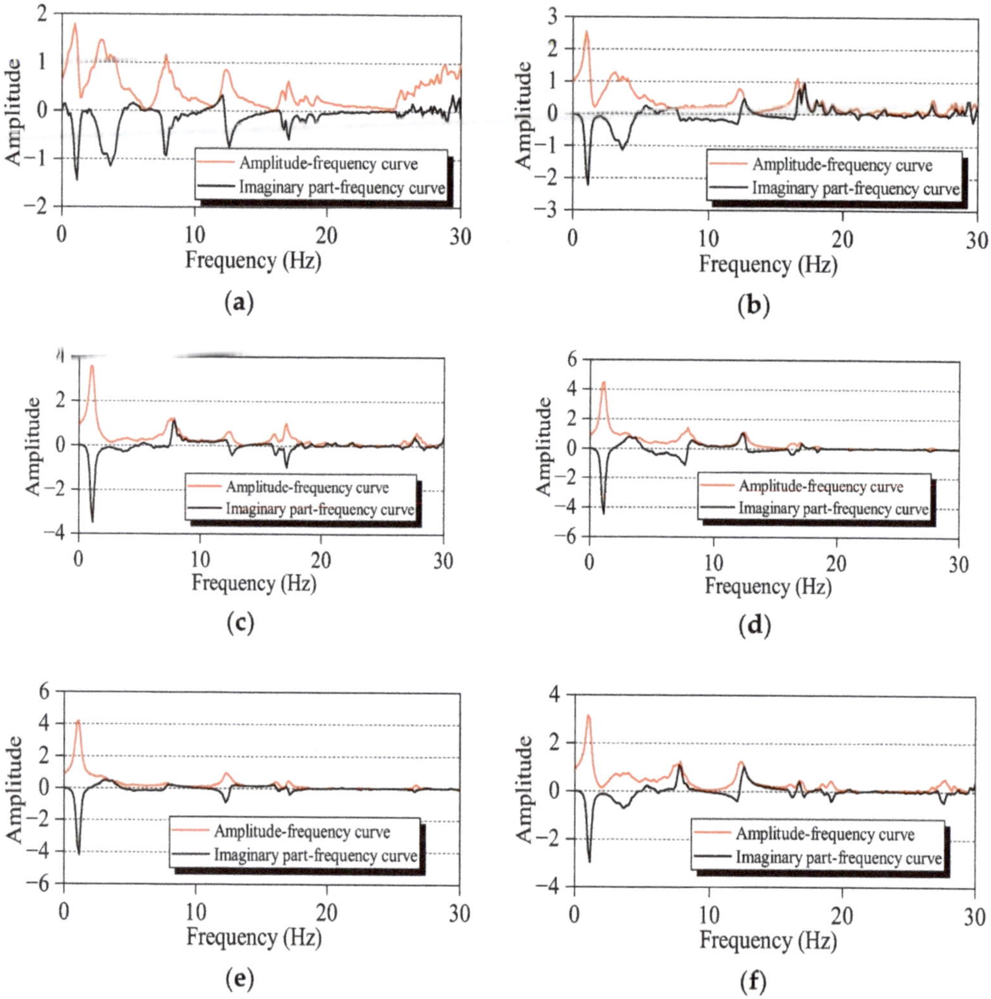

Figure 6. Amplitude and imaginary parts of the transfer function of the superstructure by WN05: (**a**) Basement; (**b**) First story; (**c**) Second story; (**d**) Third story; (**e**) Fourth story; (**f**) Fifth story.

Table 6. Frequencies and damping ratios of the first five modes.

Dynamic Property	Mode				
	1	2	3	4	5
Frequency (Hz)	1.1209	3.1053	7.8180	12.3631	16.7636
Damping ratio (%)	16.51	8.18	6.60	1.86	3.60

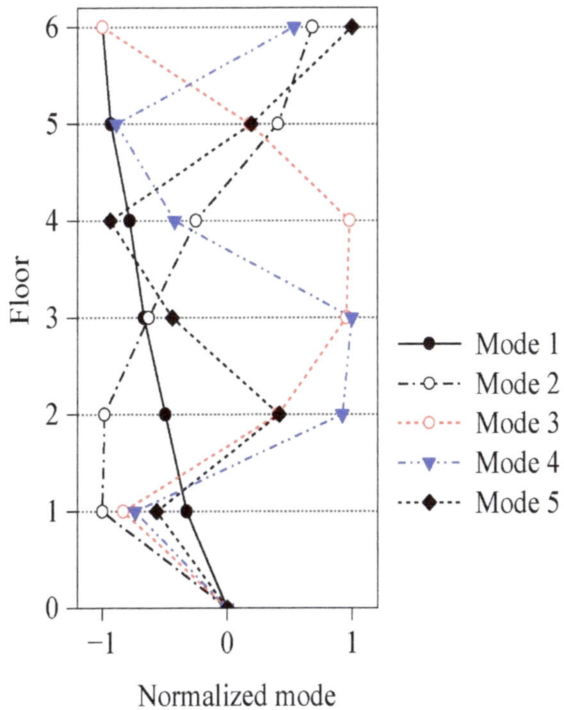

Figure 7. Mode shapes of the isolated structure.

4.2. Seismic Responses of the Frame Structures

Figure 8 displays the maximum inter-story drifts of the superstructure. It can be clearly seen from the figure that NF ground motions caused larger inter-story drifts than FF ground motions, especially when the input ground motion PGA exceeded 0.4 g. When the input PGA is 0.40 g, corresponding to the rare earthquake of intensity 8, the maximum inter-story drift of the superstructure is 0.72%, which is smaller than the specified value of 2% for the elastoplastic story drift in the current Chinese seismic code. When the input ground motion PGA was 0.60 g, the inter-story drifts caused by pulse-like NF ground motion were obviously larger than that caused by FF ground motion and NF ground motion without a pulse. The Kobe_NP ground motion caused the largest inter-story drift (1.32%) of the superstructure, followed by the ChiChi_NP pulse-like ground motion (0.959%). When the input ground motion PGA was 0.80 g, inter-story drifts of the superstructure caused by pulse-like NF ground motion were much larger than that caused by FF ground motion and NF ground motion without a pulse. The inter-story drift caused by ChiChi_NP ground motion was 1.79%. In contrast, the inter-story drift of the second and first floor of the superstructure caused by Kobe_NP ground motion was 2.79% and 3.15%, respectively. The measurement errors of the inter-story drifts of the third to the fifth floors were caused by the uplift of the LNRs. It can be drawn from the test results that the seismic responses of pulse-like NF ground motions are maximum, while those of the FF ground motions are minimum. The reason is that the principal frequency of the structure is closest to the main frequency content of the pulse-like NF ground motions and is furthest from that of the FF ground motions.

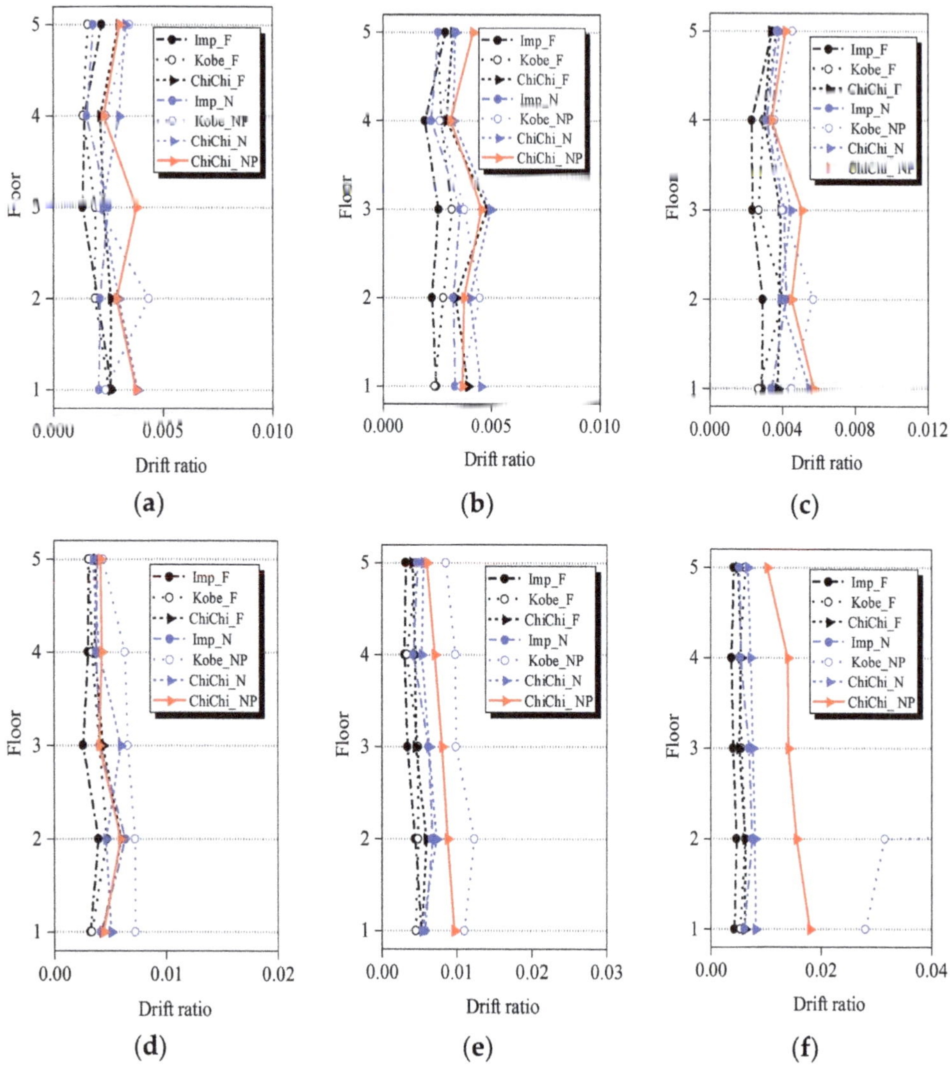

Figure 8. Maximum inter-story drifts of the superstructure: (**a**) PGA of 0.10 g; (**b**) PGA of 0.20 g; (**c**) PGA of 0.30 g; (**d**) PGA of 0.40 g; (**e**) PGA of 0.60 g; (**f**) PGA of 0.80 g.

Table 7 illustrates the amplification factors of the roof acceleration. The acceleration magnification was not obvious when the isolated structure was subjected to Imp_Fground motion, and the amplification factor was smaller than one when the input PGA was larger than 0.30 g. The amplification factor decreased gradually with the increased input PGA of the NF ground motions. The amplification factor increased when subjected to Kobe_NP and ChiChi_NP ground motions with a PGA of 0.80 g because of the impact of the substructure on the isolation bearing.

Table 7. Amplification factors of the roof acceleration.

Ground Motion	PGA (g)					
	0.10	0.20	0.30	0.40	0.60	0.80
Imp_F	1.09	1.02	0.87	0.92	0.91	0.87
Kobe_F	1.51	1.25	1.25	1.26	1.24	1.20
ChiChi_F	1.65	1.68	1.79	1.70	1.58	1.46
Imp_N	1.86	1.69	1.67	1.66	1.64	1.38
Kobe_N	1.46	1.45	1.15	1.10	1.05	1.57
ChiChi_N	1.80	1.72	1.49	1.36	1.19	1.15
ChiChi_NP	1.69	1.38	1.19	1.13	1.00	1.06

4.3. Performance Evaluation of the Isolation System under Earthquakes

The base shear shown in Figure 9 was calculated as the sum of the shear forces of the four isolation bearings divided by the total weight of the substructure. The presented base shear was normalized by the total weight, W, which is equal to 180 kN. The normalized base shear increased with the PGA of the input. When PGA was smaller than 0.6 g, the normalized base shear was smaller than one. The normalized base shear exceeded one when the isolation structure was subjected to Kobe-NP and ChiChi_NP ground motions, with a PGA of 0.8 g. The NF ground motions significantly affected the normalized base shear compared with FF ground motions. The Kobe_NP ground motion yields a considerably large normalized base shear for the isolation system.

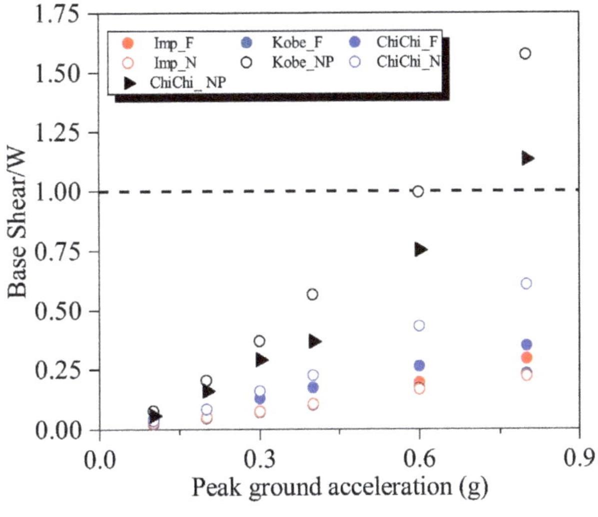

Figure 9. Base shear variation with the PGA of the ground motion inputs.

The overturning moment of the superstructure induced a substantial variation of the axial load on the north or south bearings. In a few cases, tension was afforded in the LNRs. Figures 10 and 11 present the histories of axial force on the bearings for the isolated moment frame structure subjected to the Kobe_NP_0.60 g and ChiChi_NP_0.60 g, respectively. The bottom graph in the figures displays the history of the total axial forces on all four bearings. When the ground motion PGA was 0.6 g, the total axial force on all four bearings slightly varied, indicating that there was little vertical acceleration on the model. In contrast, the total axial force in the two north and south bearings experienced large variations of the order of 100%. This considerable variation in the total axial force of four bearings indicates that a large overturning moment is present. As shown in Figure 10, the excitation PGA is 0.8 g, and the total axial force on all four bearings exhibited large variations due to the impact on the load cells after the uplift of the entire structure.

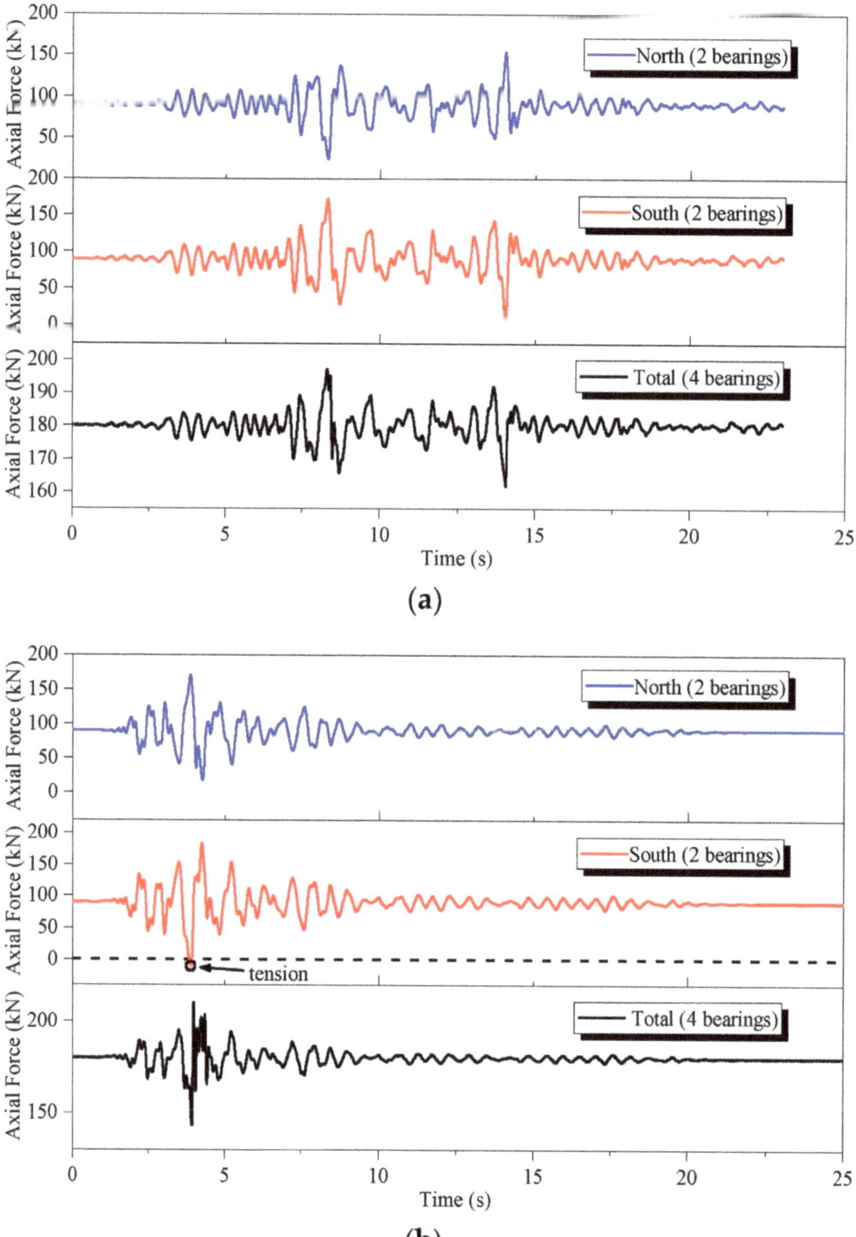

Figure 10. History of the record axial force of the LNRs subjected to: (**a**) ChiChi_NP_0.60 g; (**b**) Kobe_NP_0.60 g.

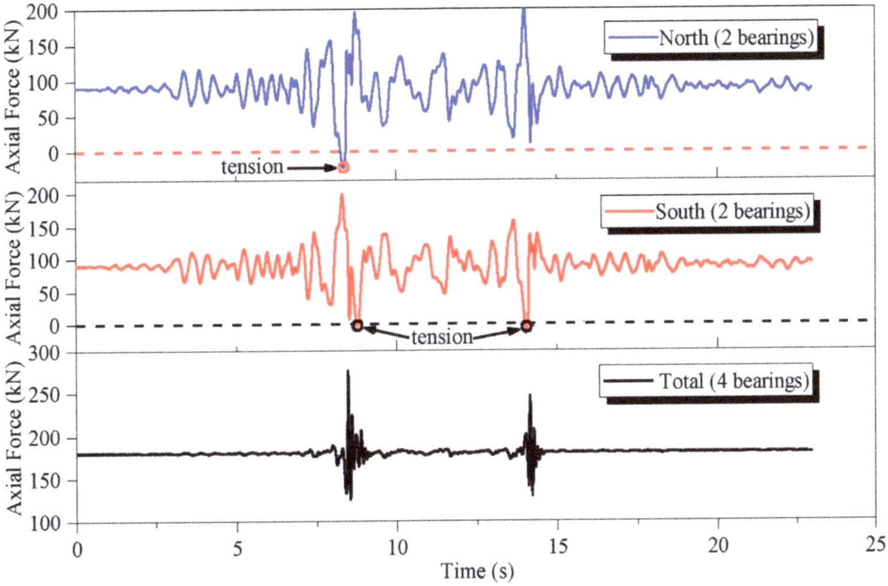

(a)

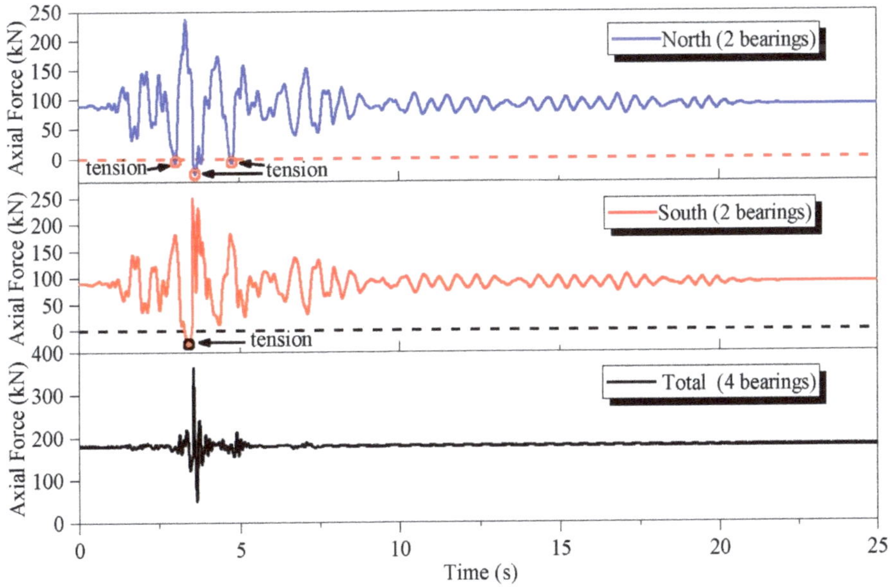

(b)

Figure 11. History of the record axial force of the LNRs subjected to: (**a**) ChiChi_NP_0.80 g; (**b**) Kobe_NP_0.80 g.

As shown in Figure 10b, the two south low damping rubber bearings experienced slight tension (a value less than zero) under Kobe_NP_0.60 g. In Figure 11, all four bearings experienced tension. The uplift of the entire structure affected the load cells, and the total axial force on all four bearings exhibited large variations. While for the FF ground motions, the total axial force on all four bearings exhibited little variation. The rubber bearings did not experience tension.

The maximum displacement of all four bearings under different earthquake levels are listed in Table 8. The displacements of the bearings caused by FF ground motions were smaller than those caused by NF ground motions. The Kobe_NP and ChiChi_NP ground motions caused very large displacements of the bearings. When the input PGA is 0.40 g, corresponding to the rare earthquake of intensity 8, the maximum displacement of the isolation bearing is 51.0 mm, which is smaller than the design displacement of 80 mm. When the input PGA is 0.60 g, corresponding to the extremely rare earthquake of intensity 8, the maximum displacement of the isolation bearing is 83.61 mm, which exceeds the design displacement of 80 mm. The displacement of all four bearings slightly exceeded the design displacement of 80 mm under Kobe_NP ground motion excitation with an input PGA of 0.6 g, and the displacement exceeded 100 mm when PGA was increased to 0.8 g. The time history of the northeast bearing under the Kobe_NP and ChiChi_NP ground motions are shown in Figures 12 and 13, respectively.

Table 8. Displacement of the bearings (unit: mm).

Bearings	Ground Motion	PGA(g)					
		0.1	0.2	0.3	0.4	0.6	0.8
Bearing northeast	Imp_F	2.0	3.2	5.4	7.2	16.3	25.7
	Kobe_F	1.7	3.0	5.4	8.0	13.8	19.8
	ChiChi_F	3.2	6.3	10.2	14.5	23.2	32.3
	Imp_N	2.0	3.4	5.4	8.3	13.4	19.0
	Kobe_NP	5.6	17.0	32.3	51.0	83.6 *	120.3 *
	ChiChi_N	3.9	9.7	13.6	19.8	39.0	56.5
	ChiChi_NP	4.4	10.8	23.4	33.0	63.3	90.5 *
Bearing southeast	Imp_F	1.8	3.6	5.5	6.9	15.6	26.2
	Kobe_F	2.3	3.7	6.0	8.0	14.2	19.3
	ChiChi_F	3.6	6.0	10.9	14.1	24.3	33.6
	Imp_N	2.1	4.2	5.8	7.4	14.2	19.1
	Kobe_NP	6.3	16.6	32.3	51.7	82.7 *	118.8 *
	ChiChi_N	4.7	9.9	13.3	20.2	38.9	57.0
	ChiChi_NP	5.3	12.1	24.0	33.0	64.4	90.4 *
Bearing northwest	Imp_F	3.2	4.1	6.7	7.8	15.8	34.0
	Kobe_F	2.4	3.7	7.1	7.6	14.2	25.6
	ChiChi_F	4.4	6.8	11.9	14.7	23.1	33.3
	Imp_N	2.5	3.7	6.3	8.1	14.4	24.0
	Kobe_NP	5.4	15.2	30.4	50.2	82.4 *	104.8 *
	ChiChi_N	3.1	8.0	13.1	18.3	37.4	60.0
	ChiChi_NP	3.9	13.0	26.0	30.8	65.7	98.4 *
Bearing southwest	Imp_F	1.9	3.2	5.1	7.1	15.3	24.7
	Kobe_F	1.7	3.0	5.6	8.2	14.8	21.2
	ChiChi_F	3.1	6.3	10.8	14.7	22.5	31.7
	Imp_N	2.1	3.8	5.8	8.3	14.5	20.5
	Kobe_NP	5.7	16.6	32.0	51.7	82.4 *	-
	ChiChi_N	4.3	9.2	13.6	20.6	39.1	55.7
	ChiChi_NP	3.5	11.4	24.9	32.7	63.6	90.9 *

* denotes that the displacement of the bearing exceeds the design displacement, and—denotes unrecorded displacement.

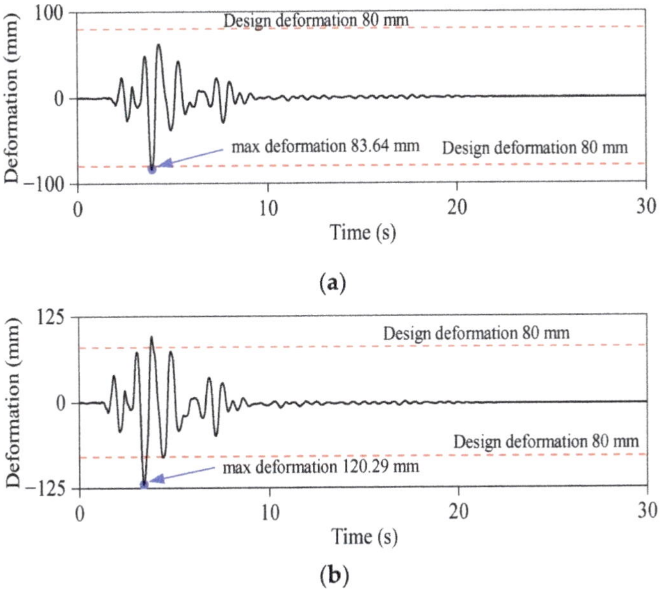

Figure 12. History of the displacement of the LNR subjected to: (**a**) Kobe_NP_0.60 g; (**b**) Kobe_NP_0.80 g.

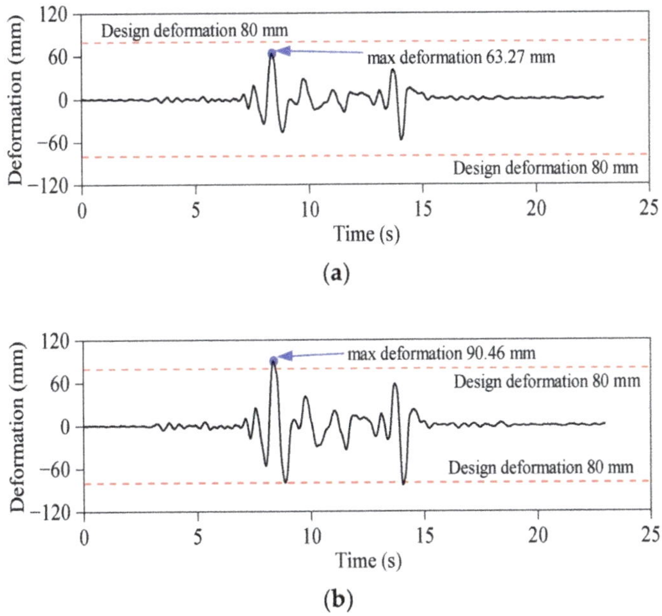

Figure 13. History of displacement of the LNR subjected to: (**a**) Kobe_NP_0.60 g; (**b**) Kobe_NP_0.80 g.

Figure 14 displays the maximum displacements of the viscous dampers subjected to excitations with different earthquake levels. The displacement response of the four viscous dampers affords similar distributions under different earthquake levels. Moreover, displacements caused by NF ground motions were significantly larger than those caused by

FF ground motions, except for the Imp_N ground motion. The response spectrum shows that in the neighborhood of the fundamental period of the isolated structure, the response spectrum value of the Imp_N ground motion is smaller than that of other NF ground motions and is close to that of FF ground motions. The damper displacement exceeds the design displacement of the damper (75 mm) under Kobe_NP ground motion with a PGA of 0.8 g. The displacement of the dampers increased with the PGA value of the excitation, and the growth rate of the displacement increased with the input PGA.

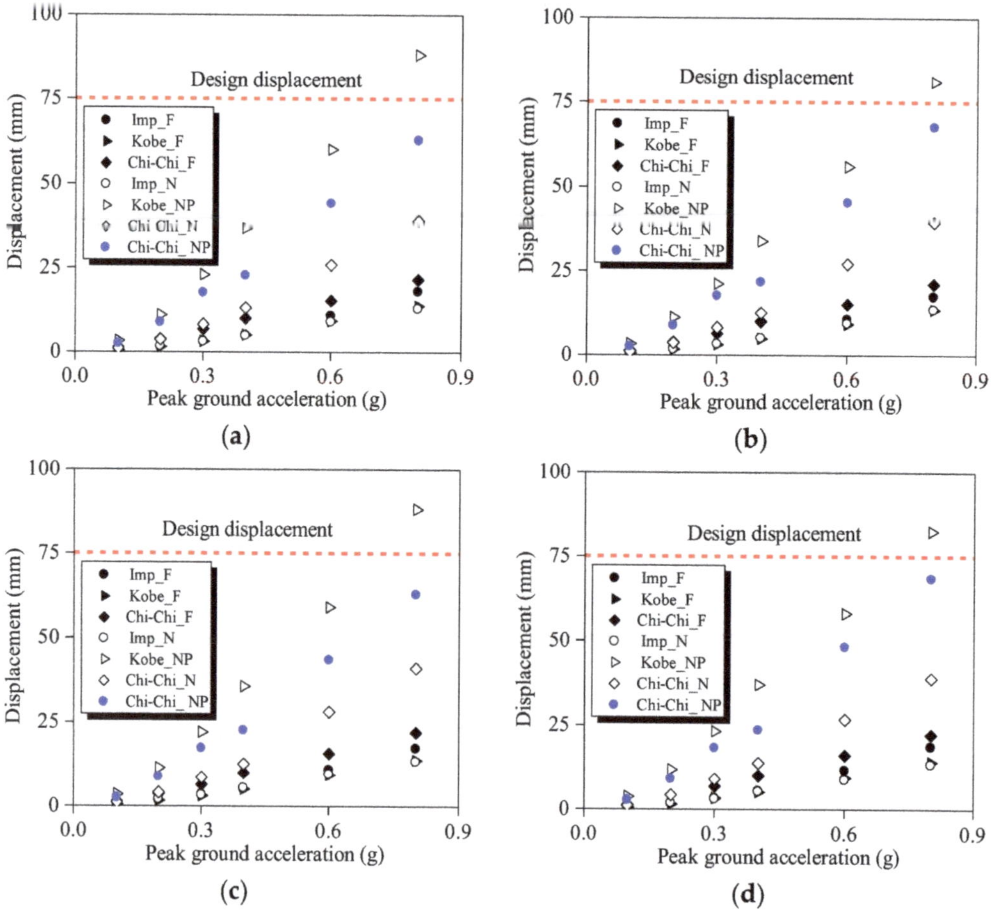

Figure 14. Maximum damper displacement in different earthquake levels: (**a**) northeast location; (**b**) southeast location; (**c**) northwest location; (**d**) southwest location.

Figure 15 displays the residual displacement of the viscous dampers. The maximum value of the residual displacement of the NLVDs was smaller than 0.5 mm until the input PGA exceeded 0.4 g. For the two dampers located in the northeast and southeast, maximum residual displacement was induced when the isolated structure was subjected to Kobe_NP_0.80 g. For the two dampers located in the northwest and southwest, maximum residual displacement was induced when the isolated structure was subjected to ChiChi_F_0.80 g. Residual displacements of the NLVDs were very small during earthquakes.

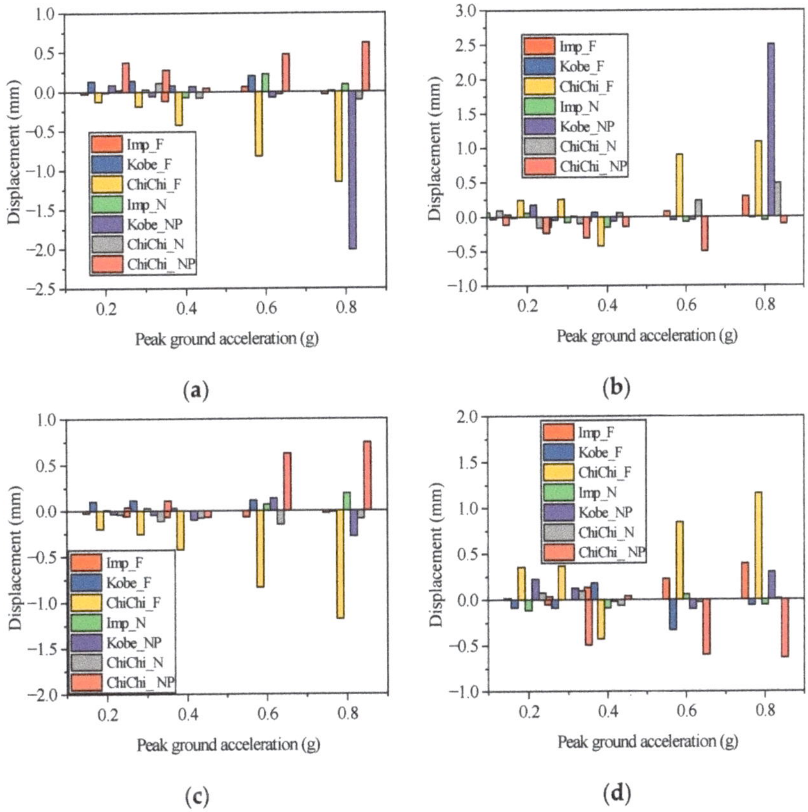

Figure 15. Residual displacement of the viscous dampers: (**a**) Northeast location; (**b**) Southeast location; (**c**) Northwest location; (**d**) Southwest location.

Peak values of the axial forces of the viscous damper are listed in Table 9. The maximum damper force occurred in the damper located in the northeast when the isolated structure was subjected to Kobe_NP_080 g. The maximum damper force was 25.46 kN, which is smaller than the design maximum damper force of 30 kN. Figure 16 displays the hysteretic curves of the four NLVDs under Kobe_NP ground motion with the input PGA of 0.80 g. The shape of the hysteretic curve is full, and it indicates that the NLVDs have a good energy dissipation effect.

Table 9. Maximum axial force of the NLVDs (unit: kN).

NLVD Location	PGA (g)					
	0.10	**0.20**	**0.30**	**0.40**	**0.60**	**0.80**
northeast	7.54	11.34	14.31	16.62	20.32	25.46
southeast	7.13	10.92	13.64	16.08	19.50	23.88
northwest	7.61	11.13	13.70	16.14	20.07	23.35
southwest	7.71	11.69	14.45	16.74	20.18	24.56

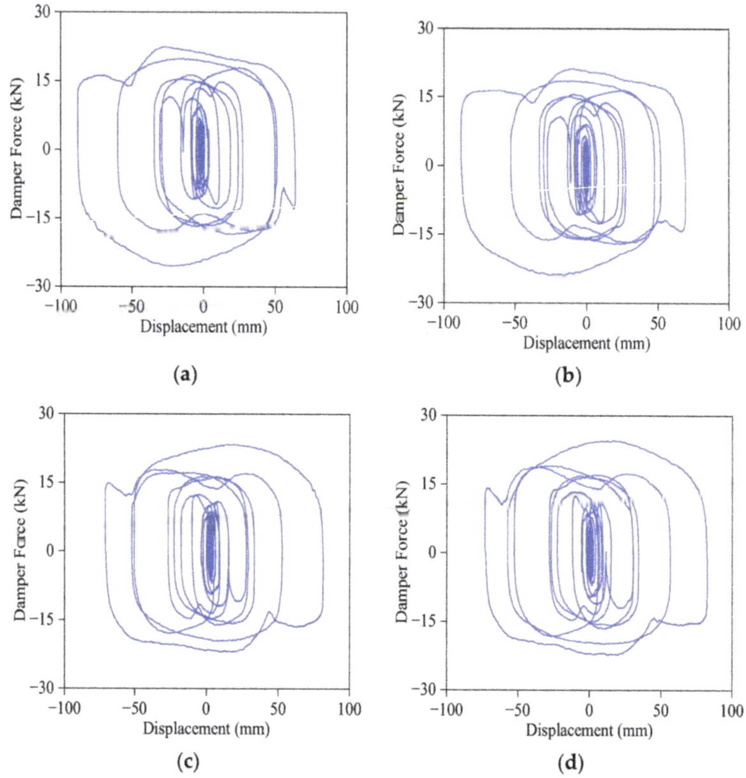

Figure 16. Hysteretic curve of the viscous dampers under Kobe_NP ground motion with a PGA of 0.80 g: (**a**) Northeast location; (**b**) Southeast location; (**c**) Northwest location; (**d**) Southwest location.

5. Conclusions

The shaking table test was conducted for a five-story building model in the moment frame in conjunction with isolation systems. The isolation systems utilized were LNRs with additional NLVDs. Seven ground motions, including three FF ground motions and four NF ground motions, were employed for the shaking table test. Compared with the existing studies, the novelty of the current study is that it experimentally investigates the seismic performance of isolation systems with passive control devices and superstructures subjected to NF ground motions. The main conclusions of the paper are as follows:

1. NF ground motions caused larger inter-story drifts than FF ground motions, especially when the input PGA exceeded 0.4 g. The Kobe_NP NF ground motion with pulse period T_p neighboring the fundamental period of the isolated structure caused the largest inter-story drift and floor acceleration of the superstructure.
2. NF ground motions cause larger base shear force of the isolation system compared with FF ground motions. Pulse-like NF ground motions, e.g., Kobe_NP and ChiChi_NP ground motions, caused a considerably large base shear for the isolation system. The base shear exceeded the total weight of the isolated structure. The axial load of the isolation bearings experienced tension under the pulse-like NF ground motion excitations because of a large overturning moment. The large overturning moment caused an uplift of the LNR when the isolated system was subjected to pulse-like NF ground motions with a large PGA value.
3. The displacements of the LNRs caused by FF ground motions were smaller than those caused by NF ground motions. The displacement of all four bearings slightly

exceeded the design displacement of 80 mm under Kobe_NP pulse-like NF ground motion excitation with a PGA of 0.6 g, and the displacement exceeded 100 mm when the PGA was increased to 0.8 g.

4. Nonlinear viscous dampers performed well during the earthquake simulations; the maximum displacement and damper force were within the design value. The residual displacements of the viscous dampers were very small under NF ground motions. NLVDs protected the LNRs very under NF ground motions, but the severe responses of the LNRs could be caused by NF ground motion with pulse period T_p neighboring the fundamental period of the isolated structure.

In this paper, the seismic performance of base-isolated structures with viscous dampers was studied, and the performance of isolation bearing and viscous damper was also evaluated. Although viscous dampers have performed well in reducing isolation structure under near-field earthquakes, the large deformation of the isolation layer was caused for near-field earthquakes with the pulse period neighboring the period of the isolated structure. The combination of dampers with different mechanisms or new dampers should be further used to control the deformation of the isolation layer under large near-field earthquakes without affecting the isolation effect under moderate earthquakes. The corresponding calculation analysis and experimental study should be carried out in the future.

Author Contributions: Conceptualization, Q.L. and B.F.; methodology, Q.L. and W.L.; validation, B.F. and L.L.; formal analysis, Q.L.; investigation, Q.L.; resources, W.Q. and M.Z.; data curation, Q.L. and Z.Z.; writing—original draft preparation, Q.L.; writing—review and editing, B.F. and B.Z.; visualization, W.Q. All authors have read and agreed to the published version of the manuscript.

Funding: This study was supported by the National Natural Science Foundation of China (Grant Nos. 52008216, 51908129, 51908048); the Opening Funds of State Key Laboratory of Building Safety and Built Environment and National Engineering Research Center of Building Technology (Grant No. BSBE2019-01); the Scientific Research Fund of Institute of Engineering Mechanics, China Earthquake Administration (Grant No. 2020D22); the Basic Public Welfare Research Project of Zhejiang Province, China (Grant No. LGG21E080007); and the Natural Science Foundation of Zhejiang province, China (Grant No. LY20A020005).

Data Availability Statement: Some or all data that support the findings of this study are available from the corresponding author upon reasonable request.

Acknowledgments: Thanks are due to Kaiqi Yu for assistance with the experiments and to Dayang Wu and Peng Zhang for valuable discussion.

Conflicts of Interest: The authors declare no conflict of interest.

References

1. Rong, Q. Optimum parameters of a five-story building supported by lead-rubber bearings under near-fault ground motions. *J. Low Freq. Noise, Vib. Act. Control* **2020**, *39*, 98–113. [CrossRef]
2. Bhandari, M.; Bharti, S.D.; Shrimali, M.K.; Datta, T.K. Seismic Fragility Analysis of Base-Isolated Building Frames Excited by Near- and Far-Field Earthquakes. *J. Perform. Constr. Facil.* **2019**, *33*, 1–16. [CrossRef]
3. Hall, J.F.; Heaton, T.H.; Halling, M.W.; Wald, D. Near-Source Ground Motion and its Effects on Flexible Buildings. *Earthq. Spectra* **1995**, *11*, 569–605. [CrossRef]
4. Jangid, R.S.; Kelly, J.M. Base isolation for near-fault motions. *Earthq. Eng. Struct. Dyn.* **2001**, *30*, 691–707. [CrossRef]
5. Jangid, R. Optimum lead–rubber isolation bearings for near-fault motions. *Eng. Struct.* **2007**, *29*, 2503–2513. [CrossRef]
6. Providakis, C. Pushover analysis of base-isolated steel–concrete composite structures under near-fault excitations. *Soil Dyn. Earthq. Eng.* **2008**, *28*, 293–304. [CrossRef]
7. Mazza, F.; Vulcano, A. Nonlinear response of RC framed buildings with isolation and supplemental damping at the base subjected to near- fault earthquakes. *J. Earthq. Eng.* **2009**, *13*, 690–715. [CrossRef]
8. Bhagat, S.; Wijeyewickrema, A.C. Seismic response evaluation of base-isolated reinforced concrete buildings under bidirectional excitation. *Earthq. Eng. Eng. Vib.* **2017**, *16*, 365–382. [CrossRef]
9. Bhagat, S.; Wijeyewickrema, A.C. Seismic collapse probability considering pounding and financial loss estimation of base isolated reinforced concrete buildings. *J. Earthq. Tsunami* **2018**, *12*, 1–21. [CrossRef]
10. Nagarajaiah, S.; Ferrell, K. Stability of elastomeric seismic isolation bearings. *J. Struct. Eng.* **1999**, *125*, 946–954. [CrossRef]

11. Vestroni, F.; Vulcano, A.; Di Pasquale, G. Earthquake response analysis of a nonlinear model of a base-isolated structure. In Proceedings of the International Meeting on Earthquake Protection of Buildings, Ancona, Italy, 6–8 June 1991.
12. Mazza, F.; Vulcano, A. Seismic response of buildings with isolation & supplemental dissipation at the base. In Proceedings of the 5th World Congress on Joints, Bearings and Seismic Systems for Concrete Structures, Rome, Italy, 7–11 October 2001.
13. Kelly, J.M. The role of damping in seismic isolation. *Earthq. Eng. Struct. Dyn.* **1999**, *28*, 3–20. [CrossRef]
14. Ahmet, H.D.; Huseyin, B. Effects of the isolation parameters on the seismic response of steel frames. *Earthq. Struct.* **2018**, *15*, 319–334.
15. Kun, Y.; Yan, X.; Liang, H. A direct displacement-based design procedure for base-isolated building structures with lead rubber bearings (LRBs). *Eng. Struct.* **2019**, *197*, 1–9.
16. Ayoub, S.; Gholamreza, G.A.; Mohammad, S. Effects of ductility and connection design on seismic responses of base-isolated steel moment-resisting frames. *Soil Dyn. Earthq. Eng.* **2021**, *143*, 1–19.
17. Alataby, E.K.; Kadhim, J.A.; Ahmed, M.A. Study on the Effect of Distribution of Viscous Damper for Steel Frame Structure. *J. Phys. Conf. Ser.* **2021**, *1973*, 012220. [CrossRef]
18. Makris, N.; Chang, S.P. Effect of viscous, viscoplastic and friction damping on the response of seismic isolated structures. *Earthq. Eng. Struct. Dyn.* **2000**, *29*, 85–107. [CrossRef]
19. Palazzo, B.; Petti, L. Combined control strategy: Base isolation and tuned mass damping. *ISET J. Earthq. Technol.* **1999**, *36*, 121–137.
20. Zhang, Y.; Iwan, W.D. Protecting base isolated structures from near-field ground motion by tuned interaction damper. *J. Eng. Mech.* **2002**, *128*, 287–295. [CrossRef]

buildings

MDPI

Article

Study on the Constitutive Model of Concrete Confined by Multi-Spiral Composite Stirrups

Kun Yang *, Tao Yu, Guiliang Ma, Jiaxiang Zhao and Shanshan Sun

School of Civil Engineering, Chang'an University, Xi'an 710061, China
* Correspondence: yangkun8224@chd.edu.cn

Abstract: Through axial compression tests, the influence of three stirrup indexes (space, form, and strength) on the confining performance of multi-spiral composite stirrups in square reinforced concrete (RC) columns were analyzed, and the square RC columns with traditional well-shaped composite stirrups were used as the reference group. The results show that the multi-spiral composite stirrups had a positive contribution to the important indexes (bearing capacity and ductility) of the square RC columns due to its multiple restraint mechanism on core concrete. In terms of constraint effect, the five-spiral composite stirrup is the best, followed by the four-spiral composite stirrup, and the last is the traditional well-shaped composite stirrup. The section of the concrete square column is divided into highly constrained, partially constrained and unconstrained regions and the constraint mechanism of multi-spiral composite stirrups is discussed. The formulas for calculating the peak stress, peak strain, and ultimate compressive strain of the constrained column are presented, and the relative error between the theoretical values and the tested values is small. The constitutive model of concrete constrained by multi-spiral composite stirrups is established and compared with other constitutive models. The results show that the proposed model fits well with the experimental curves.

Keywords: multi-spiral stirrup; strength; ductility; constitutive model; confined concrete

Citation: Yang, K.; Yu, T.; Ma, G.; Zhao, J.; Sun, S. Study on the Constitutive Model of Concrete Confined by Multi-Spiral Composite Stirrups. *Buildings* **2022**, *12*, 2179. https://doi.org/10.3390/buildings12122179

Academic Editor: Abdelhafid Khelidj

Received: 14 November 2022
Accepted: 6 December 2022
Published: 9 December 2022

Copyright: © 2022 by the authors. Licensee MDPI, Basel, Switzerland. This article is an open access article distributed under the terms and conditions of the Creative Commons Attribution (CC BY) license (https://creativecommons.org/licenses/by/4.0/).

1. Introduction

China is one of the countries most affected by earthquake disasters. The earthquake damage shows that the reinforced concrete (RC) columns in the building have been damaged to different degrees, especially the columns located at the bottom of the whole structure are often the most seriously damaged at the column end [1,2]. The failure of the column end may cause the rapid degradation of the bearing capacity or even the continuous collapse of the column. The lateral constraint of the RC column can not only help it support more vertical load, but also increase the deformation capacity of the column. General constraint methods can be divided into active and passive constraints. In related research, passive restraint methods such as square hoops, spiral stirrups [3], steel pipes [4], fiber materials [5], or their combinations [6–11] are used to provide effective lateral restraint for concrete. Studies have proved that using stirrup to constrain concrete can enhance the mechanical properties of columns, thereby preventing or delaying serious damage to columns [12]. Among them, the traditional spiral stirrup is widely used in a cylinder because of its good constraint effect, and its confining effect on a cylinder is generally better than that on rectangular stirrups columns (including square columns).

Scholars abroad and at home have conducted a series of investigations on the performance of concrete confined with stirrups. For the purpose of obtaining the exact strength of constrained concrete, Skeikh and Uzumeri [13] fully considered the influence of stirrup spacing and lengthways reinforcement arrangement, and analyzed the effectively constrained area in the core concrete area. Based on that, they obtained the stress–strain expression of constrained concrete. Mander et al. [14,15] provided a stress–strain model for constrained concrete considering the effects of strain rate and cyclic loading. The model fully considered the arch function between rectangular stirrups and lengthways reinforcement and defined the concept of effectively constrained area based on this model. It is suitable for various restraint forms such as traditional rectangular hoops, spiral stirrups, etc. Cusson et al. [16,17] believed that the axial compressive strength of restrained concrete columns would arrive at its peak point before the stirrup stress, and proposed an innovative method to calculate the actual stirrup stress under the peak stress of the constrained concrete. Saatcioglu and Razvi [18] established a constitutive model curve composed of ascending parabola segments and descending straight segments based on the concept of equivalent constrained stress. Later, Bing et al. [19] combined the model of Mander and Saatcioglu and proposed a fitting equation for the ultimate longitudinal strain of high-strength concrete. This formula is suitable for the fracture of the first rectangular hoop or spiral stirrup. Through the axial compression test of high-strength multi-spiral stirrup columns, Rong et al. [20] proposed a new energy absorption method and established a relationship between energy absorption and lateral confinement in concrete columns. Then, the passive strain model, the boundary point model, and the constraint coefficient model of the passive restrained concrete are proposed. Li et al. [21,22] proposed an analytical method for the axial performance of high-strength multi-spiral reinforced concrete columns based on strain compatibility and presented an iterative method for calculating the actual stress of spiral stirrup when concrete is located at peak stress. Through the axial compression experiment of restrained high-strength concrete, Shi and Yang et al. [23,24] believed that the real stress of a high-strength stirrup should be used when considering the restrained stress of a high-strength stirrup, so as to ensure accuracy in judging the effect of strength improvement on concrete columns. Ouyang et al. [25] unified three strength levels of concrete into their calculation formula for a compressive stress–strain curve. According to the experimental results of 42 concrete cylinders confined by a spiral stirrup, Zheng et al. [26] investigated the relationship between stirrup tensile strain under peak compressive stress and three stirrup factors (volume stirrup, yield strength, and spacing). In the end, the constitutive model for this type of column was derived. In order to study the mechanical properties and size-effect behavior of a large RC column confined by stirrups under axial compression loads, Jin et al. [27] carried out corresponding tests and established a constitutive model of an RC column confined by stirrups with size-effect. Li et al. [28] tested a total of six RC columns confined by composite spiral stirrups under eccentric loading with different stirrup spacing and force eccentricities. The experimental analyses explained that the dual confinement of the concrete consisting of core confinement by spiral stirrups and surrounding confinement by rectangular hoops had significant effects on failure modes. Finally, a formula for calculating the eccentric loading capacity of composite spiral stirrups confined concrete columns was proposed.

In practical engineering, a rectangular (or square) column is more applicable than a circular column. In order to solve the problem that spiral stirrups do not adapt to rectangular (or square) columns, Yin [29,30] developed a series of new restraint types of rectangular RC columns and proved that spiral stirrups can also show good restraint effects in those columns through a number of experimental researches. Subsequently, in terms of the axial compression test of large-size SRC (steel reinforced concrete) columns and RC (reinforced concrete) columns equipped with five-spiral stirrups, Yin et al. [31] believed that the new five-spiral SRC column has excellent strength and ductility under the premise of the same total longitudinal steel. In addition, the five-spiral stirrups can be processed automatically, which can shorten the construction period in engineering practice and bring economic benefits. Then, Yin et al. [32,33] carried out a large number of axial compression and transverse cyclic load tests on concrete column specimens equipped with four different stirrup forms (traditional well-shape, four-spiral, five-spiral, and spiral with cross cable ties). The test results show that each spiral stirrup contributes to the overall pressure, which makes the degradation of strength more slowly after the peak stress point. In addition, the axial pressure of each constraint region in the section depends on the geometric area of the spiral stirrup. Hung et al. [34] proposed a pier system with cross multi-spiral composite stirrups. It is proved that this pier system not only contributes to the enhancement of strength and ductility but also has advantages in cost-effectiveness and constructability. The experiment conducted by Chen et al. [35] indicated that the strength and ductility of columns are increased by 33% and 145% under axial compression, and by 7.14% and 21.4% under eccentric compression, respectively. When the overlap area of the stirrup increases to 10%, the above two analysis indexes increase by 2.56% and 11.25%, respectively. Weng et al. [36] used five-spiral stirrups in concrete rectangular columns and proved the effectiveness of five-spiral stirrups in such applications. Shih et al. [37] came up with a new type of interlocking spiral stirrup for a rectangular column, in which a circular spiral stirrup is interlocked with a star-shaped spiral stirrup to improve its restraint effect. Wu et al. [38] proposed the section constraint form of six-interlocking spiral transverse RC columns and proved through tests that the ductility of such specimens was significantly higher than the requirements of the design specifications. Tang et al. [39] used the finite element method (FEM) to study the varying law of stress distribution in sections of concrete columns with stirrup constraints. The analysis results show that the stress and area of the core section of the column increase obviously. The small spiral stirrup located at four corners can compensate for the stress reduction because of the existence of plain concrete. Finally, a more accurate calculation formula for loading capacity for this type of column is given. Using the simplified elastic finite element method and experiment results, Wang et al. [40] found that the constrained area is closely related to the radius ratio of different spiral stirrups and positively related to the bearing capacity of the column. At present, the application of circular spiral stirrup in rectangular RC columns (or squares) needs further study.

In summary, based on the axial compression test of square RC columns confined by multi-spiral composite stirrups, the constraint mechanism of this type of stirrup and its corresponding constitutive model is studied in this paper.

2. Research Significance

A traditional spiral stirrup is only suitable for a circular column, but the research on multi-spiral composite stirrups breaks through this limitation and applies spiral stirrup to the square column. At present, the application of multi-spiral composite stirrups is not common in many countries, and there is a lack of applied research for corresponding standards. The design of carrying capacity for this type of column is not reflected in the national standards. Therefore, based on the corresponding experimental and theoretical analysis, this paper discusses the curve of stress–strain relationship and the calculation formula of related parameters, so as to provide a reference for the study of mechanical properties of this type of column and its engineering application.

3. Experiment

3.1. Test Program

In this test [41], 8 RC column specimens were designed and fabricated, including 7 specimens with multi-spiral composite stirrup and 1 specimen with traditional well-type stirrup. The size of the square column was 260 × 260 × 750 (mm), and the thickness of the concrete protective layer was 10mm. The design of the specimen mainly considered three variables: stirrup strength, stirrup spacing, and stirrup form. The specific size, stirrup form, and corresponding geometric model of the specimen are shown in Figure 1, and the design parameters are shown in Table 1. In Table 1, stirrup types include A (quadruple spiral composite stirrup), B (five-spiral composite stirrup), and C (traditional well composite stirrup). The ratio of the volume hoop ρ_v refers to the ratio of the stirrup volume to the corresponding concrete volume within a stirrup spacing. f_{cu} is the cube compressive strength of concrete measured by using a 150 mm × 150 mm × 150 mm cube as the sample of standard. f_c is the axial compressive strength of concrete measured by using 150 mm × 150 mm × 300 mm prisms as the sample of standard. Longitudinal compression reinforcement is made of HPB235 steel with a diameter of 8 mm. The strength grades and mechanical properties of the three stirrups are shown in Table 2.

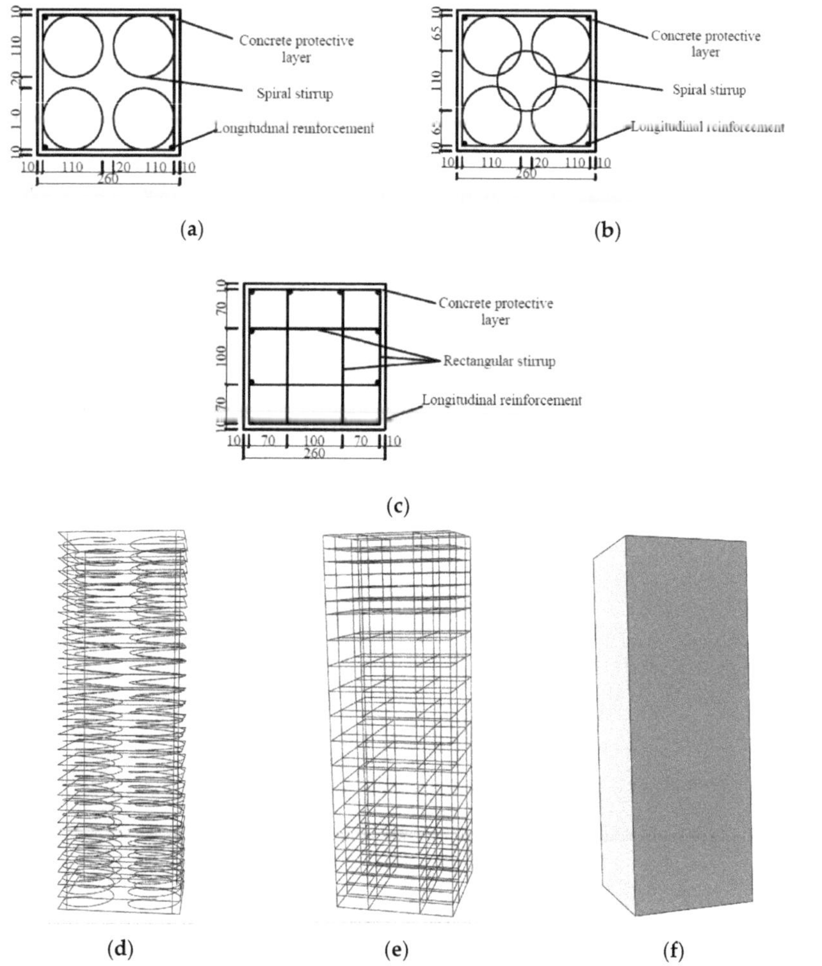

Figure 1. Specimen size and stirrup form. (**a**) four-spiral composite stirrups; (**b**) five-spiral composite stirrups; (**c**) traditional well-shaped composite stirrups; (**d**) 3d diagram of four-spiral composite stirrups; (**e**) 3d diagram of traditional well-shaped composite stirrups; (**f**) A geometric model of confined concrete column.

Table 1. Design parameters of specimen.

Specimen No.	Stirrup Form	Stirrup Yield Strength /MPa	Stirrup Spacing /mm	Ratio of Volume Hoop ρ_v /%	f_{cu} /MPa	f_c /MPa
A-1	A	685	30	6.31	37.3	24.9
A-2	A	685	50	3.96	30.8	20.6
A-3	A	685	70	2.79	30.8	20.6
A-4	A	412	50	3.96	37.3	24.9
A-5	A	919	50	3.96	37.3	24.9
B-1	B	412	50	4.53	37.3	24.9
B-2	B	919	50	4.53	30.8	20.6
C-1	C	919	50	3.96	30.8	20.6

Table 2. Mechanical properties of reinforcement.

Type	Diameter /mm	Yield Strength f_{yv}/MPa	Elastic Modulus E/10^5
HRB400	8	412	2.0
HTRB630	8	685	2.0
1000 MPa	8	919	2.0

The production process of the specimens includes: binding the steel cage; sticking the strain gauge on the steel bar; building formwork on the outside of the steel cage; pouring concrete into the cage along the long side of the formwork; vibrating concrete until it compacts; conserving for 28 days under natural conditions; removing the concrete formwork; attaching strain gauge to the concrete surface. Figure 2 shows the main process of specimen production.

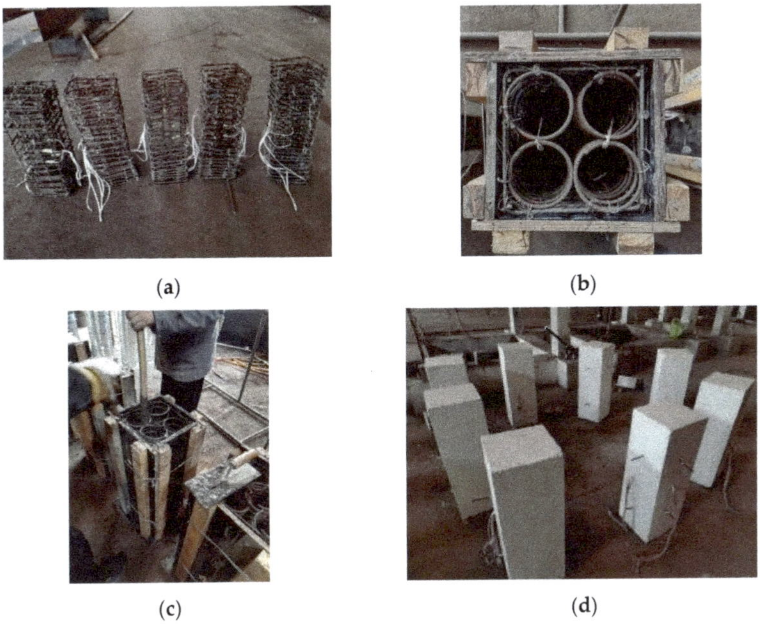

Figure 2. The main production process of constrained concrete column specimen. (**a**) Binding the steel cage; (**b**) Building formwork on the outside of the steel cage; (**c**) Vibrating concrete. (**d**) Attaching strain gauge to concrete surface.

The test was carried out on the 2000 KN hydraulic test machine in the Laboratory. The distribution of devices and measured points is shown in Figure 3. The height range of 450 mm in the middle of the specimen was the test area, and two displacement meters were arranged on the surface of the adjacent sides of the specimen to measure the axial deformation of the specimen. The stirrup strain was measured by attaching a resistance plate to the two rings of the stirrup in the middle of the specimen. This investigation chose the loading way of displacement control, and the loading was stopped when the load value drops to 75% of the peak load.

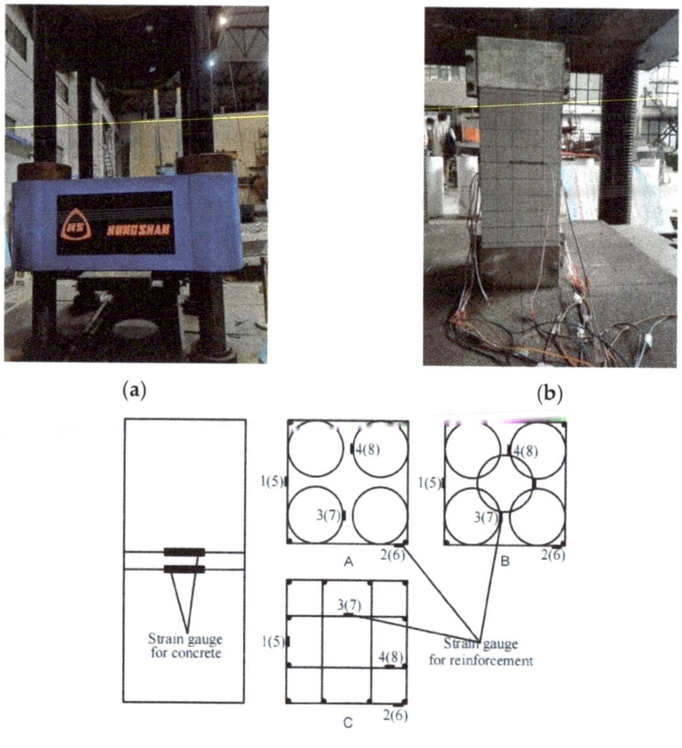

Figure 3. Distribution of loading device and measuring point. (**a**) 2000 KN hydraulic test machine; (**b**) Tested specimen; (**c**) Distribution of strain gauge.

3.2. Analysis of Strength and Ductility

Table 3 shows the specific data of the test results, where f_{cc} and ε_{cc} represent the peak stress and peak strain of the constrained RC column specimen, respectively. f_{co} and ε_{co} are the peak stress and peak strain of the unconstrained RC column specimens, respectively. ε_u is the strain of the specimen when the stress of the concrete column specimen decreases to 85% of the peak stress. The increasing multiple of peak stress K_f ($K_f = f_{cc}/f_{co}$), increasing multiple of peak strain K_ε ($K_\varepsilon = \varepsilon_{cc}/\varepsilon_{co}$) and ductility ratio μ_ε ($\mu_\varepsilon = \varepsilon_u/\varepsilon_{cc}$) can be used as evaluated indicators of the strength and ductility of confined concrete. Comparing specimens A-1 (s = 30 mm), A-2 (s = 50 mm), and A-3 (s = 70 mm), it can be seen that the greater the stirrup spacing, the lower the ascending of peak stress and strain, and the lower the ductility ratio. The increased times of peak stress and strain of specimen A-5 (f_{yv} = 919 MPa) are 1.15 and 1.66 times that of specimen A-4 (f_{yv} = 412 MPa), respectively, which means that the higher the stirrups strength, the greater the peak stress and strain. By comparing specimens A-4 and B-1, as well as specimens B-2 and C-1, it can be seen that the strength and ductility of the five-spiral composite stirrup-restrained concrete specimens are significantly better than those of the four-spiral composite stirrup-restrained concrete specimens and slightly better than those of the traditional well-shaped stirrup restrained concrete specimens. The ductility ratio of specimens B-1 and B-2 is the highest (1.98 and 1.97, respectively), which means that the ductility of concrete specimens constrained by a five-spiral composite stirrup is the best in all specimens.

Table 3. Test result.

Specimen No.	f_{cc}/MPa	ε_{cc}	ε_u	μ_ε	K_f	K_ε
A-1	56.9	0.02803	0.05473	1.95	2.29	14.02
A-2	33.4	0.01921	0.03343	1.74	1.62	9.61
A-3	26.3	0.01686	0.02472	1.47	1.28	8.43
A-4	43.6	0.01408	0.02270	1.61	1.75	7.04
A-5	50.2	0.02340	0.03841	1.64	2.02	11.7
B-1	50.0	0.02471	0.04892	1.98	2.01	12.36
B-2	34.3	0.03508	0.06894	1.97	1.67	17.54
C-1	28.4	0.03521	0.06105	1.73	1.38	17.61

In summary, the strength and ductility of concrete columns confined by multi-spiral composite stirrups are higher than those confined by traditional well-shaped composite stirrups.

3.3. Analysis of Influencing Factors

3.3.1. Stirrup Spacing

When the stirrup form and strength are the same, the influence of stirrup spacing on the stress–strain curve of the specimen is shown in Figure 4. When the stirrup spacing is 30 mm, the strength of specimen A-1 is significantly increased, and the declining phase of the stress-strain curve is gentle. When the stirrup spacing is 70 mm, the peak strength of specimen A-3 increases slightly, and the declining phase of the stress–strain curve is slightly steep. When the stirrup spacing is 50 mm, the strength and ductility of specimen A-2 are between the former two. Therefore, with the diminution of stirrup spacing, that is, the increase of the ratio of volume stirrup, the carrying capacity and ductility of the specimens are significantly improved, and the deformation ability is also better.

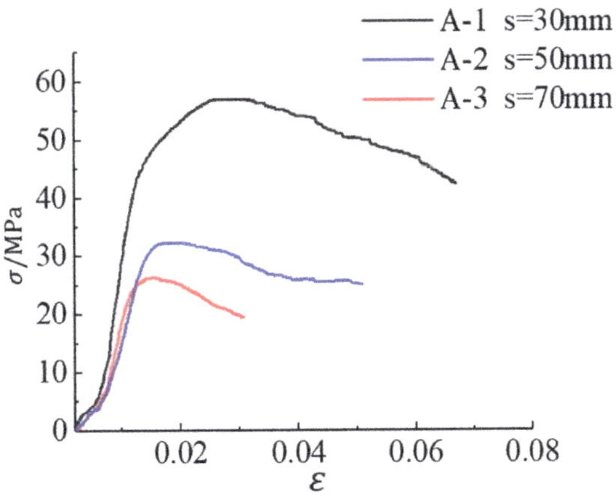

Figure 4. Influence of stirrup spacing.

3.3.2. Stirrup Form

Figure 5 shows the effect of different stirrup types on the stress-strain curves of specimens when the yield strength of the stirrup is 919 MPa and 412 MPa, respectively. According to Figure 5a, when the stirrup strength, spacing, and concrete strength are the same, the restraint effect of the five-spiral composite stirrup form (specimen B-2) is better than that of the well-shaped composite stirrup form (specimen C-1), and the peak strength of the former is higher and the declining phase is more gentle. This is because the small concrete cylinder constrained by the inner spiral hoop is in a state of three-way compression itself, and at the same time is constrained by the outer rectangular hoop. In addition, due to human factors during concrete pouring, the concrete strength of the four-spiral composite stirrup specimen A-5 (f_c = 28.3 MPa) was higher than that of the specimens B-2 and C-1 (f_c = 23.4 MPa). Therefore, the peak strength of the specimen was the highest, but the declining phase of the curve was slightly steeper and the ductility was poor. Figure 5b indicated that the peak strength and strain of specimen B-1 constrained by a five-spiral composite stirrup are significantly larger than those of specimen A-4 constrained by a four-spiral composite stirrup, and the peak plateau of specimen B-1 is longer and the declining phase of the curve is more gentle, that is, the constraint effect of the five-spiral composite stirrup form is better than that of the four-spiral composite stirrup form. This is because the five-spiral composite stirrup interlocks with each other and has a higher degree of constraint.

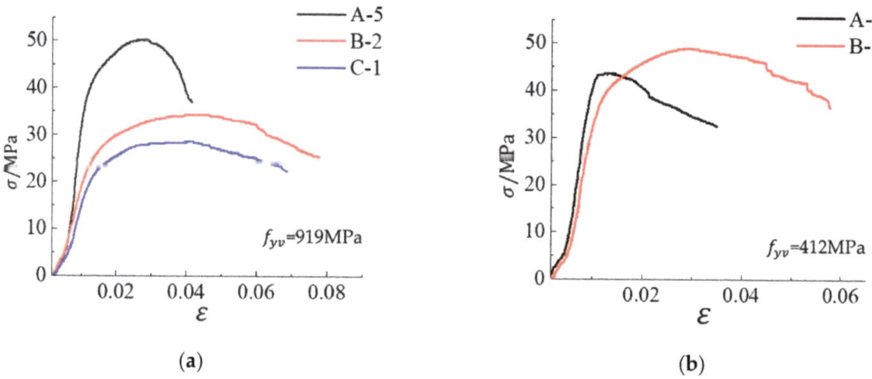

Figure 5. Influence of stirrup form. (**a**) Stress–strain curves of specimens with different stirrup forms when the stirrup strength is 919 MPa; (**b**) Stress–strain curves of specimens with different stirrup forms when the stirrup strength is 412 MPa.

3.3.3. Stirrup Strength

Figure 6 indicates the influence of stirrup strength on the stress–strain curve of the specimen. The peak stress and strain of specimen A-5 (f_{yv} = 919 MPa) are higher than those of specimen A-4 (f_{yv} = 412 MPa), and the declining phase of the stress–strain curve of the former specimen is more gentle. So, the strength and ductility of concrete specimens with high-strength stirrup constraints are better than those with ordinary-strength stirrup constraints. In addition, due to the casting problem, the concrete strength of specimen A-2 is relatively low, so the peak strength of its curve in Figure 6 is the lowest, but the descending section is the most gentle.

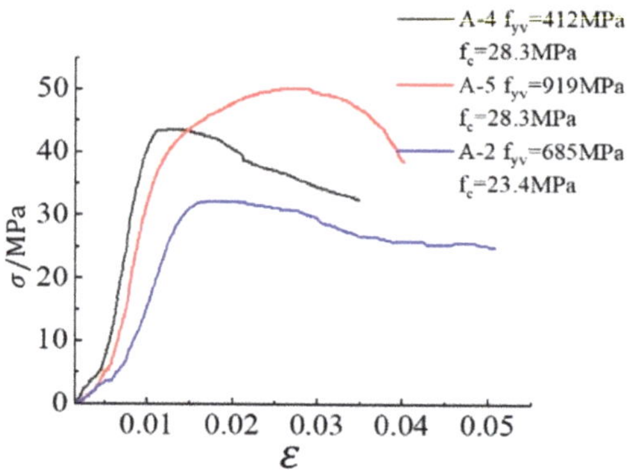

Figure 6. Influence of stirrup strength (Specimen of four-spiral composite stirrup).

In summary, the space, form, and strength of a stirrup are important factors affecting the compression performance of concrete columns confined by stirrups. The smaller the stirrup space and the higher the strength, the better the bearing capacity and ductility of the column. And the restraint form of multi-spiral composite stirrups is better than that of traditional well-shaped composite stirrups.

4. Calculation of Stress and Strain

4.1. Analysis of Constraint Mechanism

A square RC column confined by four-spiral composite stirrups is taken as an example for analysis. The distribution of constraint stress in its cross-section under axial pressure is shown in Figure 7. The rectangular hoop acts as a transverse constraint on the core concrete surrounded by it, and the distribution of constraint stress is arched along the edge length direction. The four concrete cylinders located inside the four spiral stirrups are confined by both the outer rectangular hoop and the inner spiral stirrups. In addition, the outer rectangular hoop connects the four concrete cylinders together, while allowing the RC cylinders to also provide lateral constraints on the core concrete they enclose.

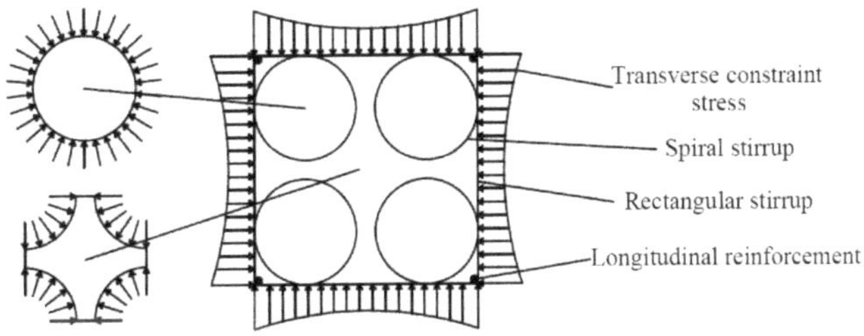

Figure 7. Distribution of transverse constraint stress of square column with four spiral stirrups.

According to the different degrees of constraint, the cross-section of the square column is divided into three parts: highly constrained region, partially constrained region, and unconstrained region, as shown in Figure 8. The highly constrained region is the area surrounded by four inner spiral hoops. The concrete in this area is constrained by both the outer rectangular hoop and inner spiral hoops. The partially constrained region is the area outside the circular spiral stirrup and inside the rectangular stirrup. The concrete in this area is constrained by the outer rectangular hoop and four concrete cylinders. The unconstrained region is the concrete protective layer.

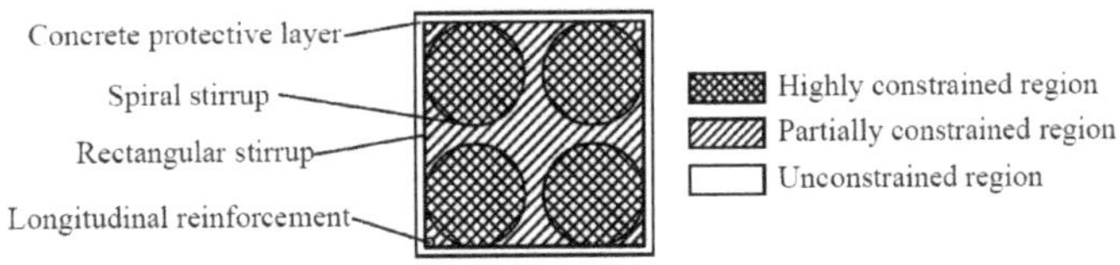

Figure 8. Division of constrained region of square column with four spiral stirrups.

4.2. Effective Lateral Restraint Stress

Mader et al. [15] proposed the effective constraint area A_e of the circular stirrup and rectangular stirrup (Equations (1) and (2)), and defined the ratio of effective constraint area A_e to net core area A_{cc} as the effective constraint coefficient k_e. The effective constraint coefficients of circular spiral and rectangular stirrup were calculated in Equations (3) and (4).

$$A_e = \frac{\pi}{4}{d_s}^2\left(1 - \frac{s'}{2d_s}\right)^2 \quad (1)$$

$$A_e = \left(b_c d_c - \sum_{i=1}^{n} \frac{(w_i')^2}{6}\right)\left(1 - \frac{s'}{2b_c}\right)\left(1 - \frac{s'}{2d_c}\right) \tag{2}$$

$$k_{e1} = \frac{\left(1 - \frac{s'}{2d_s}\right)}{1 - \rho_{cc}} \tag{3}$$

$$k_{e2} = \frac{\left(1 - \sum_{i=1}^{n} \frac{(w_i')^2}{6b_c d_c}\right)\left(1 - \frac{s'}{2d_s}\right)\left(1 - \frac{s'}{2d_s}\right)}{1 - \rho_{cc}} \tag{4}$$

where s' is the minimum distance between adjacent stirrup surfaces and surfaces. d_s is the diameter between the centers of spiral stirrups. b_c and d_c are the core dimensions of the rectangular stirrup along the x and y orientations of the center line, respectively. w_i' is the net distance between the ith adjacent lengthways bars. ρ_{cc} is the ratio of the area of lengthways reinforcement to the area of core concrete. The meanings of the above symbols are shown in Figure 9.

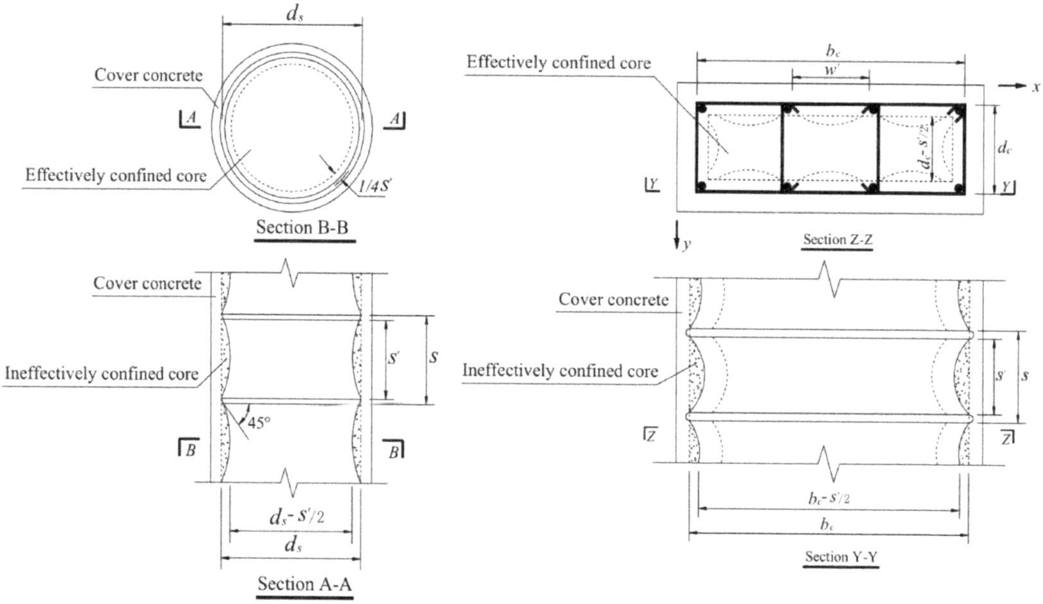

Figure 9. Effectively restrained area of transverse reinforcement.

The effective lateral constraint stresses f_{l1} and f_{l2} imposed on concrete by inner spiral hoops and outer rectangular hoops are calculated according to Equations (5) and (6), respectively.

$$f_{l1} = \frac{1}{2} k_{e1} \rho_s f_{yh1} \tag{5}$$

$$f_{l2} = \frac{1}{2} k_{e2} (\rho_x + \rho_y) f_{yh2} \tag{6}$$

k_{e1} and k_{e2} in the formula can be calculated according to Equations (3) and (4). ρ_s is the volume ratio of the circular spiral stirrup to the internally constrained concrete; ρ_x and ρ_y are the volume ratios of the rectangular stirrup and its internally constrained concrete in the x and y orientations, respectively. f_{yh1} and f_{yh2} are stirrup stresses when the confined concrete column under axial load reaches its peak stress, respectively.

For the concrete square column constrained with multi-spiral composite stirrup studied in this paper, its effective constraint stress f_l is calculated according to Equation (7) [42], where γ is the ratio of the region of constrained concrete to the region of core concrete in the circular spiral stirrup region.

$$f_l = (f_{l1} + f_{l2})\gamma + f_{l2}(1-\gamma) \tag{7}$$

4.3. Peak Stress

The William–Warnke five-parameter model [43] is used as the failure criterion model of concrete under triaxial compression. Equation (8) represents the compression meridian, and σ_{oct} and τ_{oct} represent the octahedral normal stress and tangential stress, respectively.

$$\frac{\tau_{oct}}{f_{co}} = b_0 + b_1\frac{\sigma_{oct}}{f_{co}} + b_2\left(\frac{\sigma_{oct}}{f_{co}}\right)^2 \tag{8}$$

It is assumed that the transverse constraint stresses of concrete confined by multi-spiral stirrups are equal, where both two principal stresses are transverse constraint stresses and the third stress is compressive strength subjected to maximum load. Then the three principal stresses of constrained concrete can be expressed by Equations (9) and (10).

$$\sigma_1 = \sigma_2 = -f_l \tag{9}$$

$$\sigma_3 = -f_{cc} \tag{10}$$

The stress is transformed from a Cartesian coordinate system to an octahedral coordinate system to modify the stress, as shown in Equations (11) and (12).

$$\sigma_{oct} = \frac{(\sigma_1+\sigma_2+\sigma_3)}{3} = \frac{-f_{cc}}{3} - \frac{2}{3}f_l \tag{11}$$

$$\tau_{oct} = \sqrt{\frac{(\sigma_1^2+\sigma_2^2+\sigma_3^2)}{3} - \sigma_{oct}^2} = \frac{\sqrt{2}}{3}(f_{cc}-f_l) \tag{12}$$

By substituting Equations (11) and (12) into Equation (8), the formula for calculating the peak stress of constrained concrete f_{cc} can be obtained as follows:

$$f_{cc} = f_{co}\left(\frac{3\left(\sqrt{2}+b_1\right)}{2b_2} + \sqrt{\left(\frac{3\left(\sqrt{2}+b_1\right)}{2b_2}\right)^2 - \frac{9b_0}{b_2} - \frac{9\sqrt{2}f_l}{b_2 f_{co}} - 2\frac{f_l}{f_{co}}}\right) \tag{13}$$

where f_l is the effective constraint stress; f_{co} is the peak stress of unconstrained concrete; b_0, b_1, and b_2 are the coefficients. By regression analysis of the experimental data in this investigation, $b_0 = 0.111$, $b_1 = -1.02$, and $b_2 = -0.304$ are obtained, and their values are put into Equation (13) to obtain the formula (Equation (14)) for calculating the peak stress of constrained concrete suitable for this paper.

$$f_{cc} = f_{co}\left(-1.944 + 2.663\sqrt{1+5.9\frac{f_l}{f_{co}}} - 2\frac{f_l}{f_{co}}\right) \tag{14}$$

Table 4 indicates the experimental value of the peak stress of the square RC column with multi-spiral composite stirrups tested in this paper and the theoretical value calculated according to the above equations. Through comparative analysis, it can be seen that the calculated value is generally higher than the tested value. The error ω is defined as the ratio of the calculated value minus the tested value to the tested value. The maximum ω is 6.844%, the minimum ω is 0.797%, and the average ω is 3.910%. Therefore, Formula (14) has a high accuracy when it is used to calculate the peak stress of concrete confined by multi-spiral composite stirrups.

Table 4. Tested and calculated values of peak stress.

Specimen No.	Tested Value/MPa	Calculated Value/MPa	$\omega = \frac{(\text{Calculated value} - \text{Tested value})}{\text{Tested value}}/\%$
A-1	56.9	58.9	3.515
A-2	33.4	36.0	7.784
A-3	26.3	28.1	6.844
A-4	43.6	44.7	2.523
A-5	50.2	50.6	0.797
B-1	50.0	49.8	−4.000
B-2	34.3	35.9	4.665
C-1	28.4	31.0	9.155
Average value of ω			3.910

4.4. Peak Strain

According to the formula presented by Bing et al. and combined with the regression analysis of the experimental data in this investigation, the calculation formula of peak strain ε_{cc} is as follows:

$$\frac{\varepsilon_{cc}}{\varepsilon_{co}} = 1 + 18.92\left(\frac{f_l}{f_{co}}\right)^{0.58} \tag{15}$$

Table 5 shows the experimental values of the peak strain of concrete square columns with multi-spiral composite stirrup tested in this paper and the theoretical values calculated according to the above equation. Through comparative analysis, it can be seen that the maximum ω is 9.510%, the minimum ω is 0.260%, and the average ω is 5.621%. Therefore, Equation (15) can be used to calculate the peak strain of concrete with multi-spiral composite stirrup constraints.

Table 5. Tested and calculated values of peak strain.

Specimen No.	Tested Value/MPa	Calculated Value/MPa	$\omega = \frac{(\text{Calculated value} - \text{Tested value})}{\text{Tested value}}/\%$
A-1	0.02803	0.02760	−1.534
A-2	0.01921	0.01926	0.260
A-3	0.01686	0.01529	−9.312
A-4	0.01408	0.01503	6.747
A-5	0.02340	0.02280	−2.564
B-1	0.02471	0.02236	−9.510
B-2	0.03508	0.03185	−9.208
C-1	0.03521	0.03201	−9.088
Average value of ω			5.621

4.5. Ultimate Compressive Strain

In this paper, the strain value corresponding to the axial load of the constrained concrete square column falling to 85% of the peak value is taken as the ultimate compressive strain ε_{cu}. Through regression analysis of the test data, the calculation formula of the ultimate compressive strain ε_{cu} can be obtained as follows:

$$\frac{\varepsilon_{cu}}{\varepsilon_{co}} = 2 + 41.81\left(\frac{f_l}{f_{co}}\right)^{0.76} \tag{16}$$

Table 6 shows the test values of the ultimate compressive strain of the square RC column confined by multi-spiral composite stirrups tested in this paper and the theoretical values calculated according to the above equation. Through comparative analysis, it can be seen that the maximum ω is 17.743%, the minimum ω is 1.376%, and the average ω is 8.484%. For specimen B-1, although the maximum error ω between the test value and the calculated value was more than 10%, the calculation was safe because the calculated value was less than the test value, so the corresponding formula could be used.

Table 6. Tested and calculated values of ultimate compressive strain.

Specimen No.	Tested Value/MPa	Calculated Value/MPa	$\omega = \frac{(\text{Calculated value} - \text{Tested value})}{\text{Tested value}} /\%$
A-1	0.05473	0.05410	−1.151
A-2	0.03343	0.03389	1.376
A-3	0.02472	0.02522	2.023
A-4	0.02270	0.02468	8.722
A-5	0.03841	0.04216	9.763
B-1	0.04892	0.04024	−17.743
B-2	0.06894	0.06530	−5.280
C-1	0.06105	0.06570	7.617
Average value of ω			8.484

4.6. Stirrup Stress

The test in this paper shows that when the stress of the concrete column confined by the multi-spiral composite stirrups reaches the peak stress, the high-strength stirrup does not yield completely. If the yield strength of the stirrup is used to calculate the effective lateral constraint stress, the calculation result will be too large. Therefore, the actual stress of the transverse stirrup when the column reaches the peak stress should be used to calculate the effective lateral constraint stress. Cusson gave a simplified formula to calculate the actual stress f_{yh} of high-strength stirrup under peak stress, as follows:

$$\varepsilon_{hc} = 0.5\varepsilon_{cc}\left(1 - \frac{f_l}{f_{cc}}\right) \tag{17}$$

At the same time, the iterative process of calculating the actual strain ε_{hc} and stress f_{yh} of the high-strength stirrup is given:

1. Let $f_{yh} = f_y$ and substitute them into Equations (5) and (6) to find the effective lateral constraint stresses f_{l1} and f_{l2} imposed on concrete by inner spiral hoops and outer rectangular hoops;
2. Substitute f_{l1} and f_{l2} into Equation (7) to calculate the effective transverse constraint stress f_l of high-strength stirrup on core concrete;
3. Substitute f_l into Equations (14) and (15), respectively, to calculate the peak stress f_{cc} and peak strain ε_{cc};
4. By substituting f_l, f_{cc} and ε_{cc} into Equation (17), the strain ε_{hc} of the high-strength stirrup is obtained;
5. The stress f_{yh} of the high-strength stirrup is gained by the stress-strain relation of high-strength stirrup;
6. Only when the $f_{yh} < f_y$, the calculated value of f_{yh} is re-substituted into step1 to recalculate the relevant parameters;
7. Repeat steps 2 to 6 until convergence.

In summary, based on the analysis of the constraint mechanism, the effective lateral constraint coefficient f_l is introduced to establish the formulas for calculating the peak stress f_{cc}, peak strain ε_{cc}, and ultimate compressive strain ε_{cu} of concrete confined by multi-spiral composite stirrups. In addition to this, the iterative formulas for calculating the actual stress and strain of stirrups are given.

5. Constitutive Model

5.1. Establishment of Constitutive Model

At the beginning of loading, the passive constraint of stirrups and compressive strain of concrete are small, and the existing constitutive models of constrained concrete at home and abroad have little difference in the rising section. However, after the peak stress, due to the great difference in the effect of different constraints, different constraints of concrete constitutive models will differ greatly in the vicinity of the peak stress and its declining section. In this paper, the corresponding constitutive model is proposed for concrete confined by multi-spiral composite stirrups, as shown in Equations (18)–(20). The stress-strain relation curve is shown in Figure 10. The curve equation of the Bing constitutive model is adopted in the ascending section. The declining section is a straight line whose slope is determined by the peak point and the point where the stress drops to 0.85 f_{cc}. The stress is assumed to remain constant after decreasing to 0.4 f_{cc}.

$$0 < \varepsilon_c \leq \varepsilon_{co} : \quad f_c = E_c\varepsilon_c + \frac{(f_{co} - E_c\varepsilon_c)}{\varepsilon_{co}^2}\varepsilon_c^2 \tag{18}$$

$$\varepsilon_{co} < \varepsilon_c \leq \varepsilon_{cc} : f_c = f_{cc} - \frac{(\varepsilon_c - \varepsilon_{cc})^2}{(\varepsilon_{cc} - \varepsilon_{co})^2}(f_{cc} - f_{co}) \tag{19}$$

$$\varepsilon_{cc} < \varepsilon_c : \quad f_c = f_{cc}\left(1 - \frac{0.15(\varepsilon_c - \varepsilon_{cc})}{(\varepsilon_{85} - \varepsilon_{cc})}\right) \tag{20}$$

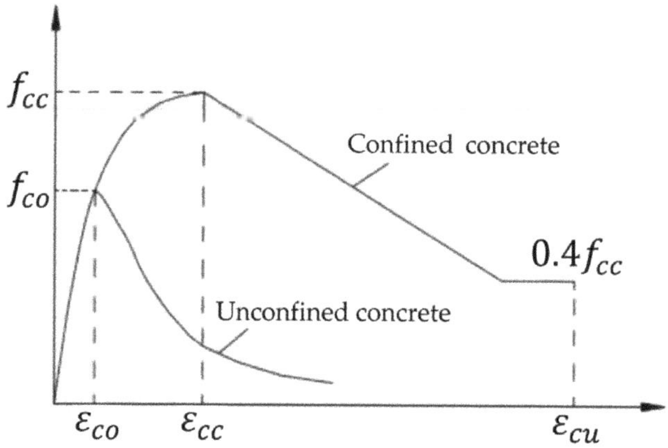

Figure 10. Stress–strain curve of confined concrete.

In the formula, ε_{cc} and f_{cc} are calculated by Equations (14) and (15), respectively. ε_{85} represents the strain when it drops to 85% of the peak stress, and can be calculated by Equation (16).

5.2. Comparative Analysis of Constitutive Models

For the specimens tested in this paper, the Mander model [15], the Saatcioglu model [18], and the constitutive model presented by this paper were respectively used to calculate the corresponding stress–strain curves and compared with the test curves in Figure 11. The differences between the rising section of the stress–strain curve calculated by each model and the test curve are small. However, the peak stress, strain, and declining section of the curves are quite different, and the calculation results of the Mander model and Saatcioglu model are higher. The main reason is that the yield strength of the high-strength stirrup is used in the calculation of the peak point, while the actual stress of the stirrup does not reach the yield. The calculation curve of the presented model fits well with the test curve, so the model can be used to predict the stress–strain relation curve of concrete confined by multi-spiral composite stirrups.

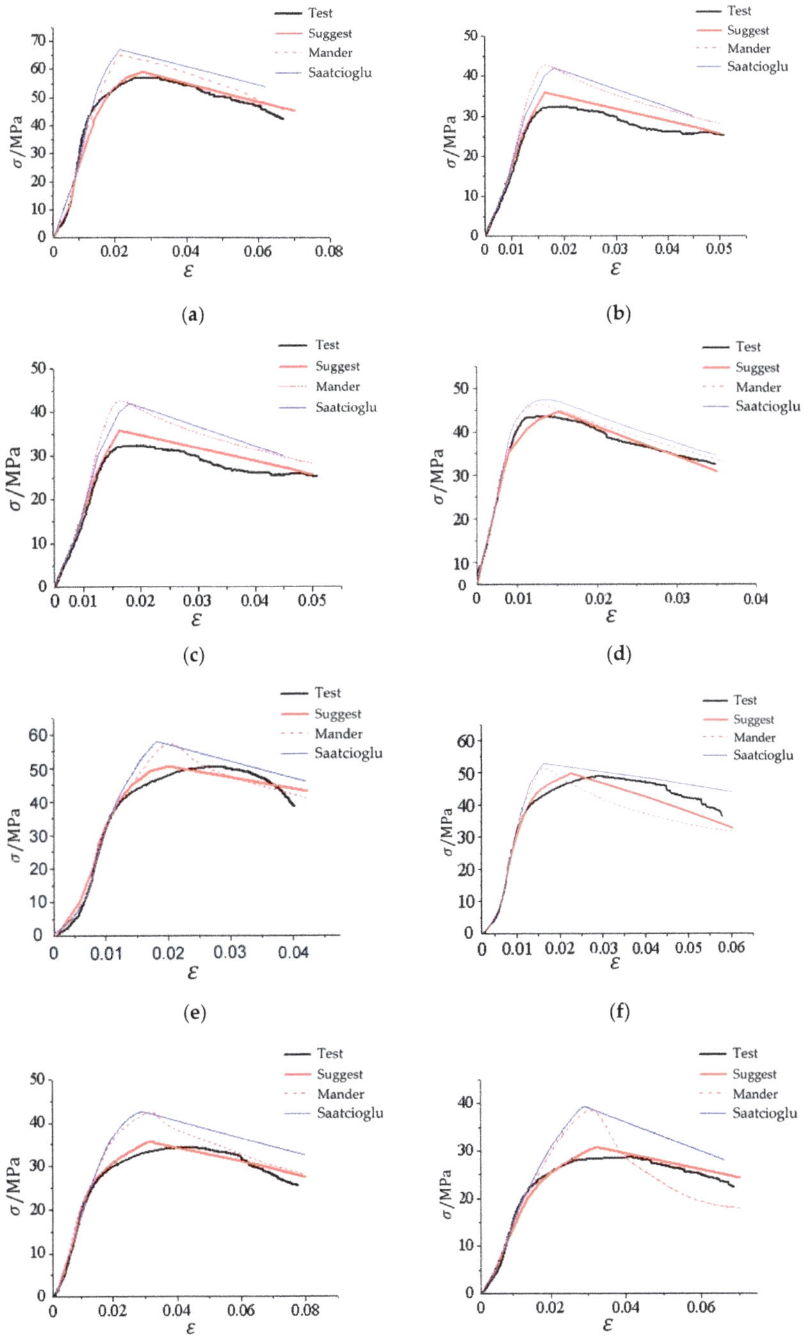

Figure 11. Comparison of stress-strain curves of specimens. (**a**) Specimen A-1; (**b**) Specimen A-2; (**c**) Specimen A-3; (**d**) Specimen A-4; (**e**) Specimen A-5; (**f**) Specimen B-1; (**g**) Specimen B-2; (**h**) Specimen C-1.

In summary, based on the theoretical analysis and experimental data, the stress-strain curves and corresponding piecewise equations of concrete confined by multi-spiral stirrups are established. The results can be used as a reference for the numerical simulation and engineering design of such components.

6. Conclusions

Based on the axial compression test of seven multi-spiral composite stirrups square RC columns and 1 traditional well-shaped composite stirrups square RC column, the following conclusions are obtained:

1. With the decrease of stirrup spacing or the increase of stirrup strength, the carrying capacity and ductility of the specimens are significantly improved, and the deformation ability is also better. The strength and ductility of the square RC column specimen confined by five-spiral composite stirrups are obviously better than those of the corresponding specimens with four-spiral composite stirrups, and slightly better than those of the corresponding specimen with traditional well-shaped composite stirrups.
2. The restraint mechanism of multi-spiral composite stirrups was analyzed. According to the different degrees of constraint, the cross-section of the square RC column is divided into three areas: the highly constrained region (the area surrounded by the circular spiral stirrups), the partially constrained region (the area outside the circular spiral stirrups and the area within the rectangular hoop) and the unconstrained region (the protective layer).
3. Combined with theoretical analysis and experimental data regression, the formulas of effective confinement stress, peak stress, peak strain, and ultimate strain of concrete confined by multi-spiral composite stirrups are proposed and compared with the experimental results. The results show that the error between the calculated and experimental values is less than 5%, which means the calculation formula presented in this paper has high accuracy.
4. A constitutive model of concrete confined by multi-spiral composite stirrups is proposed and compared with several typical constitutive models of constrained concrete. The results indicate that the presented constitutive model fits well with the experimental curve and can predict the axial compression performance of this type of constrained concrete well.

7. Discussion

According to the test results of this paper, the strength and ductility of concrete columns can be significantly improved by considering the form of cross-section reinforcement restrained by four or five spiral stirrups in practical engineering. In particular, the formulas of effective confining stress, peak stress, peak strain, and ultimate strain proposed in this paper can be used in the design of multi-spiral composite stirrup-constrained concrete columns. Due to the influence of the number of specimens and production errors, it is not possible to fully analyze the influence of each parameter on the constraint effect of specimens. Therefore, the finite element method will be used to simulate the mechanical properties of such components, and the rationality of the constitutive model proposed in this paper will be compared and analyzed.

8. Recommendations

In this paper, experimental research and theoretical analysis are carried out on the axial compression performance and constitutive model of confined concrete short columns confined by multi-spiral composite stirrups, but the analysis of the influencing factors is not comprehensive, such as the size effect, concrete strength, stirrup form optimization, and so on. In addition, further experimental studies are needed on the optimal design of axial compression, the partial compression performance, and the seismic performance of the concrete columns confined by multi-spiral composite stirrups, which can provide a theoretical basis for the practical engineering application of this type of column.

Author Contributions: Conceptualization, K.Y.; methodology, K.Y. and S.S.; software, J.Z.; validation, G.M. and J.Z.; formal analysis, T.Y.; investigation, T.Y. and G.M.; resources, K.Y.; data curation, G.M.; writing—original draft preparation, T.Y.; writing—review and editing, K.Y. and S.S.; visualization, T.Y.; supervision, K.Y.; project administration, T.Y. and S.S.; funding acquisition, K.Y. All authors have read and agreed to the published version of the manuscript.

Funding: This research was funded by National Natural Science Foundation of China, grant number 51708035; by Fundamental Research Funds of the Central Universities, grant number 300102289202; and by Natural Science Basic Research Plan in Shaanxi Province of China, grant number 2021JM-177.

Data Availability Statement: In all datasets presented in this study are included in the article.

Conflicts of Interest: The authors declare no conflict of interest.

References

1. Ye, L.P.; Lu, X.Z.; Zhao, S.C. Analysis on seismic damage of buildings in the Wenchuan earthquake. *J. Build. Struct.* **2008**, *29*, 1–9. [CrossRef]
2. Li, H.N.; Xiao, S.Y.; Huo, L.S. Damage investigation and analysis of engineering structures in the Wenchuan earthquake. *J. Build. Struct.* **2008**, *29*, 10–19. [CrossRef]
3. Chen, Z.; Chen, J.; Jiang, X.; Mo, L. Experimental research and finite element analysis on seismic behavior of square reinforced concrete columns with four interlocking spirals. *Structures* **2022**, *39*, 1–16. [CrossRef]
4. Yang, L.G.; Wang, Y.Y.; Liew, J.Y.R. Compression-Bending Strength Model for Corrugated Steel Tube Confined Reinforced Concrete Section. *J. Struct. Eng.* **2021**, *147*, 04021187. [CrossRef]
5. Qaidi, S.; Al-Kamaki, Y.S.S.; Al-Mahaidi, R.; Mohammed, A.S.; Ahmed, H.U.; Zaid, O.; Althoey, F.; Ahmed, J.; Isleem, H.F.; Bennetts, I. Investigation of the effectiveness of CFRP strengthening of concrete made with recycled waste PET fine plastic aggregate. *PLoS ONE* **2022**, *17*, e0269664. [CrossRef]
6. Ahmed, M.; Sheikh, M.N.; Hadi, M.N.S.; Liang, Q.Q. Nonlinear analysis of square spiral-confined reinforced concrete-filled steel tubular short columns incorporating novel confinement model and interaction local buckling. *Eng. Struct.* **2022**, *274*, 115168. [CrossRef]
7. Zhao, D.; Zhang, J.; Lu, L.; Liang, H.; Ma, Z. The Strength in Axial Compression of Aluminum Alloy Tube Confined Concrete Columns with a Circular Hollow Section: Experimental Results. *Buildings* **2022**, *12*, 699. [CrossRef]
8. Wei, Y.; Xu, Y.; Wang, G.; Cheng, X.; Li, G. Influence of the Cross-Sectional Shape and Corner Radius on the Compressive Behaviour of Concrete Columns Confined by FRP and Stirrups. *Polymers* **2022**, *14*, 341. [CrossRef]
9. Hao, M.J.; Zheng, W.Z.; Chang, W. Compression behavior of reinforced concrete columns jacketed with multi-spiral transverse reinforcement. *Struct. Concr.* **2022**, *26*, 2942–2967. [CrossRef]
10. Chen, Z.P.; Zhou, J.; Jing, C.G.; Tan, Q.H. Mechanical behavior of spiral stirrup reinforced concrete filled square steel tubular columns under compression. *Eng. Struct.* **2021**, *226*, 19. [CrossRef]
11. Eid, R.; Cohen, A.; Guma, R.; Ifrach, E.; Levi, N.; Zvi, A. High-Strength Concrete Circular Columns with TRC-TSR Dual Internal Confinement. *Buildings* **2019**, *9*, 218. [CrossRef]
12. Wang, Q.X.; Zhao, G.F.; Lin, L.Y. Experimental study on ductility of high strength concrete columns. *J. Build. Struct.* **1995**, *16*, 22–31. [CrossRef]
13. Sheikh, S.A.; Uzumeri, S.M. Analytical model for concrete confinement in tied columns. *J. Struct. Eng.* **1982**, *108*, 2703–2722. [CrossRef]
14. Mander, J.B. Seismic Design of Bridge Piers. Ph.D. Thesis, University of Canterbury, Christchurch, New Zealand, 1983. [CrossRef]
15. Mander, J.B.; Priestley, J.N.; Park, R. Theoretical stress-strain model for confined concrete. *J. Struct. Eng.* **1988**, *114*, 1804–1826. [CrossRef]
16. Cusson, D.; Paultre, P. High-strength concrete columns confined by rectangular ties. *J. Struct. Eng.* **1994**, *120*, 783–804. [CrossRef]
17. Cusson, D.; Paultre, P. Stress-strain model for confined high-strength concrete. *J. Struct. Eng.* **1995**, *121*, 468–477. [CrossRef]
18. Saatcioglu, M.; Razvi, S.R. Strength and ductility of confined concrete. *J. Struct. Eng.* **1992**, *118*, 1590–1607. [CrossRef]
19. Bing, L.; Park, R.; Tanaka, H. Stress-strain behavior of high-strength concrete confined by ultra-high- and normal-strength transverse reinforcements. *J. Struct. Eng.* **2001**, *98*, 395–406.
20. Rong, C.; Shi, Q. Analysis constitutive models for actively and passively confined concrete. *Compos. Struct.* **2021**, *256*, 113009. [CrossRef]
21. Li, Y.Z.; Cao, S.Y.; Jing, D.H. Axial compressive behaviour of RC columns with high-strength MTS transverse reinforcement. *Mag. Concr. Res.* **2017**, *69*, 436–452. [CrossRef]
22. Li, Y.Z.; Cao, S.Y.; Jing, D.H. Analytical compressive stress-strain model for concrete confined with high-strength multiple-tied-spiral transverse reinforcement. *Struct. Des. Tall Spec. Build.* **2018**, *27*, e1416. [CrossRef]
23. Shi, Q.X.; Yang, K.; Liu, W.Y.; Zhang, X.H.; Jiang, W.S. Experimental study on mechanical behavior of high strength concrete confined by high-strength stirrups under concentric loading. *Eng. Mech.* **2012**, *29*, 141–149.

24. Yang, K.; Shi, Q.X.; Zhao, J.H.; Jiang, W.S.; Meng, H. Study on the constitutive model of high-strength concrete confined by high-strength stirrups. *China Civ. Eng. J.* **2013**, *46*, 34–41. [CrossRef]
25. Ouyang, X.; Wu, Z.M.; Shan, B.; Chen, Q.; Shi, C.J. A critical review on compressive behavior and empirical constitutive models of concrete. *Constr. Build. Mater.* **2022**, *323*, 126572. [CrossRef]
26. Zheng, W.Z.; Zhang, J.; Wang, G.; Wang, Y. Experimental Study on Axial Compression Behavior of High- strength Concrete Columns Confined by Spiral Stirrups. *China J. Highw. Transp.* **2022**, *35*, 22–35. [CrossRef]
27. Jin, L.; Li, P.; Du, X.L. Compressive stress-strain model for stirrup-confined concrete columns considering the effect of structural size. *J. Civ. Environ. Eng.* **2020**, *42*, 81–89. [CrossRef]
28. Li, M.H.; Zhou, W.; Liu, M.J.; Liu, G.Y.; Zhang, S. Experimental investigation of reinforced concrete columns with composite spiral stirrups under eccentric loading. *J. Harbin Inst. Technol.* **2019**, *51*, 113–120. [CrossRef]
29. Yin, Y.L. Researches and developments of alternative confinements for rectangular concrete columns (I). *China Civ. Eng. J.* **2004**, *37*, 1–10. [CrossRef]
30. Yin, Y.L. Researches and developments of alternative confinements for rectangular concrete columns (II). *China Civ. Eng. J.* **2004**, *37*, 1–12. [CrossRef]
31. Yin, Y.L.; Weng, Z.Q.; Wang, R.Z.; Liang, J.Y. Axial compressive behavior of precast SRC columns with multi-spirals. *Eng. Sci.* **2006**, *8*, 16–30.
32. Yin, Y.L.; Wu, T.L.; Liu, T.C.; Sheikh, S.A.; Wang, R. Interlocking spiral confinement for rectangular columns. *Concr. Int.* **2011**, *33*, 38–45.
33. Yin, Y.L.; Wang, J.C.; Wang, P.H. Development of multi-spiral confinements in rectangular columns for construction automation. *J. Chin. Inst. Eng.* **2012**, *35*, 309–320. [CrossRef]
34. Hung, H.H.; Wang, P.H.; Yin, Y.L.; Wang, J.C.; Chang, K.C. Large-scale cyclic loading test on a multi-spiral stirrup bridge pier constructed by automated method. In Proceedings of the 15th World Conference on Earthquake Engineering, Lisbon, Portugal, 24–28 September 2012.
35. Chen, Y.; Feng, J.; Yin, S. Compressive behavior of reinforced concrete columns confined by multi-spiral hoops. *Comput. Concr.* **2012**, *9*, 359–373. [CrossRef]
36. Weng, C.C.; Yin, Y.L.; Wang, J.C.; Liang, C.Y. Seismic cyclic loading test of SRC columns confined with 5-spirals. *Sci. China Series E-Tech. Sci.* **2008**, *51*, 529–555. [CrossRef]
37. Shih, T.H.; Chen, C.C.; Weng, C.C.; Yin, Y.L.; Wang, J.C. Axial strength and ductility of square composite columns with two interlocking spirals. *J. Constr. Steel Res.* **2013**, *90*, 184–192. [CrossRef]
38. Wu, T.L.; Ou, Y.C.; Yin, Y.L.; Wang, J.C.; Wang, P.H.; Ngo, S.H. Behavior of oblong and rectangular bridge columns with conventional tie and multi-spiral transverse reinforcement under combined axial and flexural loads. *J. Chin. Inst. Eng.* **2013**, *36*, 980–993. [CrossRef]
39. Tang, Q.; Li, Y.; Lu, X.Z.; Yan, W.M. Study on axial compression capacity of multi-spiral hoops confined concrete columns. *Eng. Mech.* **2018**, *35*, 166–171. [CrossRef]
40. Wang, P.-H.; Chang, K.-C.; Yin, S.Y.-L.; Wang, J.-C.; Ou, Y.-C. Simplified Finite-Element Analysis Method for Axial Compression Behavior of Rectangular Concrete Columns with Interlocking Multispiral Reinforcements. *J. Struct. Eng.* **2020**, *146*, 04019176. [CrossRef]
41. Yang, K.; Guo, S.H.; Wang, Y.K.; Liu, R.; Hu, Y.Y. Experimental study on axial compression performance of square concrete column confined with new multi-spiral composite stirrup. *J. Archit. Civ. Eng.* **2021**.
42. Kim, J.K.; Park, C.K. The behaviour of concrete columns with interlocking spirals. *Eng. Struct.* **1999**, *21*, 945–953. [CrossRef]
43. Willam, K.J. Constitutive model for the triaxial behaviour of concrete. *Proc. Intl. Assoc. Bridge Structl. Engrs.* **1975**, *19*, 1–30.

buildings

MDPI

Article

Effects of Vertical Ground Motion on Pedestrian-Induced Vibrations of Footbridges: Numerical Analysis and Machine Learning-Based Prediction

Xinxin Wei [2], Bo Fu [1,*], Wenyan Wu [1] and Xinrui Liu [1]

[1] School of Civil Engineering, Chang'an University, Xi'an 710061, China
[2] Department of Civil and Environmental Engineering, Ruhr-Universität Bochum, Universitätsstraße 150, D-44780 Bochum, Germany
* Correspondence: 90_bofu@chd.edu.cn

Abstract: Current codes and guidelines for the dynamic design of footbridges often only specify the pedestrian-induced excitations. However, earthquakes may occur during the passing stage of pedestrians in earthquake-prone regions. In addition, modern footbridges tend to be slender and are sensitive to vertical ground motions. Therefore, we investigate the effects of vertical ground motion on pedestrian-induced vibrations of footbridges. A total of 138 footbridges with different materials, dimensions, and structural types are considered as the target structures. The classical social force model combined with the pedestrian-induced load is used to simulate crowd loads for the scenarios with six typical pedestrian densities. Furthermore, 59 vertical ground motions with four seismic intensities are taken as the seismic inputs. An amplification factor is introduced to quantify the amplification effects of vertical ground motion on human-induced vibrations of footbridges. Four machine learning (ML) algorithms are used to predict the amplification factor. The feature importance indicates that the scaled peak ground acceleration, the pedestrian density, and the bridge span are the three most important parameters influencing the amplification factor. Finally, the vibration serviceability of the footbridge subjected to both crowd load and vertical ground motion is assessed.

Keywords: footbridge; vibration; serviceability; crowd load; earthquake; machine learning

Citation: Wei, X.; Fu, B.; Wu, W.; Liu, X. Effects of Vertical Ground Motion on Pedestrian-Induced Vibrations of Footbridges: Numerical Analysis and Machine Learning-Based Prediction. *Buildings* **2022**, *12*, 2138. https://doi.org/10.3390/buildings12122138

Academic Editor: Humberto Varum

Received: 26 October 2022
Accepted: 2 December 2022
Published: 5 December 2022

Copyright: © 2022 by the authors. Licensee MDPI, Basel, Switzerland. This article is an open access article distributed under the terms and conditions of the Creative Commons Attribution (CC BY) license (https://creativecommons.org/licenses/by/4.0/).

1. Introduction

Pedestrians are the main users of footbridges. Therefore, pedestrian-induced footbridge vibration has drawn much attention from researchers in the past two decades, especially since the London Millennium Bridge incident that was induced by crowds [1]. Till now, researchers have made great contributions on human-induced loads [2–5], vibration serviceability evaluation [6–12], pedestrian–structure interaction [13–20], and pedestrian-induced vibration control [21–31]. The pedestrian-induced footbridge vibration falls into the serviceability category. Excessive vibration may cause pedestrian uncomfortableness and even endanger the bridge's safety. Correspondingly, several specifications have been issued, e.g., Sétra (2006) [32], ISO (2007) [33], and HiVoSS (2008) [34], regarding the serviceability design of footbridges. It is notable that current specifications for the dynamic design of footbridges only consider the relevant pedestrian-induced excitations.

When located in earthquake-prone regions, it is also possible that footbridges are subjected to not only pedestrian-induced loads but also ground motions. Simply, earthquakes can occur during the crowd passing process. In fact, the earthquake action is non-negligible for footbridges in earthquake-prone regions. In addition, modern footbridges tend to be slender, which makes these footbridges also sensitive to vertical earthquake loads. After the London Millennium Bridge incident, great efforts have been made to avoid lateral vibrations of footbridges. For instance, the UK National Annex to Eurocode 1 (2008) [35] specifies lock-in stability boundaries to avoid unstable lateral responses under crowd loads. However, vertical crowd-induced vibration is also an important issue and cannot be ignored. Therefore, this study mainly focuses on footbridge vibrations in the vertical direction, which is subjected to much larger loads than other directions. There are also studies investigating the influence of vertical ground motions on different types of structures, e.g., long-span cable-stayed bridges [36], segmental post-tensioned bridges [37], vehicle–bridge systems [38,39], long-span steel structures [40,41], long-span latticed arch-type structures [42], masonry structures [43], underground subway stations [44], undersea shield tunnels [45], and multi-directional base isolation systems [46]. The aforementioned studies indicate that vertical ground motion has a significant impact on structural performance. To the best knowledge of the authors, however, there is no existing research considering the combined effects caused by human-induced loads and vertical ground motions. It should be noted that the occurrence of an earthquake may induce crowd panic and increase the vibration. Therefore, it is quite meaningful to investigate the influence of vertical ground motion on crowd-induced vibration. Table 1 summarizes some representative references related to the topic of the current study.

To fill the gap, this study conducts a series of time-history analyses for 138 footbridges with various dimensions, materials, and structural types subjected to the combined actions of crowd-induced loads and vertical ground motions. The crowd-induced loads consider the crowd scenarios with six typical pedestrian densities ranging from 0.1 to 1.5 pedestrians/m^2. In total, there are 59 vertical ground motions with four different intensities for the seismic inputs. Thus, the total amount of calculation cases is 195,408 (=138 $\times$ 6 $\times$ 59 $\times$ 4). Furthermore, the amplification factor, which is defined as the ratio of the maximum acceleration induced by the combined actions to the maximum acceleration induced by crowds only, is calculated for each case and used as a guide for the serviceability design of footbridges subjected to both crowds and vertical ground motions. As there exists a strong nonlinear relationship between the inputs and the output (amplification factor), machine-learning (ML) techniques [47–52], which are quite suitable for solving nonlinear regression problems, are used to predict the amplification factor. Two individual-type ML algorithms, i.e., decision tree (DT) [53] and artificial neural network (ANN) [54], and two ensemble ML algorithms [55], i.e., random forest (RF) [56] and gradient boosting regression tree (GBRT) [51,52], are adopted to construct the predictive models. Ten parameters, including four structure-related parameters, pedestrians' density, and five earthquake-related parameters, are taken as the input variables, while the amplification factor is taken as the output variable. By using ML techniques, it relates the multiple factors (the structure, pedestrian, and earthquake parameters) to the amplification factor. Therefore, the total peak response of the structure due to combined loads can be obtained by multiplying the amplification factor with the pedestrian-induced vibration amplitudes. For a specific structure, the pedestrian-induced vibration levels can be conveniently measured by real-world measurements or predicted by numerical models. Finally, the vibration serviceability of the footbridge subjected to both crowd load and vertical ground motion is assessed.

The remaining parts of the paper are organized as follows. Section 2 introduces the simulation method and results of the footbridges under crowd loads. In Section 3, a total of 59 vertical ground motions with four intensities are selected. Section 4 investigates the influence of vertical ground motion on crowd-induced vibrations of footbridges. The amplification factor is then defined in Section 5. Four ML algorithms are adopted to predict the amplification factor based on a database containing 171,572 datasets. The vibration serviceability is evaluated in Section 6. Finally, several important conclusions are presented in Section 7.

Table 1. Summary of related previous research.

Reference Number	Authors (Year)	What Was Performed	Main Findings
1. Pedestrian-induced loads and vibrations			
1.1 Human-induced loads			
[2]	Ingólfsson and Georgakis (2011)	A new stochastic load model was proposed to simulate the frequency and amplitude-dependent pedestrian-induced lateral forces.	The prediction of the critical number of pedestrians is consistent with the incident on the London Millennium Bridge.
[3]	Racic and Brownjohn (2012)	A mathematical model was developed to create synthetic narrow-band lateral forces induced by pedestrians.	The model can be used to assess the dynamic performance in everyday design practice.
[4]	Bruno and Corbetta (2017)	A new multi-scale model was developed to simulate uncertainties in pedestrian traffic.	The variability of traffic random variables is larger than structural properties ones.
[5]	Casciati et al. (2017)	A time-variant stochastic field model was proposed to model the walking forces induced by a small group of pedestrians.	The developed model can consider different idealizations of human-induced excitation and can be used in a serviceability limit state design.
1.2 Vibration serviceability evaluation			
[6]	Bruno and Venuti (2010)	A simplified serviceability assessment method for footbridges under lateral crowd loading was proposed.	The proposed method can reflect the actual walking behaviour of pedestrians by using the speed–density and frequency–speed relationship.
[7]	Živanović (2012)	A comprehensive experimental dataset of a box-girder footbridge that is lively in the vertical direction was provided.	Walking frequency, step length, and pedestrian speed in normal traffic obey a normal distribution, while pedestrian arrival time follows a Poisson distribution.
[8]	Setareh (2016)	Three important issues regarding vibration serviceability were investigated on a slender steel footbridge.	When the crowd speed closes to the first-mode resonance frequency of the footbridge, the enhancement factor for the group effect becomes closer to the group size.
[9]	Bedon (2019)	A preliminary dynamic characterization of an existing suspension glass footbridge was presented using on-site vibration tests and refined Finite Element methods.	A combination of multiple aspects has a significant influence on the structural performances and modal dynamic estimations.
[10]	Feng et al. (2019)	The acceleration of 21 pedestrian bridges in Beijing were recorded under different service conditions.	The fundamental frequency and acceleration are the two most important controlling factors in vibration serviceability design.
[11]	Fu and Wei (2021)	A two-stage ML-based analysis method for the human-induced vibration of a concrete footbridge was proposed.	The elastic modulus of concrete can markedly affect the human-induced vibration of concrete footbridges
[12]	Gong et al. (2022)	The vibration serviceability of two recent long-span footbridges in China was comprehensively assessed with six current specifications.	The total structural responses considering the contributions of closely spaced multiple modes are significantly larger than those using the specifications based on the single dominating mode only.

Table 1. *Cont.*

Reference Number	Authors (Year)	What Was Performed	Main Findings
1.3 Pedestrian–structure interaction			
[13]	Morbiato et al. (2011)	The pedestrian structure interaction was considered by developing a non-linear double pendulum model.	When synchronization occurs, pedestrian motion becomes in-phase quadrature with a quarter-of-period before the bridge motion.
[14]	Carroll et al. (2012)	A discrete element theory (DET) based method was proposed to simulate the crowd–bridge interaction.	The proposed method can predict emergent crowd behaviour better than earlier hydrodynamic models.
[15]	Jiménez-Alonso et al. (2016)	A biomechanical crowd–structure interaction model was developed.	The proposed model can accurately reflect the change in the dynamic properties of the structure induced by pedestrian flows.
[16]	Shahabpoor et al. (2017)	A vibration serviceability assessment method was proposed according to the actual vibration level experienced by each pedestrian.	The method can accurately estimate the structural responses compared to current design guidelines.
[17]	Toso et al. (2017)	A fully synchronized force model for walking pedestrians was proposed and compared with a simple force-only model and experimental vibration data was recorded in a real composite footbridge.	The proposed model can improve the simple force-only model and this may obtain a more realistic simulation of the dynamic structural behaviour.
[18]	Mulas et al. (2018)	The footbridge-walking pedestrian coupled equation of motion in the vertical direction was analytically derived using Lagrange's equation and a discrete modelling framework.	The numerical simulations exhibit significant variability in the response due to relatively small variations in the loading scenarios.
[19]	Setareh and Gan (2018)	The human–structure interaction on the dynamic behaviour of a slender two-span steel footbridge was studied.	The contribution of the wood decking to the structural stiffness is limited while their mass can be included.
[20]	Ahmadi et al. (2019)	The influence of human–structure interaction on the structural response of a lively lightweight GFRP footbridge was studied.	The bridge vibration has a significant impact on walking force, and to a lesser extent on the dynamics of the human–structure system.
1.4 Pedestrian-induced vibration control			
[22]	Li et al. (2010)	The multiple tuned mass damper (MTMD) designed by a random optimization procedure was adopted to reduce the crowd-induced vibration of a footbridge.	The proposed MTMD is more effective than the traditional MTMD in terms of reduction efficiency and reducing the off-tuning effect of MTMD.
[24]	Venuti and Bruno (2013)	A new strategy of using walkway shaping was developed to mitigate the human-induced lateral vibrations on footbridges.	The new strategy is less expensive and more durable than traditional structural countermeasures based on increasing stiffness and damping, respectively.
[27]	Venuti and Anna (2018)	A crowd flow control strategy by installing obstacles located along the footbridge span was proposed to control the human-induced vertical vibrations of footbridges.	The maximum reduction of 31% can be achieved if the obstacles are placed to generate local bottlenecks along the footbridge.
[31]	Gong et al. (2021)	The effectiveness of installing TMD on mitigating the pedestrian-induced vibration on a typical glass suspension footbridge in China was studied.	The commonly used TMD can effectively reduce the vibration levels of the footbridge.

Table 1. *Cont.*

Reference Number	Authors (Year)	What Was Performed	Main Findings
2. Seismic performance of long-span structures subjected to vertical earthquakes			
[36]	Shrestha (2015)	The effect of the near-fault vertical ground motions on the seismic response of a long-span cable-stayed bridge was numerically studied.	Vertical displacement of the bridge deck at mid-span is sensitive to vertical ground motion.
[40]	Xiang et al. (2017)	The seismic response of steel structures subjected to vertical seismic excitation was studied by using an idealized model and inelastic displacement ratio.	The inelastic displacement ratio-based method can estimate the seismic responses of steel structures subjected to severe vertical ground motions.
[41]	Fayaz and Zareian (2019)	The influences of the vertical component of near-fault ground motions on special moment-resisting steel frames and special concentrically braced frame-braced steel frames were studied.	The current seismic load combinations in ASCE 7 are inadequate to consider the influences of the vertical near-fault ground motions.
[42]	Qu et al. (2019)	An improved multidimensional modal pushover approach with two-stage analyses was developed for seismic assessment of latticed arches subjected to lateral and vertical ground motions.	The developed method has good agreement with those of time-history analysis and is superior to the existing methods in terms of accuracy.

The main contributions of the study can be summarized as three aspects. Firstly, the structural responses of footbridges subjected to the combination of crowd loads and vertical earthquakes are analysed. Secondly, a huge amount of time-history analysis is conducted to consider the influences of structure-related, crowd-related, and earthquake-related parameters on the structural responses. Thirdly, four ML models are used to predict the amplification factor.

The driving ideas traced from the literature review and organization of the paper are depicted in Figure 1.

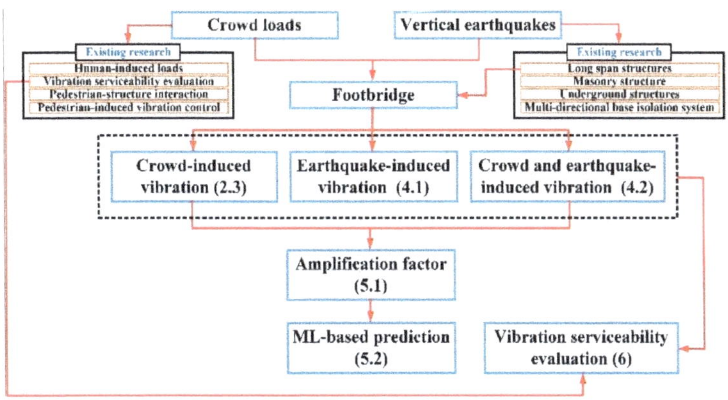

Figure 1. Driving ideas traced from the literature review and organization of the paper.

2. Simulation of Footbridge Vibration under Crowd Loads

2.1. Analytical Model of Footbridge

Wei et al. (2019) [57] comprehensively reviewed 138 footbridges, which were mostly built after 1991 and reported in the literature, e.g., 73 footbridges were also evaluated by [58]. In this study, an analytical footbridge model is constructed based on the typical characteristics of real-world footbridges as summarized by Wei et al. (2019) [57] and the popular guidelines, e.g., Sétra and HiVoSS. Typical footbridge characteristics are summarized as follows.

- Material: Different materials are applied in footbridge construction. As shown in Figure 2a, conventional construction materials include steel, concrete, steel–concrete composites, timber, and aluminium. Almost half of the footbridges are made of steel (67/138, i.e., 48.6%). Furthermore, the proportion of concrete footbridges is over a quarter (38/138, i.e., 27.5%). New constructional materials such as FRP (14/138, i.e., 10.1%) are also increasingly applied. Based on available data, the conventional footbridges (1200 kg/m^2) can be approximately 8.6 times heavier than FRP footbridges (140 kg/m^2), in terms of the physical mass per square meter.
- Dimension: Very few bridge decks have variant widths along the spans, with almost all bridge decks being typical rectangles. The rectangular decks vary in the main spans and widths of bridge decks (Figure 3a). For those bridges with variable widths along the spans, the corresponding mean widths are considered in Figure 3a. As presented in Figure 3a, the spans and widths are within the ranges of [4.8, 230] m and [0.78, 13.4] m, respectively. In particular, most spans and widths are correspondingly smaller than 50 m and 5 m, respectively. Furthermore, no obvious trend is found between the width–span relationships.
- Structural type: To satisfy engineering and realistic needs, different types are selected in bridge construction (Figure 2b). Most footbridges are typical bridge types, e.g., girder (25.4%), truss/truss-girder (20.3%), arch (10.9%), cable-stayed (9.4%), suspension (5.1%), and stress-ribbon (2.9%). The remaining bridge types are unknown due to unavailable information from the literature [57,58]. The boundary conditions of the reported footbridges are basically simply supported. Simply supported is not only the simplest boundary condition, but also the basic element for other more complex boundary conditions [59]. This is also in accordance with the common practice that, in the calculations of human-induced vibrations for footbridges, it often applies a simply supported beam model with sinusoidal mode shapes as the analytical model [60–63]; when experimental data with good quality are available, a good match between the calculated and measured responses can often be obtained, e.g., with the help of model updating techniques [64].
- Fundamental natural frequency: Figure 3b shows the fundamental natural frequencies of the vertical modes for the bridges. Most of the frequencies are below 5 Hz and may fall into the frequency range of human-induced excitations [32,34]. Furthermore, the fundamental natural frequency $f_{1,v}$ (unit: Hz) basically follows a fitted numerical relationship with the main span L (unit: m) as [57]:

$$f_{1,v} = \frac{100.5}{L} \tag{1}$$

- Damping ratio: The damping ratios fall within the range of [0.14%, 7.9%]. Based on the estimated non-exceedance probability, less than 50% of the footbridges have damping ratios higher than 1.0%. Most (92%) damping ratios are lower than 3%.

To model real-world footbridges as realistically and simply as possible, an analytical model is proposed. The model considers different construction materials, bridge span lengths and widths, natural frequencies, damping ratios, modal masses, etc. Different boundary conditions and cross-sectional properties can also be considered when necessary.

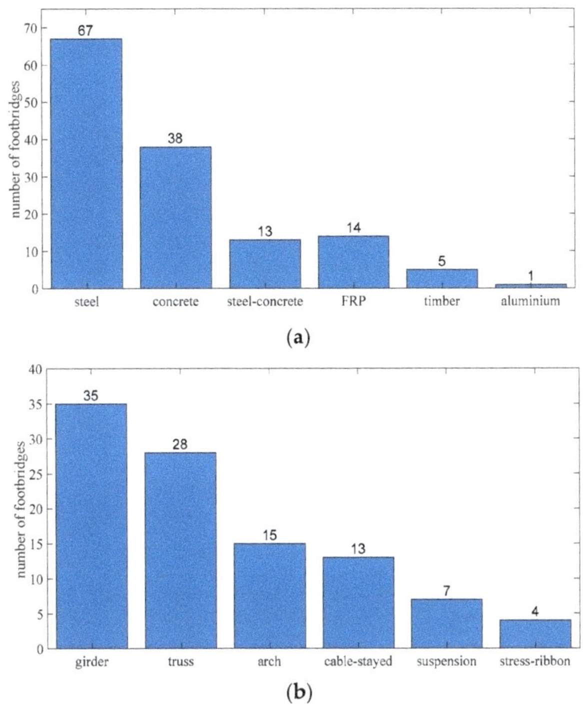

Figure 2. Number of footbridges. (**a**) Construction materials; (**b**) bridge types.

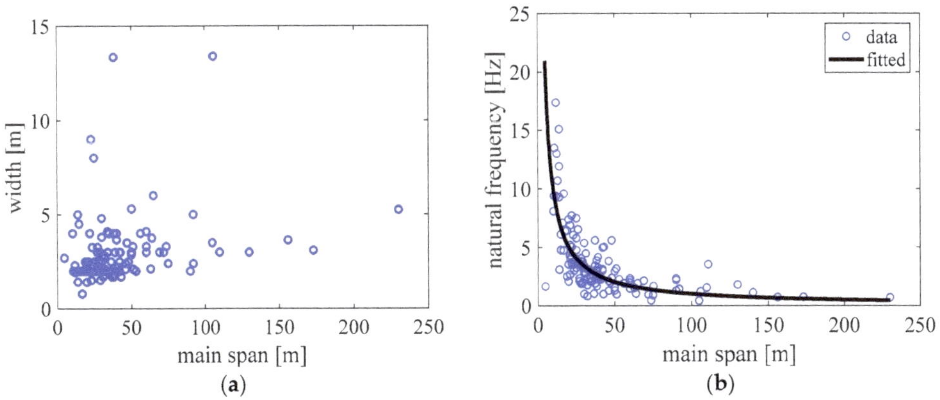

Figure 3. Plots of (**a**) main widths and (**b**) fundamental natural frequencies of the vertical modes over the main spans of constructed footbridges.

Therefore, in the current investigations, the basic assumptions of the proposed analytical model are:

1. Bridge type, boundary conditions, and mode shapes: The simply supported beam-like footbridge with sinusoidal mode shapes is considered as the basic analytical model [65]. The applied analytical model of the footbridge is idealized as a simply supported beam in the vertical (Z) direction. The bridge deck has a rectangular walking surface in the XY plane, with X the longitudinal direction and Y the lateral direction.

2. Bridge deck span lengths and widths: The considered bridge decks are typical rectangles with different widths and lengths as summarized by the real-world footbridges in Figure 3.
3. Natural frequencies, damping ratios, and modal masses: The fundamental natural frequencies, as shown in Figure 4, are calculated based on the span length, according to Equation (1). In Figure 4, the solid line is the mean value of the frequencies, while the two dashed lines represent mean ± St.D. (standard deviation). The damping ratios are random values within the range of [0.14%, 7.9%]. However, in this study, damping ratios are assumed to be identical if the bridge is made of the same material. Typical (average) damping ratios for different materials are 0.4% (steel), 1.3% (concrete), 0.6% (steel–concrete), 1.5% (timber), 1.1% (aluminium), and 2.5% (FRP), according to the real-world footbridges [57] and HiVoSS guidelines. Therefore, the aforementioned six damping ratios are used in the following analytical analysis. The modal masses of the fundamental mode can be set as half of the total masses of the footbridges, which are mainly governed by the construction material density, cross-sectional properties, and bridge length and width. In accordance with the ratio of the physical mass per square meter [57] for conventional and FRP footbridges, the modal masses of conventional footbridges are considered as 8.6 times higher than FRP footbridges. Specially, the modal mass for the fundamental vertical mode is considered as:

$$M_{1,v} = \frac{m \cdot L \cdot W}{2} \tag{2}$$

with m the physical mass per square meter, i.e., $m = 1200 \text{ kg/m}^2$ for conventional footbridges and 140 kg/m^2 for FRP footbridges [57], while W is the bridge width.

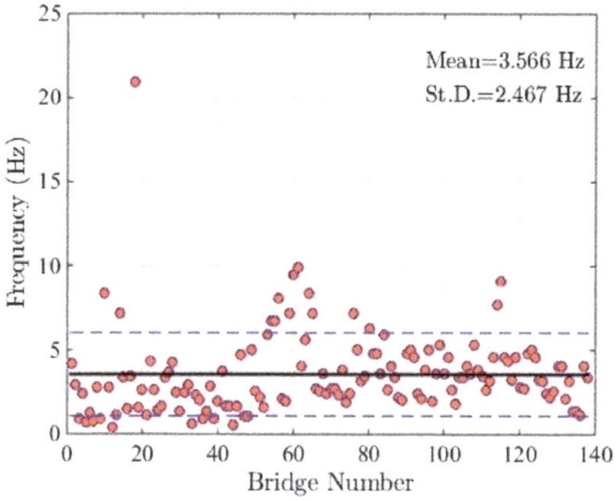

Figure 4. Natural frequencies of the first bending mode for the analytical structures.

According to modal analysis of the simply supported beam [65], the mode shape, natural frequency, and modal mass of the nth vertical mode are expressed as Equations (3)–(5), respectively. It should be noted that the current study assumes that the structure has constant natural frequencies and damping ratios. In the future study, it is more realistic to use variable natural frequencies and damping ratios induced by long-term effects such as prestressing losses [66–68].

$$\varnothing_{n,v}(x) = \sin\left(\frac{n\pi x}{L}\right) \tag{3}$$

$$f_{n,v} = n^2 \cdot f_{1,v} = n^2 \cdot \frac{100.5}{L} \tag{4}$$

$$M_{n,v} = M_{1,v} = \frac{m \cdot L \cdot W}{2} \tag{5}$$

2.2. Model of Crowd-Induced Loads under Evacuation

2.2.1. Crowd-Induced Load

During the crowd evacuation on a footbridge, each person excites the structure. The human-induced load of a person in a crowd is not the same as the case when he/she is in free walking status. In particular, his/her behaviour is affected by others and the surroundings [61,62,69]. Thus, it is realistic to consider the inter- and intra-subject variabilities in pedestrian behaviour and the induced forces. To model the pedestrian evacuation behaviour, a microscopic crowd evacuation model is required.

Since its development in 1995 [69], the social force model has been widely applied to simulate pedestrian dynamics in many applications, such as in transport stations, buildings, and other urban public area scenarios [70]. Despite its simplicity of mathematical formulation, the model demonstrates a good ability of pedestrian dynamics reproduction. Till now, the model has been applied not only for crowd evacuation in normal situations [69], but also for unusual situations when people are in panic mode [71], e.g., in earthquakes [72]. Thus, in this study, the social force model is utilized to model the crowd evacuation behaviour during an earthquake. The crowd's evacuation behaviour is guided by physical and psychological interactions with other persons and borders (obstacles). These physical and psychological interactions are considered as physical and psychological forces (accelerations), respectively. Based on Newtonian mechanics, for a random pedestrian α with a mass of m_α, the relationship between displacements (time-variant location $\vec{r}_\alpha(t)$), velocities (time-variant velocity $\vec{v}_\alpha(t)$), and accelerations (time-variant acceleration $\vec{a}_\alpha(t)$) are coupled as:

$$\frac{d\vec{r}_\alpha(t)}{dt} = \vec{v}_\alpha(t) \quad (6)$$

$$\frac{d\vec{v}_\alpha(t)}{dt} = \vec{a}_\alpha(t) = \frac{\vec{F}_\alpha(t)}{m_\alpha} \quad (7)$$

The solutions of these coupled equations output the real-time walking behaviour, i.e., the realistic evacuation behaviour of each person in earthquakes. By including pedestrian-induced forces following the time-variant pedestrian locations and velocities, the crowd-induced loads under an evacuation scenario are obtained. The harmonic load model in terms of Fourier series from Bachmann and Ammann (1987) [73] was applied. To be concise, detailed information on crowd behaviour modelling and crowd-induced load formulation is referred to in [11,61,62,69,71].

2.2.2. Parameter Settings

This subsection presents the parametric settings for a case of the structure with main span length $L = 50$ m and width $W = 3$ m. Table 2 summarizes the six representative crowds with different pedestrian densities ρ_{crowd}, from 0.1 (very weak traffic) to 1.5 (exceptionally dense traffic) pedestrians/m^2, as defined in HiVoSS.

Table 2. Parameters of considered crowds.

Pedestrian Density (Pedestrians/m^2)	Number of Persons (-)	Arrival Time of First Person (s)	Arrival Time of Last Person (s)	Expected Speed (m/s)	Expected Passing Time (s)
0.1	15	3.56	34.86	1.34	37.32
0.2	30	3.32	37.32	1.34	37.32
0.5	75	2.26	35.96	1.30	38.50
0.8	120	0.80	42.90	1.17	42.90
1.0	150	0.38	46.96	1.06	47.26
1.5	225	0.22	61.74	0.81	61.99

For each simulated scenario, it assumes that when people are evacuating on the bridge from one end ($x = 0$) to another end ($x = L$), the earthquake occurs at a random time instant t_{eq}. The arrival times on the bridge of the pedestrians are assumed to follow a Poisson distribution [7,61]. The arrival times of the first and last persons are listed in Table 2. For instance, for 0.1 pedestrians/m^2, the first pedestrian arrives on the structure at a time instant $t_1 = 3.56$ s. The arrival time of the last person is $t_{25} = 34.86$ s, where the subscript 25 denotes the number of pedestrians for 0.1 pedestrians/m^2. For 1.5 pedestrians/m^2, the first pedestrian arrives on the structure at a time instant $t_1 = 0.22$ s. The arrival time of the last person is $t_{375} = 61.71$ s, where the subscript 375 denotes the number of pedestrians for 1.5 pedestrians/m^2. The pedestrians arrive on the bridge with random positions, i.e., with a random value of a coordinate in the Y direction in the range of $[r_\alpha, W - r_\alpha]$. r_α is the radius of a random pedestrian α and thus the range meets the minimum requirement of the pedestrian body to avoid a collision with the borders. As suggested by [61,65], r_α is assigned as 0.3 m.

The initial desired speeds of the crowd can be described as following a normal distribution: $N(1.34, 0.26)$ m/s [69]. In a pedestrian crowd in daily-life conditions, the mean walking speed decreases with an increase in crowd density, according to experimental observations by [74]. Detailed formulations on the relationship between the speed of movement and crowd density is referred to in [75]. Based on their results, the expected mean walking speeds and average passing times of the crowd are determined for walking crowds in normal situations. For example, as shown in Table 2, mean walking speed is expected to be approximately 1.34 m/s and 0.81 m/s for densities of 0.1 and 1.5 pedestrians/m^2, respectively. Correspondingly, it can be predicted that the crowd needs approximately 37.32 s and 61.99 s to pass the bridge, respectively. Due to a lack of real-world walking speed data of pedestrian evacuation in earthquakes, the average passing times needed in normal situations for each relevant density case are considered as the time span in the simulations. The time steps in the crowd simulations are adopted as 0.02 s, in accordance with the time step of the recorded earthquake accelerations.

For any other footbridges with different main span lengths and widths, similar procedures can be taken as the illustrative example. For different span lengths and widths, the corresponding arrival times of pedestrians may be different.

2.3. Dynamic Response of Bridge under Crowd Loads

In this subsection, the structural responses due to crowd loads are calculated. The social force model with the parametric values used in the illustrative example is applied to realistically simulate crowd behaviour. The induced vibrations by the pedestrian crowd are also determined.

2.3.1. Simulated Crowd Behaviour

Figure 5 shows the mean speed of the dynamic crowd on the bridge. For both densities, in the first approximately 5 s, the mean walking speed experiences abrupt changes. This results from the fact that it needs large adjustments in walking parameters at the entrance of the bridge, where newly arrived persons start evacuations and need more sufficient adjustments (see Figure 5). For most of the time instants after the initial stage, the mean speed of the crowd fluctuates at a lower value than the desired mean speed. This demonstrates that after the initial stage, walking speeds are partially restricted in the crowd. Late-arriving persons tend to maintain similar walking speeds as the pedestrians ahead. These 'traffic jam' effects most probably occur in a very crowded situation. After the fluctuation stage, the mean walking speed in a high pedestrian density crowd decreases gradually with increasing pedestrian numbers on the bridge. For the low-density case, the mean speed can even show an increasing trend after the abrupt fluctuation stage, depending on the initial desired walking speed of the persons. This reflects the fact that 'conflicts' among pedestrians do not occur very often in a low-density crowd. Thus, late-arriving faster pedestrians can maintain their walking speeds for a longer time. Furthermore, the mean speed in the low-density case is more sensitive to the scatter in the initial desired walking speeds of newly arrived pedestrians in the crowd. It also proves that the speed–density relationship in earthquakes may be quite different from the experimental observations from daily-life conditions as discussed in [74]. The pedestrian evacuation in earthquakes may suffer from anxiety and even panic. The walking speeds in earthquakes may experience more abrupt changes.

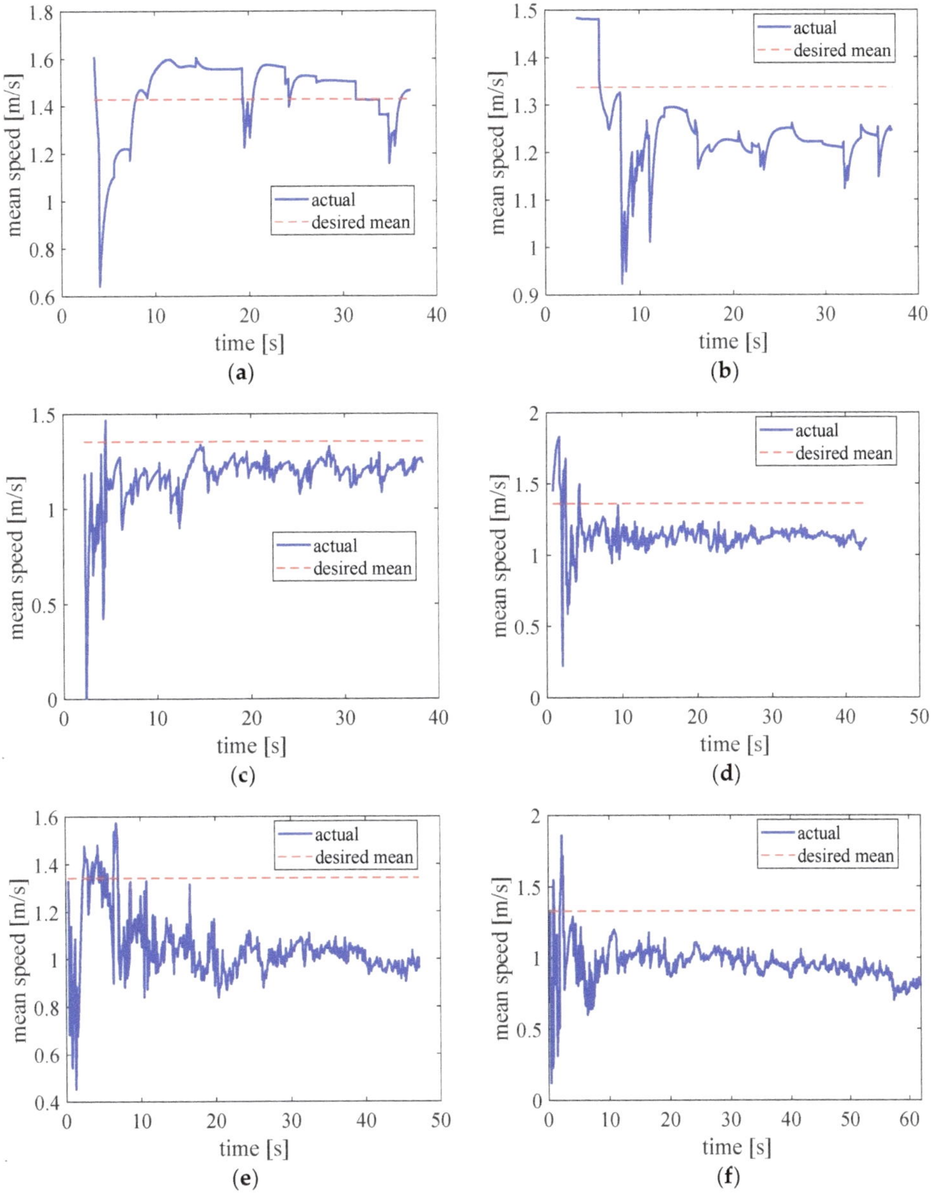

Figure 5. The mean speed of real-time evacuating persons on the bridge. (**a**) 0.1 pedestrians/m^2, (**b**) 0.2 pedestrians/m^2, (**c**) 0.5 pedestrians/m^2, (**d**) 0.8 pedestrians/m^2, (**e**) 1.0 pedestrians/m^2, (**f**) 1.5 pedestrians/m^2.

Figure 6 presents the behaviour of an evacuating pedestrian in a crowd who arrives on the bridge at 4.14 s and stops walking at 37.32 s when the simulation ends. As shown in Figure 5, more abrupt changes in the walking speed are observed during the first approximately 5 s, when the person needs to quickly adjust his/her walking parameters when entering the bridge to avoid collisions as much as possible.

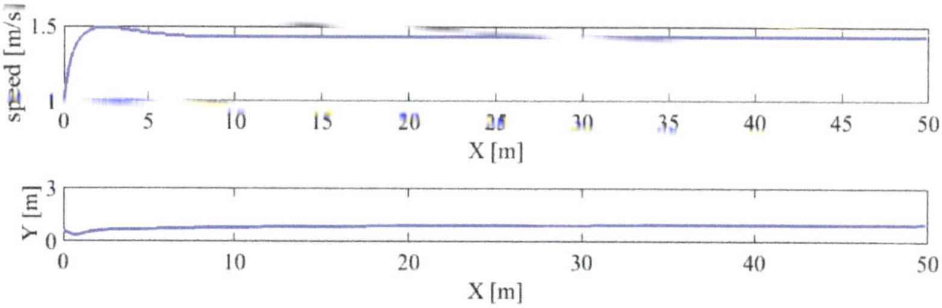

Figure 6. The walking speed (figure above) and trajectory (figure below) of a random pedestrian who arrives at 4.14 s and stops walking at 37.32 s when the simulation ends.

2.3.2. Single Pedestrian-Induced Forces and Vibrations

Figure 7 illustrates the representative person-induced walking forces in the vertical (Z) direction acting on the structure. The excited walking forces are not perfectly harmonic loads because the step frequencies are time-variant, resulting from the time-variant walking speeds. The most 'imperfect' part is at the beginning when the person arrives on the structure, which is in accordance with abrupt changes in both the walking speed and trajectory, as shown in Figure 6.

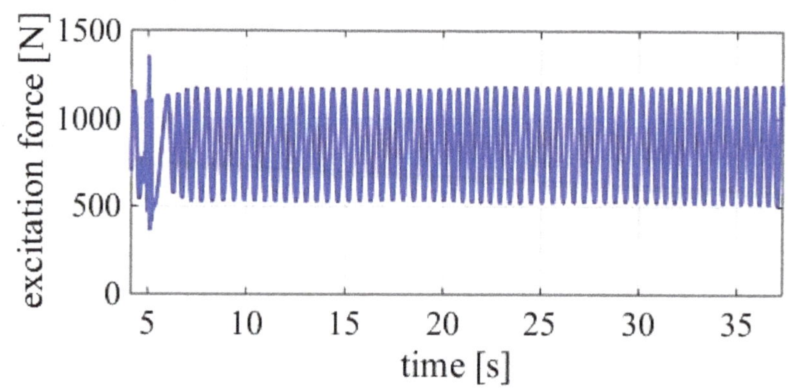

Figure 7. The time history of a representative single pedestrian-induced load in the vertical (Z) direction. The pedestrian is the one who arrives at 4.14 s and stops walking at 37.32 s when the simulation ends.

Figure 8 shows the real-time induced structural responses in the vertical (Z) direction by the representative person. The amplitude of the vertical responses is 0.08 m/s^2. The maximum acceleration amplitudes occur when the person is passing near the midspan of the structure, as the time instant when the modal load amplitudes are a maximum.

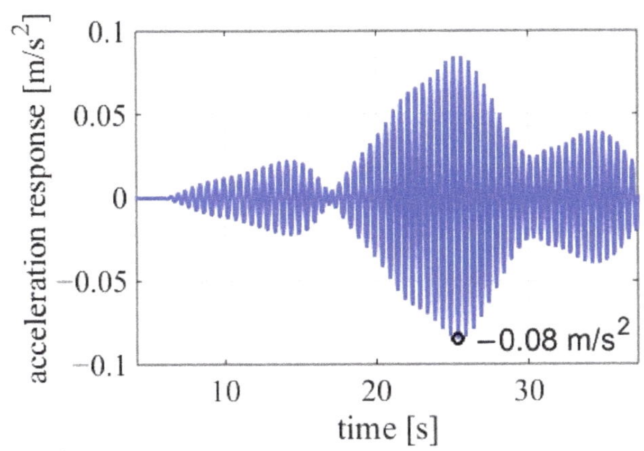

Figure 8. The time history of a representative single pedestrian-induced vibrations in the vertical (Z) direction at the midspan of the structure. The pedestrian is the one who arrives at 4.14 s and stops walking at 37.32 s when the simulation ends.

2.3.3. Crowd-Induced Loads and Vibrations

The crowd-induced loads are obtained by superimposing the force contributions of all individuals, who have different timings for arriving and leaving the bridge. For each time instant, it considers all the real-time persons on the bridge. Figure 9 depicts the time-variant crowd-induced loads in the vertical (Z) direction for different pedestrian densities. As expected, the load fluctuates and has a general increasing tendency with time due to the increasing number of pedestrians for each density case. The fluctuations in the induced load are caused by the adjustments of walking parameters of pedestrians in the crowd. The minimum and maximum load are induced by the lowest and the highest considered density of 0.1 and 1.5 pedestrians/m^2, respectively. Furthermore, the load increases nonlinearly with the density. It results from the crowd-induced load being superimposed by the force contributions of all single persons, while each pedestrian has different timings for arriving and leaving the bridge and different timings for each footfall.

Figure 10 exhibits the time history of the crowd-induced vibrations in the vertical (Z) direction for different pedestrian densities. The lowest and highest structural responses are obtained for the lowest and highest density of 0.1 and 1.5 pedestrians/m^2, respectively. The maximum acceleration response amplitude does not always increase with the density, with the turning point at a density of 0.8 pedestrians/m^2. At the case with 0.8 pedestrians/m^2, although the crowd-induced loads are higher than the lower density cases, the corresponding excitation frequency contents are off the near-resonance.

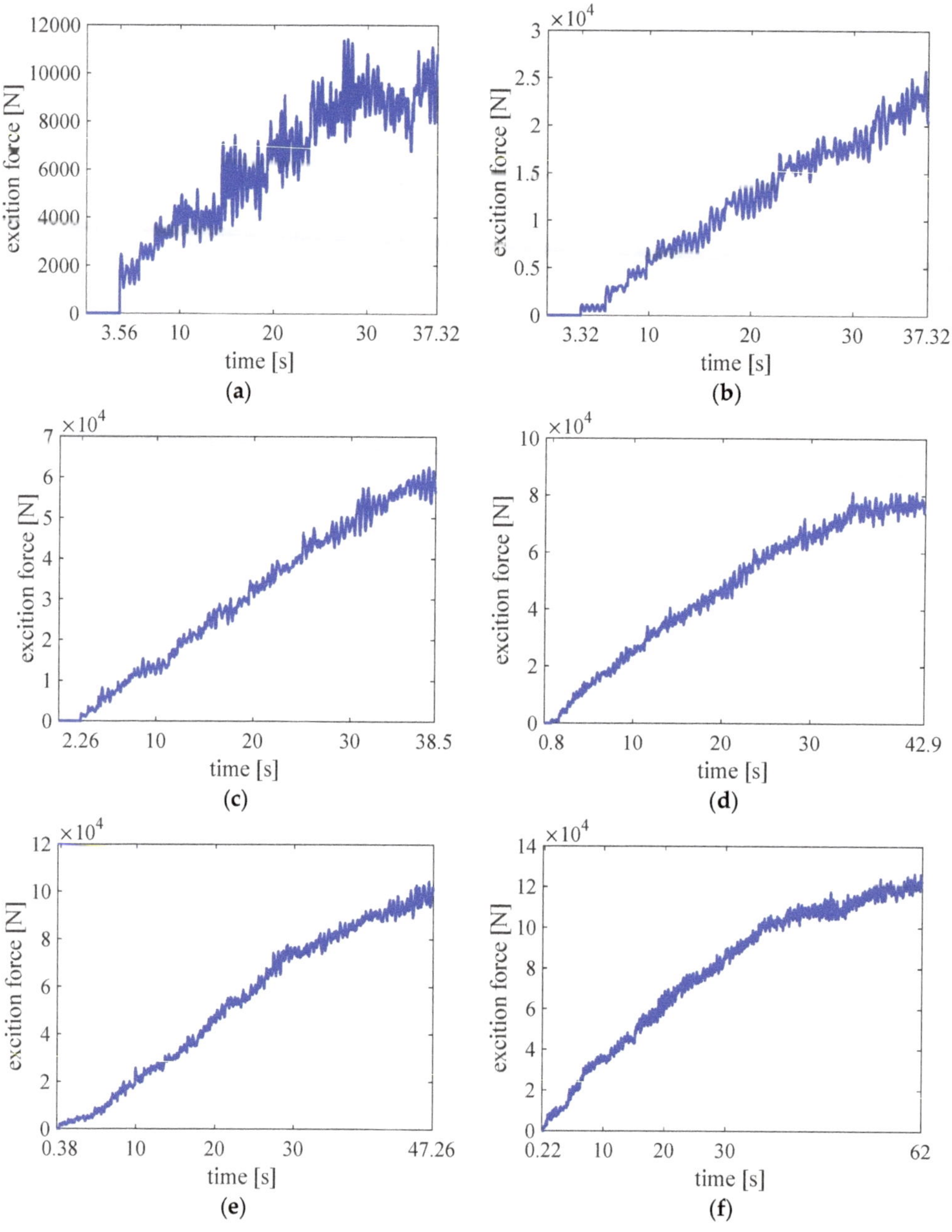

Figure 9. The time history of the crowd-induced loads in the vertical (Z) direction for different pedestrian densities. (**a**) 0.1 pedestrians/m^2, (**b**) 0.2 pedestrians/m^2, (**c**) 0.5 pedestrians/m^2, (**d**) 0.8 pedestrians/m^2, (**e**) 1.0 pedestrians/m^2, (**f**) 1.5 pedestrians/m^2.

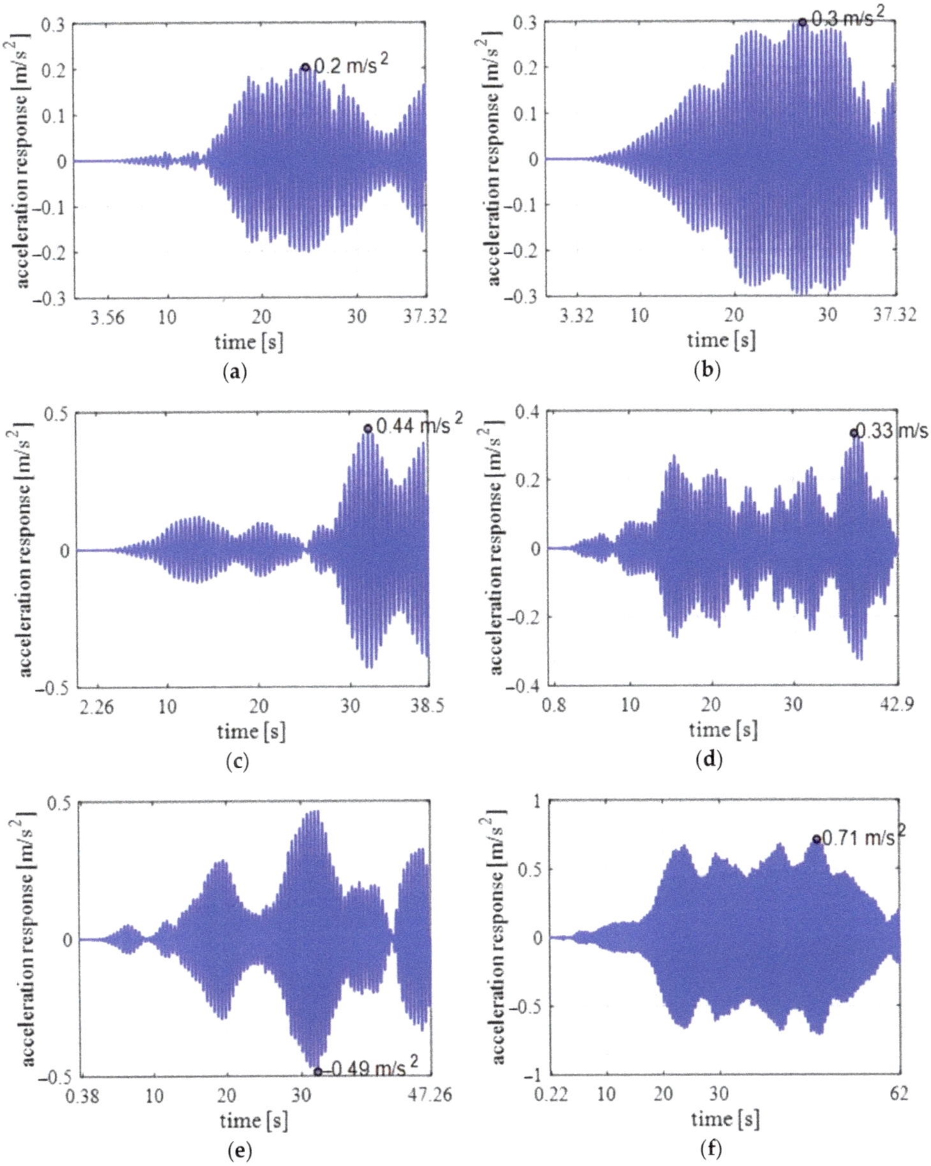

Figure 10. The time history of crowd-induced vibrations at the midspan of the structure in the vertical (Z) direction for different pedestrian densities. (**a**) 0.1 pedestrians/m^2, (**b**) 0.2 pedestrians/m^2, (**c**) 0.5 pedestrians/m^2, (**d**) 0.8 pedestrians/m^2, (**e**) 1.0 pedestrians/m^2, (**f**) 1.5 pedestrians/m^2.

3. Vertical Ground Motions

In this study, a total of 59 vertical earthquake records are collected from a publicly accessible database via the website (https://www.strongmotion.org/, accessed on 6 February 2021). The original peak ground accelerations (PGAs) of the 59 earthquake records range from 0.054 g to 2.370 g. Apart from PGA, three other intensity measures, i.e., peak ground velocity (PGV), $S_{a\text{-}1s}$ (spectral acceleration at a period of 1 s), and $S_{a\text{-}2s}$ (spectral acceleration at a period of 2 s), are also taken to characterize the ground motions. The basic information of the selected vertical ground motions is summarized in Table 3. It is assumed that the bridge is located in regions in China with seismic intensities of 6, 7, 8, and 9 [76]. When conducting elastic time-history analysis of the horizontal earthquakes, the PGAs of the minor earthquakes with a return period of 50 years should be scaled to 0.018 g, 0.035 g, 0.07 g, and 0.14 g for the four intensities, respectively. Furthermore, it is recommended that the V/H (Vertical to Horizontal) ratio should be large or equal to 2/3 (e.g., [44]). Finally, the PGAs of the selected vertical ground motions are scaled to 0.012 g, 0.023 g, 0.047 g, and 0.093 g for the four intensities, respectively. Figure 11 shows the response spectra and average spectrum of the scaled vertical ground motions.

Table 3. Vertical ground motions used in this study.

Number	Earthquake	Station	Year	Magnitude	PGA (g)	PGV (m/s)	$S_{a\text{-}1s}$ (g)	$S_{a\text{-}2s}$ (g)
1	Gazli, Uzbekistan	Karakyr	1976	6.8	1.257	0.602	0.515	0.153
2	Kobe, Japan	Nishi-Akashi	1995	6.9	0.371	0.174	0.148	0.040
3	Kobe, Japan	JR Takatori	1995	6.9	0.272	0.162	0.252	0.225
4	Northridge, USA	Beverly Hills—14145 Mulholland Drive	1994	6.7	0.319	0.201	0.311	0.057
5	Northridge, USA	Canyon Country—W Lost Cany	1994	6.7	0.286	0.189	0.194	0.299
6	Kobe, Japan	Shin–Osaka	1995	6.9	0.059	0.065	0.089	0.048
7	Izmit-Kocaeli, Turkey	Arcelik	1999	7.4	0.079	0.082	0.082	0.040
8	Landers, USA	Yermo Fire Station	1992	7.3	0.136	0.132	0.222	0.059
9	Loma Prieta, USA	Capitola	1989	6.9	0.510	0.194	0.227	0.043
10	Loma Prieta, USA	Gilroy Array #3	1989	6.9	0.369	0.448	0.410	0.369
11	Manjil, Iran	Abbar	1990	7.4	0.538	0.448	0.563	0.248
12	Cape Mendocino, USA	Rio Dell Overpass–FF	1992	7.0	0.195	0.104	0.263	0.100
13	Chi-Chi, Taiwan	CHY101	1999	7.6	0.156	0.274	0.199	0.180
14	Chi-Chi, Taiwan	TCU045	1999	7.6	0.339	0.201	0.270	0.131
15	Lytle Creek, USA	Wrightwood Park	1970	5.3	0.054	0.045	0.030	0.004
16	Livermore-02, USA	Liv.-Morgan TP	1980	5.4	0.079	0.035	0.079	0.005
17	Chi-Chi, Taiwan	CHY006	1999	7.6	0.216	0.232	0.327	0.244
18	NW China-03	Jiashi	1997	6.1	0.384	0.102	0.104	0.030
19	Kobe, Japan	Kakogawa	1995	6.9	0.158	0.107	0.257	0.055
20	Hollister-03, USA	Hollister City Hall	1974	5.1	0.068	0.030	0.020	0.011
21	Kozani, Gr-02, Greece	Chromio	1995	5.1	0.072	0.023	0.007	0.000
22	Loma Prieta, USA	SF Intern. Airport	1989	6.9	0.065	0.056	0.121	0.033
23	Loma Prieta, USA	Fremont, Mission	1989	6.9	0.083	0.092	0.178	0.024
24	Northridge, USA	Arleta—Nordhoff	1994	6.7	0.552	0.178	0.260	0.194
25	Whittier, USA	Whittier Dam	1987	5.7	0.532	0.101	0.071	0.024
26	San Fernando, USA	Pacoima Dam	1971	6.6	0.710	0.585	0.350	0.332
27	Chi-Chi, Taiwan	TCU065	1999	7.6	0.263	0.706	0.444	0.411
28	Kobe, Japan	Takarazuka	1995	6.9	0.433	0.354	0.405	0.196
29	Kobe, Japan	Takatori	1995	6.9	0.272	0.162	0.252	0.225
30	Loma Prieta, USA	Saratoga	1989	6.9	0.361	0.272	0.297	0.158
31	Northridge, USA	Rinaldi	1994	6.7	0.847	0.159	0.088	0.040
32	Northridge, USA	Newhall	1994	6.7	0.548	0.313	0.332	0.098
33	Northridge, USA	Converter	1994	6.7	0.535	0.389	0.310	0.181
34	Northridge, USA	W. Pico Canyon	1994	6.7	0.286	0.294	0.414	0.151
35	Superstition Hills, USA	Wildlife Liquef	1987	6.6	0.423	0.055	0.103	0.037
36	Tabas, Iran	Tabas	1978	7.4	0.746	0.415	0.653	0.254
37	Kobe, Japan	KJMA	1995	6.9	0.343	0.391	0.658	0.294
38	Imperial Valley-06	Bonds Corner	1979	6.5	0.355	0.127	0.218	0.068
39	Imperial Valley-06	El Centro Array #5	1979	6.5	0.479	0.469	0.182	0.195
40	Imperial Valley-06	El Centro Array #6	1979	6.5	1.644	0.581	0.439	0.246
41	Imperial Valley-06	El Centro Array #7	1979	6.5	0.472	0.279	0.323	0.230

Table 3. *Cont.*

Number	Earthquake	Station	Year	Magnitude	PGA (g)	PGV (m/s)	$S_{a\text{-}1s}$ (g)	$S_{a\text{-}2s}$ (g)
42	Imperial Valley-06	El Centro Array #8	1979	6.5	0.356	0.250	0.193	0.149
43	Imperial Valley-06	El Centro Differential Array	1979	6.5	0.464	0.275	0.183	0.123
44	Imperial Valley-06	Holtville Post Office	1979	6.5	0.209	0.149	0.067	0.074
45	Kobe, Japan	Port Island (0 m)	1995	6.9	0.562	0.718	0.505	0.670
46	Izmit-Kocaeli, Turkey	Yarimca	1999	7.4	0.241	0.325	0.327	0.497
47	Northridge, USA	Jensen Filter Plant Administrative Building	1994	6.7	0.401	0.412	0.509	0.280
48	Northridge, USA	Sylmar—Converter Sta East	1994	6.7	0.494	0.265	0.290	0.276
49	Nahanni, Canada	Site 1	1985	6.8	2.370	0.421	0.457	0.231
50	Nahanni, Canada	Site 3	1985	6.8	0.182	0.158	0.085	0.084
51	Cape Mendocino, USA	Cape Mendocino	1992	7.0	0.754	0.781	0.394	0.227
52	Northridge, USA	Jensen Filter Plant Generator Building	1994	6.7	0.760	0.329	0.511	0.201
53	Northridge, USA	Los Angeles Dam	1994	6.7	0.323	0.260	0.271	0.124
54	Northridge, USA	Pacoima Kagel Canyon	1994	6.7	0.180	0.144	0.260	0.206
55	Northridge, USA	Arleta—Nordhoff Fire Sta	1994	6.7	0.552	0.178	0.260	0.194
56	Northridge, USA	Newhall—W Pico Canyon Rd.	1994	6.7	0.286	0.294	0.414	0.151
57	Northridge, USA	Rinaldi Receiving Sta	1994	6.7	0.847	0.477	0.516	0.208
58	Northridge, USA	Sylmar—Converter Sta Valve Group 1–6	1994	6.7	0.535	0.389	0.310	0.181
59	Northridge, USA	Sylmar—Converter Sta Valve Group 7	1994	6.7	0.787	0.429	0.533	0.233

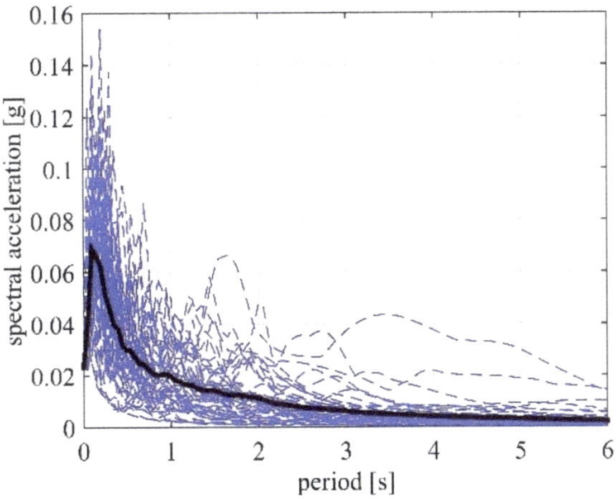

Figure 11. The response spectra (dashed lines) and average spectrum (bold solid line) of 59 scaled vertical ground motions (scaled PGA is 0.023 g).

4. Influence of Vertical Ground Motion on Crowd-Induced Vibration of Footbridge

In this section, the footbridge subjected to earthquake loads is firstly given. Next, the vibration levels are calculated for the case with both crowd loads and earthquake loads. Because of the limited space available in this paper, the numerical results of the earthquakes with intensity 7 are provided as an illustration.

4.1. Footbridge Vibration Induced by Earthquake Loads

In this subsection, the footbridge is only subjected to earthquake loads. The earthquake loads described in Section 3 are applied to calculate the induced vibrations of the illustrative structure. In the response calculation, considering the 'rich' frequency contents of the seismic inputs, the contributions from the first five vertical modes are considered. The earthquakes can occur at a random time instant t_{eq} within the relevant total simulation time span.

Figure 12 illustrates the time history of the structural acceleration responses in the vertical direction subjected to the Kobe Earthquake (intensity 7), which is assumed to occur at time instant $t_{eq} = 0$ s for the illustrative example. The maximum acceleration amplitude reached 1.81 m/s^2.

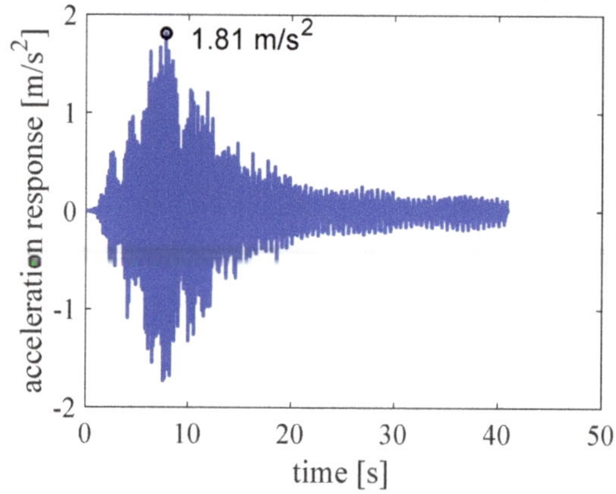

Figure 12. The time history of the structural acceleration responses in the vertical (Z) direction subjected to the Kobe Earthquake (intensity 7), which occurs at time instant $t_{eq} = 0$ s.

Figure 13 shows the empirical cumulative distribution function (CDF) based on the 59 maximum acceleration amplitudes induced by different ground accelerations which are assumed to occur at time instant $t_{eq} = 0$ s. As shown in Figure 13, most amplitudes ranged from 0.39 to 3.07 m/s^2.

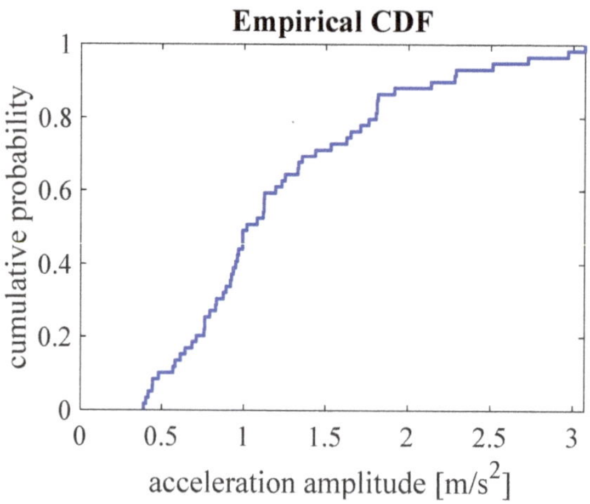

Figure 13. An empirical cumulative distribution function (CDF) plotted by the 'cdfplot' Matlab function, based on the 59 maximum acceleration amplitudes (intensity 7 as an example) induced by different ground accelerations in the vertical (Z) direction.

4.2. Footbridge Vibration Induced by Crowd Loads and Earthquake Loads

This subsection investigates the case that the footbridge is subjected to both crowd loads and earthquake loads. The induced total response is a combination of the vibrations due to crowd loads and earthquake loads. Theoretically, an earthquake can occur at any time during the crowd passing. To consider the randomness of the earthquake occurring at time instant t_{eq}, it can be considered as:

$$0\,\mathrm{s} \leq t_{\mathrm{eq}} \leq t_{\mathrm{last}} \tag{8}$$

where t_{last} is the arrival time of the last person in the crowd, e.g., for the illustrative example, $t_{\mathrm{last}} = t_{25} = 34.86$ s for 0.1 pedestrians/m^2 and $t_{\mathrm{last}} = t_{375} = 61.74$ s for 1.5 pedestrians/m^2. The time is long enough for the structural responses to reach maxima. Furthermore, a time shift of 0.02 s is adopted for each pair of two different neighbouring t_{eq}.

Figure 14 depicts the maximum amplitudes in the time history of the total structural acceleration responses in the vertical direction subjected to crowd loads and the Kobe Earthquake (intensity 7) that occur at a different time instant t_{eq}. The combined structural responses are significantly affected by the time instant t_{eq} for all density cases. The maximum amplitudes of the total responses are 2.01, 2.11, 2.25, 2.14, 2.27, and 2.52 m/s^2 for the responses induced by the earthquake and the six crowds with different densities. For the same earthquake, the maximum amplitude does not increase linearly with the density.

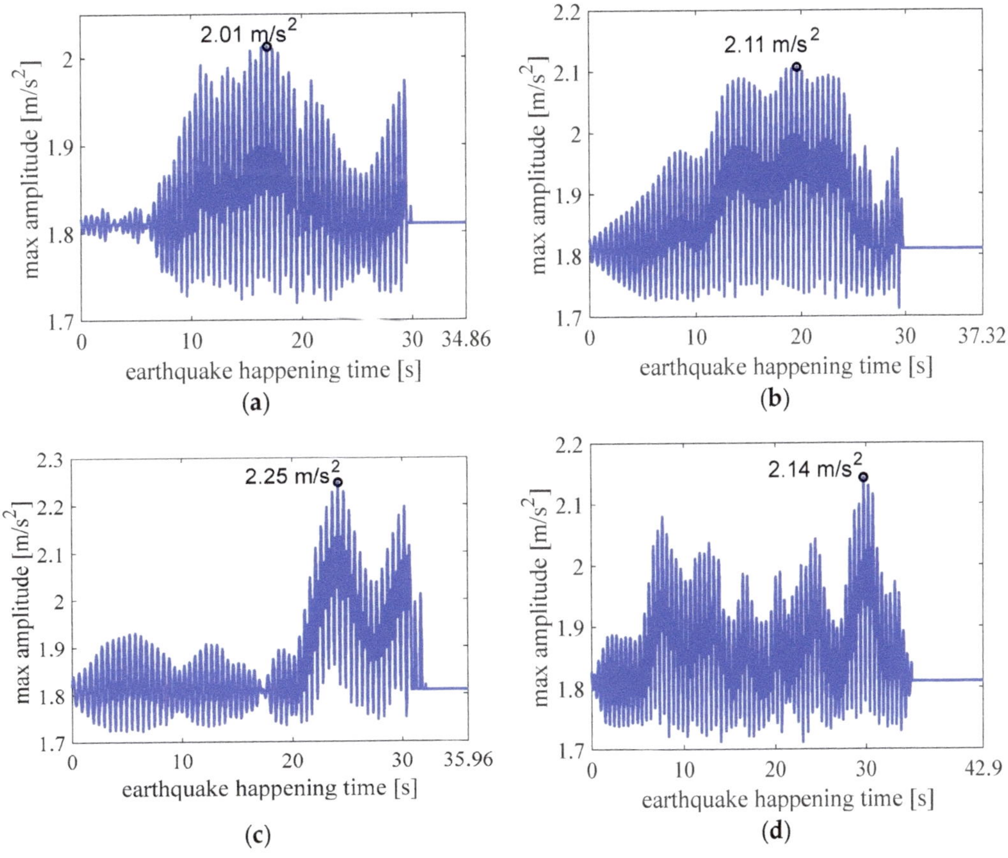

Figure 14. *Cont.*

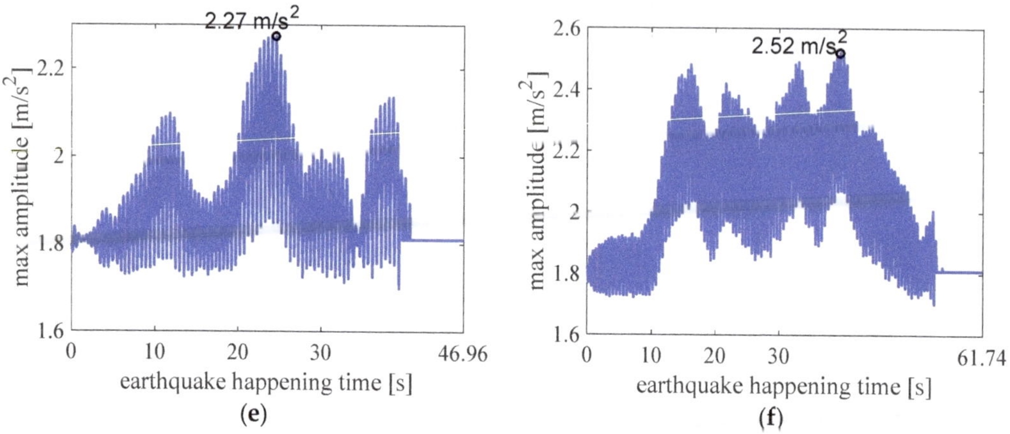

Figure 14. The maximum amplitudes in the time history of the total structural acceleration responses in the vertical (Z) direction to the crowd and the Kobe Earthquake (intensity 7 as an example) that occurs at different time instants t_{eq}. (**a**) 0.1 pedestrians/m^2, (**b**) 0.2 pedestrians/m^2, (**c**) 0.5 pedestrians/m^2, (**d**) 0.8 pedestrians/m^2, (**e**) 1.0 pedestrians/m^2, (**f**) 1.5 pedestrians/m^2.

Figure 15 presents the empirical cumulative distribution functions (CDFs) based on 7 times of 59 maximum acceleration amplitudes induced by different ground accelerations (1 time) and by both earthquake and crowd loads (6 times). The cases with both the crowd and earthquake loads basically have much higher acceleration amplitudes than the cases with crowd load or earthquake load only. As expected, the lowest and the highest acceleration amplitude curves are obtained by the cases with a low density of 0.1 pedestrians/m^2 and high density of 1.5 pedestrians/m^2, respectively. However, the amplitudes do not increase with density and there exists a valley for the case with 0.8 pedestrians/m^2. For the densities in between, the case with a lower density of 0.2 pedestrians/m^2 may induce even higher acceleration amplitudes than the case with 0.8 pedestrians/m^2.

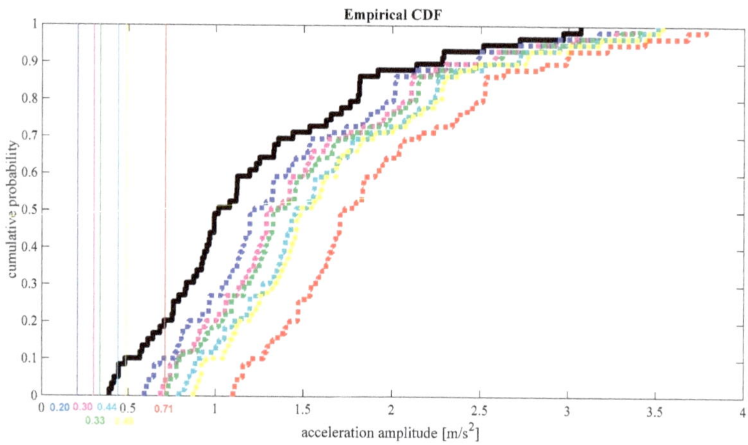

Figure 15. Empirical cumulative distribution functions (CDFs) based on 7 times of 59 maximum acceleration amplitudes (intensity 7) induced by different ground accelerations (1 black wide solid curve) and by both earthquake and crowd loads (6 dashed curves: the blue wide dashed curve for low density of 0.1 pedestrians/m^2, the red dotted wide curve for high density of 1.5 pedestrians/m^2, and other dashed curves for other densities: pink for 0.2, green for 0.8, cyan for 0.5, yellow for 1.0).

For comparison, six additional vertical straight lines are added as the induced maximum responses by the low density of 0.1 pedestrians/m^2 crowd (the blue narrow solid line of 0.20 m/s^2), the high density of 1.5 pedestrians/m^2 (red narrow solid line of 0.71 m/s^2) crowd, and four other density cases (pink for 0.2, green for 0.8, cyan for 0.5, yellow for 1.0).

5. Amplification Effects of Vertical Ground Motion

5.1. Amplification Factors of Structural Responses Due to Ground Motion

To quantify the effects of the ground motion on the structural response subjected to crowd loads and earthquake loads, an amplification factor is introduced as the ratio of maximum acceleration responses to combined loads (both crowd and earthquake) and crowd loads only. In total, there are 195,408 (= 138 × 6 × 59 × 4) calculation cases for 138 footbridges, 6 pedestrians' densities, and 59 vertical ground motions with 4 intensities. In some calculation cases, the input parameters are incomplete and have been removed. As a result, 171,572 datasets are finally selected. Table 4 shows the statistical values of amplification factors for different earthquake intensities. As shown in Table 4, the effects of higher earthquake intensity are generally larger than those of the corresponding lower earthquake intensity. It is characterized by higher mean values of the amplification factor because larger vibration responses are caused by earthquakes with higher intensity. The scatter (characterized with standard deviation) is also larger for higher earthquake intensity. This results from the fact that when the pedestrian-induced vibration levels are kept constant, the contribution of the ground motion in the structural responses is reasonably more significant for earthquakes with higher intensity. Consequently, the amplification factor is more easily affected by the randomness of the ground motion. In other words, relatively higher vibration levels induced by the earthquake can result in high mean and standard deviation values for the amplification factor. This can also be supported by the observations in Table 5. Generally, higher mean and standard deviation amplification factor values are found for lower acceleration amplitudes induced by the crowd, when the earthquake intensity is kept constant (intensity 7). It is also notable that the basic trend is slightly altered due to random characteristics of crowd loads. This is because the amplification factor is not only determined by earthquake loads but also by crowd loads.

Table 4. Statistical values of amplification factor for different earthquake intensities.

Earthquake Intensity	Maximum	Minimum	Mean	St.D.
6	165.98	1.00	5.30	6.84
7	322.29	1.00	9.42	13.29
8	645.96	1.00	17.96	26.64
9	1290.53	1.00	34.97	53.24

Table 5. Statistical values of amplification factor for different densities in the case of intensity 7.

Density (Pedestrians/m^2)	Acceleration Amplitude Induced by Crowd Loads (m/s^2)	Mean	St.D.
0.1	0.20	2.26	0.63
0.2	0.30	1.87	0.44
0.5	0.44	1.69	0.35
0.8	0.33	2.26	0.64
1.0	0.49	1.70	0.36
1.5	0.71	1.66	0.33

The amplification factor is governed by the structure, the crowd, and the earthquake, so the structural-related, crowd-related, and earthquake-related parameters are defined as inputs and the amplification factor is taken as an output. Ten parameters, including four structure-related parameters, one crowd-related parameter, and five earthquake-related parameters, are taken as the input variables, with the amplification factor taken as the only output variable. The statistical values of the input and output variables are listed in Table 6.

Table 6. Statistical values of input and output variables.

Variable Type	Parameters		Unit	Maximum	Minimum	Mean	St.D.
Input	Structure-related	L	m	230.00	4.80	38.94	28.45
		W	m	13.35	0.78	2.76	1.51
		$M_{1,v}$	kg	724,500.00	922.74	64,781.25	76,013.95
		ξ	%	2.50	0.40	0.95	0.66
Input	Crowd-related	ρ_{crowd}	pedestrians/m^2	1.50	0.10	0.67	0.47
Input	Earthquake-related	Scaled PGA	g	0.09	0.01	0.04	0.03
		Original PGA	g	2.37	0.05	0.44	0.38
		Original PGV	m/s	0.78	0.02	0.27	0.18
		Original $S_{a\text{-}1s}$	g	0.66	0.01	0.28	0.16
		Original $S_{a\text{-}2s}$	g	0.67	0.00	0.17	0.13
Output		Amplification factor		1290.53	1.00	16.91	32.73

Figure 16 plots the relationship between the amplification factor and 10 input variables. It shows that the scaled PGA, which is closely related to seismic intensity, has an obvious positive correlation with the amplification factor. With the increase of the main span L, there is a general trend that the amplification factor increases. Conversely, the amplification factor has a descending tendency with the increase of the damping ratio ξ and pedestrian density ρ_{crowd}. There are no significant correlations between the remaining parameters and the amplification factor.

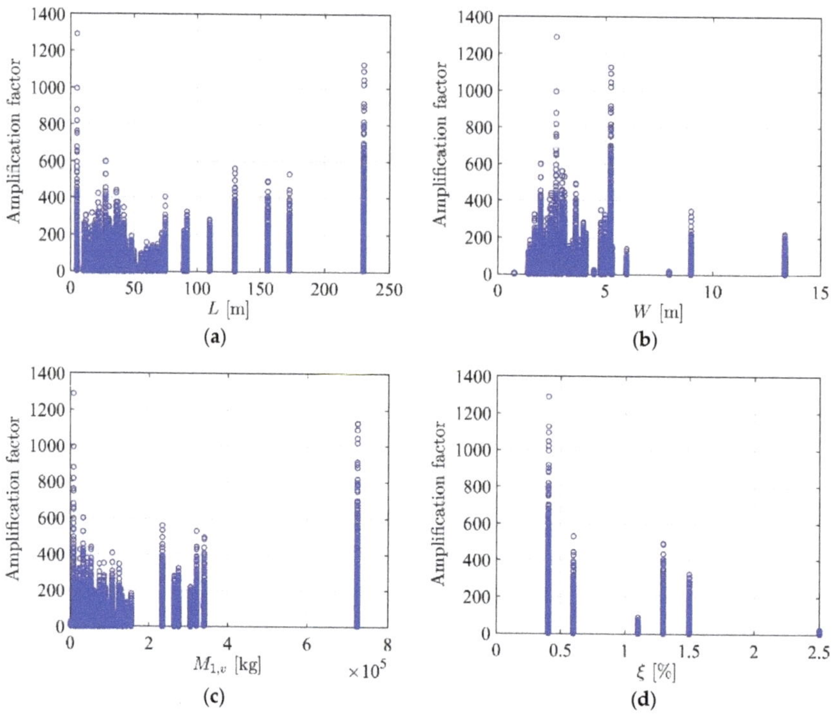

Figure 16. *Cont.*

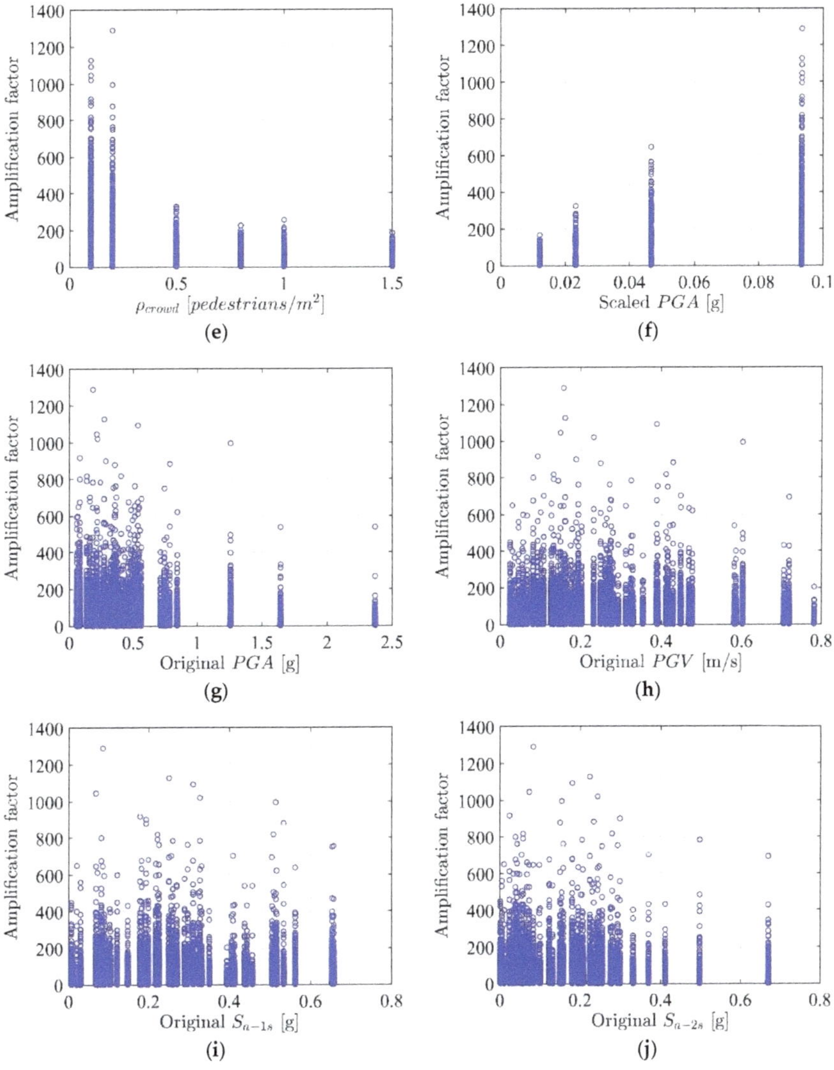

Figure 16. Relationship between the amplification factor and 10 input variables. (**a**) Input 1, (**b**) Input 2, (**c**) Input 3, (**d**) Input 4, (**e**) Input 5, (**f**) Input 6, (**g**) Input 7, (**h**) Input 8, (**i**) Input 9, (**j**) Input 10.

5.2. Machine Learning (ML)-Based Prediction of Amplification Factor

As there exists a strong nonlinear relationship between the inputs and output, machine-learning (ML) techniques, which are suitable for solving the nonlinear regression problem, are used to predict the amplification factor. Two individual-type ML algorithms, i.e., decision tree (DT) and artificial neural network (ANN), and two ensemble ML algorithms, i.e., random forest (RF) and gradient boosting regression tree (GBRT), are adopted to construct the predictive models. The characteristics of the four ML algorithms are briefly summarized as follows.

The most widely used DT algorithm is the classification and regression tree (CART). By using the CART, a characteristic space can be separated into several units. Each unit corresponds to an output. Based on the characteristic of any testing data, it can be designated into a unit and then acquire the output. The DT often has an over-fitting issue and the drawback of processing missing data. The ANN algorithm consists of a large number of neurons or processing elements arranged in different layers. The idea of the ANN originates from the biological nervous systems. A neural network becomes a vector mapper which maps input vectors to an output vector. The RF is a famous bagging-type ensemble learning algorithm based on the DT. The principle of the bagging approach is to separate the training dataset into m new training datasets and generate an independent model for each training dataset. As for the RF, m training datasets can be created by the bootstrap approach. A DT is then generated for each training dataset. The over-fitting issue can be avoided by using the RF. The GBRT is a widely used boosting-type ensemble learning algorithm. It uses a negative gradient of loss function to represent the residual error. By integrating different weaker learners, the GBRT can decrease the deviation and maintain the low variance of the weaker learners.

As mentioned in Section 5.1, a database including 171,572 datasets is used to train and test the ML algorithms. As a common practice [77], 70% and 30% of the data are used as the training and testing datasets, respectively. The Bayesian optimization method is adopted to determine the optimized hyper-parameters of the ML algorithms. The optimized parameters of the four ML algorithms are listed in Table 7.

The predictive accuracy of the ML algorithms is quantitatively evaluated by three widely used performance indices, i.e., coefficient of determination R-squared (R^2), root mean square error (RMSE), and mean absolute error (MAE):

$$R^2 = 1 - \frac{\sum_{i=1}^{N}(C_i - P_i)^2}{\sum_{i=1}^{N}\left(C_i - \overline{C}\right)^2} \tag{9}$$

$$\text{RMSE} = \sqrt{\frac{\sum_{i=1}^{N}(C_i - P_i)^2}{N}} \tag{10}$$

$$\text{MAE} = \frac{\sum_{i=1}^{N}|C_i - P_i|}{N} \tag{11}$$

where C_i and P_i are the calculated and predicted values, respectively; N is the number of datasets in the database; and $\overline{C}$ is the average calculated value. A good predictive model requires that its R^2 is close to 1 and its RMSE and MAE are small.

The performance measures of the four ML algorithms are tabulated in Table 8. Figure 17 illustrates the relationship of the predicted values and the reference values for both the training and testing datasets.

It can be concluded from Table 8 and Figure 17 that the two ensemble algorithms have a better predictive performance than the two individual ML algorithms. In terms of the performance measures of the testing dataset, the best predictive model is GBRT, whose R^2 is closest to 1 and RMSE and MAE are the smallest. Based on the GBRT model, the feature importance [49,50] is conducted to quantify the importance of different features (input variables) on the amplification factor. The relative feature importance of all input variables is plotted in Figure 18. It can be concluded from Figure 18 that the scaled PGA (earthquake-related), ρ_{crowd} (crowd-related), and L (structure-related) are the three most important features, while the influence of W on the amplification factor is less significant.

Table 7. Optimized parameters of the four ML algorithms.

ML Algorithm	Parameters
ANN	activation = 'tanh' alpha = 0.3030395941208759 hidden_layer_sizes = 493 max_iter = 496 random_state = 5 solver = 'lbfgs'
DT	criterion = 'friedman_mse' max_depth = 29 max_features = 9 min_samples_leaf = 6 min_samples_split = 12 random_state = 5
GBRT	Criterion = 'mse' learning_rate = 0.3830013954408691 loss = 'lad' max_depth = 9 max_features = 7 min_samples_leaf = 11 min_samples_split = 11 n_estimators = 285
RF	max_depth = 25 max_features = 7 min_samples_leaf = 2 min_samples_split = 5 n_estimators = 169 random_state = 5

Table 8. Performance measures of four ML algorithms.

ML Algorithm	Datasets	Performance Indices		
		R^2	RMSE	MAE
DT	Training	0.890	10.75	3.72
	Testing	0.780	15.62	5.18
ANN	Training	0.837	13.11	6.04
	Testing	0.791	15.23	6.25
RF	Training	0.942	7.83	2.54
	Testing	0.823	14.04	4.21
GBRT	Training	0.923	9.02	2.33
	Testing	0.870	12.00	3.00

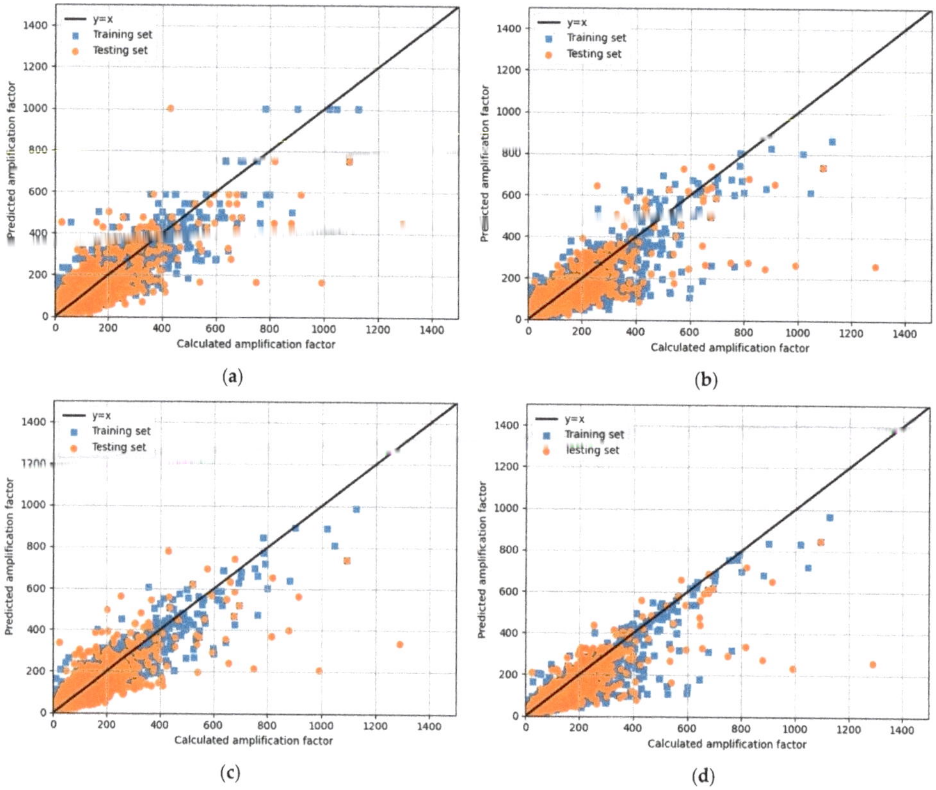

Figure 17. Prediction results of amplification factor using four ML algorithms. (**a**) DT, (**b**) ANN, (**c**) RF, (**d**) GBRT.

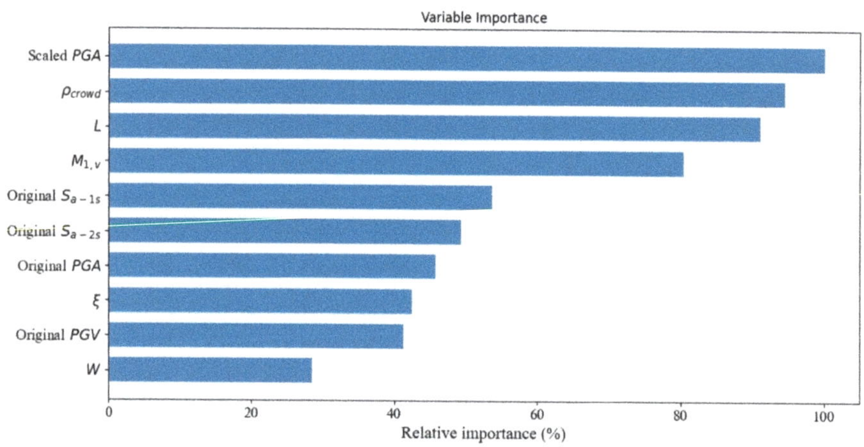

Figure 18. Relative importance of all features (input variables).

6. Vibration Serviceability Evaluation

For the case with earthquake intensity 7, the results are presented in Figure 15. It can be concluded from the empirical cumulative distribution function plot (Figure 15) that when the structure is only subjected to vertical ground motions, there is approximately

50% probability that the vibration levels fall into the range of the minimum comfort limits of 1.0–2.5 m/s^2 according to Setra and HiVoSS, with the probability of the vibration levels exceeding the human comfort limits in the vertical direction (2.5 m/s^2) being very low according to HiVoSS. For crowd load only, the acceleration level is always lower than the comfort limits. When the structure is subjected to earthquake and pedestrian walking by a low-density crowd, the exceedance probability to the human comfort limits is approximately 10%. For high-density crowd evacuation during an earthquake, the corresponding exceedance probability is approximately 20%. However, when recalling the amplification factor values for different earthquake intensities (Table 4), the mean values of the amplification factor are 1.91 (=17.96/9.42) and 3.71 (=34.97/9.42) times higher for intensities of 8 and 9, respectively. Correspondingly, the acceleration amplitudes can be nearly doubled and quadrupled, leading to the acceleration levels exceeding the comfort limits at all or most times, which is very risky for human evacuation and may even result in pedestrians falling. Thus, the serviceability of the footbridge may be impeded.

7. Conclusions

There is no existing research considering the combined effects caused by human-induced loads and vertical ground motions of footbridges. To fill the gap, this paper investigates the effects of vertical ground motion on human-induced vibrations of footbridges. A total of 138 footbridges with different materials, dimensions, and structural types are taken as the target structures. The social force model combined with the pedestrian-induced force model is applied to simulate crowd loads with six representative pedestrian densities as required by design codes. Fifty-nine vertical ground motions with four seismic intensities are adopted as the seismic inputs. The amplification factor is defined to quantify the amplification effects of vertical ground motion on human-induced vibrations of footbridges. Four ML algorithms are used to predict the amplification factor. The vibration serviceability of the footbridge subjected to both crowd load and vertical ground motion is also assessed. Several conclusions can be drawn as follows:

1. The scaled PGA has an obvious positive correlation with the amplification factor. With the increasing of the main span L, there is a general trend of the amplification factor increasing. Conversely, the amplification factor has a descending tendency with the increase of the damping ratio ξ and pedestrian density ρ_{crowd}. There is no significant correlation between the remaining parameters and the amplification factor.
2. The amplification factor is governed by structure-related, crowd-related, and earthquake-related parameters. The scaled PGA, the pedestrian density, and the bridge span are the most important parameters determining the amplification factor.
3. For the considered load scenarios in this paper, when the footbridges are only subjected to crowd loads or the vertical ground motions, there is a very small probability that the vibration levels exceed the upper limit (2.5 m/s^2) of the minimum human comfort limits in a vertical direction as suggested by current design codes. However, it is worthwhile to note that the vibration levels can be different for other cases. Furthermore, comfort limits can be also changed by, e.g., the degree of mutual synchronization of pedestrians in the crowd and their synchronization with the natural frequency of the structure, depending on the value of this frequency.
4. With both the crowd and earthquake loads considered, the acceleration levels may exceed the comfort limits. In particular, when the earthquake intensity is larger than 7, the vibration amplitudes to the combined loads may be higher than the comfort limits at all or most times, which is very risky for human evacuation and may even result in pedestrians falling. Thus, the serviceability of the footbridge may be impeded.

This study may urge footbridge designers to consider the scenario where the crowds are evacuated in earthquakes. A first estimation of the induced vibration levels to the combined loads can be obtained by considering the amplification factor for different crowd densities and earthquakes with different intensities.

In future work, more realistic evacuation scenarios can be simulated by considering possible running persons for the low crowd density cases and pedestrian–structure interactions during earthquakes. Furthermore, the simulations can be more realistic if real-world data for pedestrian evacuation in earthquakes are available.

Author Contributions: Conceptualization, X.W. and B.F.; methodology, X.W. and B.F.; validation, X.W. and B.F.; investigation, X.W. and B.F.; data curation, X.W., B.F., W.W. and X.L.; writing—original draft preparation, X.W.; writing—review and editing, B.F. and W.W.; funding acquisition, B.F. All authors have read and agreed to the published version of the manuscript.

Funding: This research was funded by the National Natural Science Foundation of China (Grant No. 51908048), the Young Talent Fund of University Association for Science and Technology in Shaanxi, China (Grant No. 20200412), and the Natural Science Foundation of Shaanxi Province (Grant No. 2021JM-179).

Informed Consent Statement: Not applicable.

Data Availability Statement: Some or all data, models, or code that support the findings of this study are available from the corresponding author upon reasonable request.

Conflicts of Interest: The authors declare no conflict of interest.

References

1. Dallard, P.; Fitzpatrick, A.J.; Flint, A.; Le Bourva, S.; Low, A.; Ridsdill, R.M.; Willford, M. The London Millennium Footbridge. *Struct. Eng.* **2001**, *79*, 17–33.
2. Ingólfsson, E.T.; Georgakis, C.T. A stochastic load model for pedestrian-induced lateral forces on footbridges. *Eng. Struct.* **2011**, *33*, 3454–3470. [CrossRef]
3. Racic, V.; Brownjohn, J.M.W. Mathematical modelling of random narrow band lateral excitation of footbridges due to pedestrians walking. *Comput. Struct.* **2012**, *90–91*, 116–130. [CrossRef]
4. Bruno, L.; Corbetta, A. Uncertainties in crowd dynamic loading of footbridges: A novel multi-scale model of pedestrian traffic. *Eng. Struct.* **2017**, *147*, 545–566. [CrossRef]
5. Casciati, F.; Casciati, S.; Faravelli, L. A contribution to the modelling of human induced excitation on pedestrian bridges. *Struct. Saf.* **2017**, *66*, 51–61. [CrossRef]
6. Bruno, L.; Venuti, F. A simplified serviceability assessment of footbridge dynamic behaviour under lateral crowd loading. *Struct. Eng. Int.* **2010**, *20*, 442–446. [CrossRef]
7. Živanović, S. Benchmark footbridge for vibration serviceability assessment under the vertical component of pedestrian load. *J. Struct. Eng.* **2012**, *138*, 1193–1202. [CrossRef]
8. Setareh, M. Vibration serviceability issues of slender footbridges. *J. Bridge Eng.* **2016**, *21*, 04016084. [CrossRef]
9. Bedon, C. Diagnostic analysis and dynamic identification of a glass suspension footbridge via on-site vibration experiments and FE numerical modelling. *Compos. Struct.* **2019**, *216*, 366–378. [CrossRef]
10. Feng, P.; Wang, Z.; Jin, F.; Zhu, S. Vibration serviceability assessment of pedestrian bridges based on comfort level. *J. Perform. Constr. Facil.* **2019**, *33*, 04019046. [CrossRef]
11. Fu, B.; Wei, X. An intelligent analysis method for human-induced vibration of concrete footbridges. *Int. J. Struct. Stab. Dyn.* **2021**, *21*, 2150013. [CrossRef]
12. Gong, M.; Shen, R.; Li, Y.; Wang, H.; Chen, W.; Wei, X. Practical suggestions for specifications in vibration serviceability of footbridges based on two recent long-span footbridges. *Struct. Eng. Int.* **2022**, *in press*. [CrossRef]
13. Morbiato, T.; Vitaliani, R.; Saetta, A. Numerical analysis of a synchronization phenomenon: Pedestrian–structure interaction. *Comput. Struct.* **2011**, *89*, 1649–1663. [CrossRef]
14. Carroll, S.; Owen, J.; Hussein, M. Modelling crowd–bridge dynamic interaction with a discretely defined crowd. *J. Sound Vib.* **2012**, *331*, 2685–2709. [CrossRef]
15. Jiménez-Alonso, J.F.; Sáez, A.; Caetano, E.; Magalhães, F. Vertical Crowd–Structure Interaction Model to Analyze the Change of the Modal Properties of a Footbridge. *J. Bridge Eng.* **2016**, *21*, C4015004. [CrossRef]
16. Shahabpoor, E.; Pavic, A.; Racic, V. Structural vibration serviceability: New design framework featuring human-structure interaction. *Eng. Struct.* **2017**, *147*, 295–311. [CrossRef]
17. Toso, M.A.; Gomes, H.M.; Vital de Brito, J.L. Crowd-structure interaction: Investigating the spatiality and synchronization of a pedestrian force model. *J. Bridge Eng.* **2017**, *133*, 510–521. [CrossRef]
18. Mulas, M.G.; Lai, E.; Lastrico, G. Coupled analysis of footbridge-pedestrian dynamic interaction. *Eng. Struct.* **2018**, *176*, 127–142. [CrossRef]
19. Setareh, M.; Gan, S. Vibration testing, analysis, and human-structure interaction studies of a slender footbridge. *J. Perform. Constr. Facil.* **2018**, *32*, 04018068. [CrossRef]

20. Ahmadi, E.; Caprani, C.; Živanović, S.; Heidarpour, A. Assessment of human-structure interaction on a lively lightweight GFRP footbridge. *Eng. Struct.* **2019**, *199*, 109687. [CrossRef]
21. Caetano, E.; Cunha, Á.; Moutinho, C.; Magalhães, F. Studies for controlling human-induced vibration of the Pedro e Inês footbridge, Portugal. Part 2: Implementation of tuned mass dampers. *Eng. Struct.* **2010**, *32*, 1082–1091. [CrossRef]
22. Li, Q.; Fan, J.; Nie, J.; Li, Q.; Chen, Y. Crowd-induced random vibration of footbridge and vibration control using multiple tuned mass dampers. *J. Sound Vib.* **2010**, *329*, 4068–4092. [CrossRef]
23. Casado, C.M.; Díaz, I.M.; de Sebastián, J.; Poncela, A.V.; Lorenzana, A. Implementation of passive and active vibration control on an in-service footbridge. *Struct. Control Health Monit.* **2013**, *25*, e2208. [CrossRef]
24. Venuti, F.; Bruno, L. Mitigation of human-induced lateral vibrations on footbridges through walkway shaping. *Eng. Struct.* **2013**, *56*, 95–104. [CrossRef]
25. Jiménez-Alonso, F.J.; Sáez, A. Robust optimum design of tuned mass dampers to mitigate pedestrian-induced vibrations using multi-objective genetic algorithms. *Struct. Eng. Int.* **2017**, *27*, 492–501. [CrossRef]
26. Moutinho, C.; Cunha, Á.; Caetano, E.; de Carvalho, J.M. Vibration control of a slender footbridge using passive and semiactive tuned mass dampers. *Struct. Control Health Monit.* **2018**, *25*, e2208. [CrossRef]
27. Venuti, F.; Anna, R. Mitigation of human-induced vertical vibrations of footbridges through crowd flow control. *Struct. Control Health Monit.* **2018**, *25*, e2266. [CrossRef]
28. Qin, S.; Zhou, Y.-L.; Kang, J. Footbridge Serviceability Analysis: From System Identification to Tuned Mass Damper Implementation. *KSCE J. Civ. Eng.* **2019**, *23*, 754–762. [CrossRef]
29. Wang, D.; Wu, C.; Zhang, Y.; Li, S. Study on vertical vibration control of long-span steel footbridge with tuned mass dampers under pedestrian excitation. *J. Constr. Steel Res.* **2019**, *154*, 84–98. [CrossRef]
30. Soria, J.M.; Díaz, I.M.; García-Palacios, J.H. Further steps towards the tuning of inertial controllers for broadband-frequency-varying structures. *Struct. Control Health Monit.* **2020**, *27*, e2461. [CrossRef]
31. Gong, M.; Li, Y.; Shen, R.; Wei, X. Glass suspension footbridge: Human-induced vibration, serviceability evaluation, and vibration mitigation. *J. Bridge Eng.* **2021**, *26*, 05021014. [CrossRef]
32. Sétra. *Assessment of Vibrational Behaviour of Footbridges under Pedestrian Loading*; AFGC: Paris, France, 2006.
33. *ISO 10137*; Bases for Design of Structures—Serviceability of Buildings and Walkways against Vibrations. International Standard (ISO): Geneva, Switzerland, 2007.
34. HiVoSS. Design of Footbridges Guideline. Human Induced Vibrations of Steel Structures. 2008. Available online: http://www.stb.rwth-aachen.de/projekte/2007/HIVOSS/download.php(RFs2-CT-2007-00033) (accessed on 6 February 2021).
35. *NA to BS EN 1991-2-2003*; UK National Annex to Eurocode 1: Actions on Structures—Part 2: Traffic Loads on Bridges. British Standards Institution (BSI): London, UK, 2008.
36. Shrestha, B. Seismic response of long span cable-stayed bridge to near-fault vertical ground motions. *KSCE J. Civ. Eng.* **2015**, *19*, 180–187. [CrossRef]
37. Khanmohammadi, M.; Kharrazi, H. Residual Capacity of Mainshock-Damaged Precast-Bonded Prestressed Segmental Bridge Deck under Vertical Earthquake Ground Motions. *J. Bridge Eng.* **2018**, *23*, 04018016. [CrossRef]
38. Jin, Z.; Pei, S.; Li, X.; Liu, H.; Qiang, S. Effect of vertical ground motion on earthquake-induced derailment of railway vehicles over simply-supported bridges. *J. Sound Vib.* **2016**, *383*, 277–294. [CrossRef]
39. Paraskeva, T.S.; Dimitrakopoulos, E.G.; Zeng, Q. Dynamic vehicle–bridge interaction under simultaneous vertical earthquake excitation. *Bull. Earthq. Eng.* **2017**, *15*, 71–95. [CrossRef]
40. Xiang, Y.; Luo, Y.; Zhu, Z.; Shen, Z. Estimating the response of steel structures subjected to vertical seismic excitation: Idealized model and inelastic displacement ratio. *Eng. Struct.* **2017**, *148*, 225–238. [CrossRef]
41. Fayaz, J.; Zareian, F. Reliability Analysis of Steel SMRF and SCBF Structures Considering the Vertical Component of Near-Fault Ground Motions. *J. Struct. Eng.* **2019**, *33*, 04019046. [CrossRef]
42. Qu, Y.; Luo, Y.; Zhu, Z.; Huang, Q. An improved multidimensional modal pushover analysis procedure for seismic evaluation of latticed arch-type structures under lateral and vertical earthquakes. *Struct. Des. Tall Spec. Build.* **2019**, *28*, e1618. [CrossRef]
43. Rinaldin, G.; Fasan, M.; Noé, S.; Amadio, C. The influence of earthquake vertical component on the seismic response of masonry structures. *Eng. Struct.* **2019**, *185*, 184–193. [CrossRef]
44. Li, W.; Chen, Q. Effect of vertical ground motions and overburden depth on the seismic responses of large underground structures. *Eng. Struct.* **2020**, *205*, 110073. [CrossRef]
45. Chen, G.; Ruan, B.; Zhao, K.; Chen, W.; Zhuang, H.; Du, X.; Khoshnevisan, S.; Juang, C.H. Nonlinear Response Characteristics of Undersea Shield Tunnel Subjected to Strong Earthquake Motions. *J. Earthq. Eng.* **2020**, *24*, 351–380. [CrossRef]
46. Pourmasoud, M.M.; Lim, J.B.P.; Hajirasouliha, I.; McCrum, D. Multi-Directional Base Isolation System for Coupled Horizontal and Vertical Seismic Excitations. *J. Earthq. Eng.* **2020**, *26*, 1145–1170. [CrossRef]
47. Salehia, H.; Burgueñoa, R. Emerging artificial intelligence methods in structural engineering. *Eng. Struct.* **2018**, *171*, 170–189. [CrossRef]
48. Wang, X.; Mazumder, R.K.; Salarieh, B.; Salman, A.M.; Shafieezadeh, A.; Li, Y. Machine Learning for Risk and Resilience Assessment in Structural Engineering: Progress and Future Trends. *J. Struct. Eng.* **2022**, *148*, 03122003. [CrossRef]
49. Feng, D.-C.; Fu, B. Shear Strength of Internal Reinforced Concrete Beam-Column Joints: Intelligent Modeling Approach and Sensitivity Analysis. *Adv. Civ. Eng.* **2020**, *2020*, 8850417. [CrossRef]

50. Wang, X.; Li, Z.; Shafieezadeh, A. Seismic response prediction and variable importance analysis of extended pile-shaft-supported bridges against lateral spreading: Exploring optimized machine learning models. *Eng. Struct.* **2021**, *236*, 112142. [CrossRef]
51. Fu, B.; Feng, D.-C. A machine learning-based time-dependent shear strength model for corroded reinforced concrete beams. *J. Build. Eng.* **2021**, *36*, 102118. [CrossRef]
52. Fu, B.; Chen, S.-Z.; Liu, X.-R.; Feng, D.-C. A probabilistic bond strength model for corroded reinforced concrete based on weighted averaging of non-fine-tuned machine learning models. *Constr. Build. Mater.* **2022**, *318*, 125767. [CrossRef]
53. Safavian, S.R.; Landgrebe, D. A survey of decision tree classifier methodology. *IEEE Trans. Syst. Man Cybern.* **2002**, *21*, 660–674. [CrossRef]
54. Schalkoff, R.J. *Artificial Neural Networks*; McGraw-Hill: New York, NY, USA, 1997.
55. Zhou, Z.-H. Ensemble learning. In *Encyclopedia Biometrics*; Springer: Boston, MA, USA, 2015; pp. 411–416.
56. Feng, D.-C.; Liu, Z.-T.; Wang, X.-D.; Jiang, Z.-M.; Liang, S.-X. Failure mode classification and bearing capacity prediction for reinforced concrete columns based on ensemble machine learning algorithm. *Adv. Eng. Inform.* **2020**, *45*, 101126. [CrossRef]
57. Wei, X.; Russell, J.; Živanović, S.; Mottram, J.T. Measured dynamic properties for FRP footbridges and their critical comparison against structures made of conventional construction materials. *Compos. Struct.* **2019**, *223*, 110956. [CrossRef]
58. Oliveira, C.S. Fundamental Frequencies of Vibration of Footbridges in Portugal: From In Situ Measurements to Numerical Modelling. *Shock Vib.* **2014**, *2014*, 925437.
59. Ahmadi, E.; Caprani, C.C.; Heidarpour, A. An equivalent moving force model for consideration of human-structure interaction. *Appl. Math. Model.* **2017**, *51*, 526–545. [CrossRef]
60. Živanović, S.; Pavic, A.; Reynolds, P. Vibration serviceability of footbridges under human-induced excitation: A literature review. *J. Sound Vib.* **2005**, *279*, 1–74. [CrossRef]
61. Wei, X.; Liu, J.-C.; Bi, S. Uncertainty quantification and propagation of crowd behaviour effects on pedestrian-induced vibrations of footbridges. *Mech. Syst. Signal Process.* **2022**, *167 Pt A*, 108557. [CrossRef]
62. Wei, X. A simplified method to account for human-human interaction in the prediction of pedestrian-induced vibrations. *Struct. Control Health Monit.* **2021**, *28*, e2753. [CrossRef]
63. Fu, B.; Wei, X.; Chen, J.; Bi, S. Shear Lag Effects on Pedestrian-Induced Vibration and TMD-Based Vibration Control of Footbridges. *Struct. Eng. Int.* **2022**. [CrossRef]
64. Friswell, M.I.; Mottershead, J.E. *Finite Element Model Updating in Structural Dynamics*; Springer: Berlin/Heidelberg, Germany, 1995.
65. Clough, R.W.; Penzien, J. *Dynamics of Structures*, 3rd ed.; McGraw-Hill: New York, NY, USA, 2003.
66. Limongelli, M.; Siegert, D.; Merliot, E.; Waeytens, J.; Bourquin, F.; Vidal, R.; Le Corvec, V.; Gueguen, I.; Cottineau, L. Damage detection in a post tensioned concrete beam—Experimental investigation. *Eng. Struct.* **2016**, *128*, 15–25. [CrossRef]
67. Noble, D.; Nogal, M.; O'Connor, A.; Pakrashi, V. The effect of prestress force magnitude and eccentricity on the natural bending frequencies of uncracked prestressed concrete beams. *J. Sound Vib.* **2016**, *365*, 22–44. [CrossRef]
68. Bonopera, M.; Liao, W.-C.; Perceka, W. Experimental–theoretical investigation of the short-term vibration response of uncracked prestressed concrete members under long-age conditions. *Structures* **2022**, *35*, 260–273. [CrossRef]
69. Helbing, D.; Molnar, P. Social force model for pedestrian dynamics. *Phys. Rev. E* **1995**, *51*, 4282–4286. [CrossRef]
70. Chen, X.; Li, H.; Miao, J.; Jiang, S.; Jiang, X. A multiagent-based model for pedestrian simulation in subway stations. *Simul. Model. Pract. Theory* **2017**, *71*, 134–148. [CrossRef]
71. Helbing, D.; Farkas, I.; Vicsek, T. Simulating dynamical features of escape panic. *Nature* **2000**, *794*, 487–490. [CrossRef]
72. Li, M.; Zhao, Y.; He, L.; Chen, W.; Xu, X. The parameter calibration and optimization of social force model for the real-life 2013 Ya'an earthquake evacuation in China. *Saf. Sci.* **2015**, *79*, 243–253. [CrossRef]
73. Bachmann, H.; Ammann, W. *Vibration in Structures-Induced by Man and Machines*; International Association for Bridge and Structural Engineering (IABSE): Zurich, Switzerland, 1987.
74. Weidmann, U. *Transporttechnik der Fussgänger*; Ivt Report No. 90; ETH: Zurich, Switzerland, 1993.
75. Bruno, L.; Venuti, F. The pedestrian speed-density relation: Modeling and application. In Proceedings of the 3rd International Conference on the Design and Dynamic Behavior of Footbridges: Footbridge 2008, Porto, Portugal, 2–4 July 2008.
76. MOHURD (Ministry of Housing and Urban-Rural Development of the People's Republic of China). *Code for Seismic Design of Urban Bridges (CJJ 166-2011)*; China Architecture & Building Press: Beijing, China, 2011. (In Chinese)
77. Friedman, J.; Hastie, T.; Tibshirani, R. *The Elements of Statistical Learning*; Springer Series in Statistics: New York, NY, USA, 2001.

buildings

MDPI

Article

Ultimate Bearing Capacity Analysis of Pile Caps with New Socket Connections

Jiping Ge [1,*], Luqi Lai [1], Side Liu [2] and Xingfei Yan [3]

1 School of Urban Construction and Safety Engineering, Shanghai Institute of Technology, Shanghai 201418, China
2 China Railway 22TH Bureau Group 3RD Engineering Corporation Limited, Xiamen 361000, China
3 Shanghai Urban Construction Design Research Institute (Group) Co., Ltd., Shanghai 200125, China
* Correspondence: bridgejiping@126.com

Abstract: Socket connection need a groove reserved in the cap to accommodate a bridge pier, which greatly weaken the vertical bearing capacity of the cap. The conventional treatment measure is to increase the thickness of the cap, and the corresponding cost will increase. The measures to enhance the vertical bearing capacity of socket caps without increasing the thickness of the cap were discussed in this paper, including a rough interface at the bottom of the pier, additional hanging bars, high-strength grouting material in the seam, and large-diameter metal corrugated pipes, etc. Based on a previous test, the finite element analysis of the vertical bearing capacity of pile caps with new socket connections was carried out. The analysis parameters included the construction method, steel bar diameter in the bottom of the cap, socket depth, thickness of the bottom plate, pile length, and friction coefficient, etc. The bearing capacity M–N relation of the full-scale model was also analyzed. Research indicated the vertical bearing capacity of the cap is mainly provided by rough interfaces, the bottom plate, and the additional hanging bars, and the contribution of the three parts was about 40%, 34%, and 26%; the vertical bearing capacity was proportional to the areas of steel bars on the cap and the thickness of the bottom plate, and was inversely proportional to the length of the pile. To obtain the vertical bearing capacity of the overall cast-in-place plan for the socket cap, the thickness of the cap needs to be increased by 27%. At last, a design formula for the calculation of the vertical bearing capacity was proposed.

Keywords: socket connection; vertical bearing capacity; vertical loading; finite element calculation; metal bellows; formula derivation

Citation: Ge, J.; Lai, L.; Liu, S.; Yan, X. Ultimate Bearing Capacity Analysis of Pile Caps with New Socket Connections. *Buildings* **2022**, *12*, 2034. https://doi.org/10.3390/buildings12112034

Academic Editor: Humberto Varum

Received: 9 October 2022
Accepted: 15 November 2022
Published: 21 November 2022

Copyright: © 2022 by the authors. Licensee MDPI, Basel, Switzerland. This article is an open access article distributed under the terms and conditions of the Creative Commons Attribution (CC BY) license (https://creativecommons.org/licenses/by/4.0/).

1. Introduction

The precast assembly technology of bridge piers is the current development direction of bridge engineering because of its advantages of rapid construction, reducing construction risks for workers, and reducing on-site labor [1], etc. There are many types of connections between the pier and the pile cap, such as grout sleeve, prestressing strands, and metal bellows [2,3]. Socket connections show good application prospects among them [4]. Due to the low material strength and lack of structural strengthening measures, the traditional embedding length requirement is longer in socket connections; the thickness of the pile cap is also larger, which leads to an increase in the cost and limits application occasions [5]. With the further study of socket connections, the method of using a high-strength grouting material, large-diameter metal bellows, rough interfaces with shear keys, and anti-punching steel reinforcement at the bottom of the groove is proposed [6]. Then, under certain conditions, the minimum socket depth can be achieved in the hollow pier and cap with the socket connection [7,8]. This is the best solution in terms of vertical bearing capacity, horizontal bearing capacity, cap thickness, and cost, etc.

Regarding the application of socket connections, there are many precedents at home and abroad. In the 1950s, the piers and columns of many railway bridge piers adopted socket connections in China. The Beijing Jishuitan Bridge Test Project, which was completed in 1992, is a highway precast bridge pier project completed earlier in China. In 2013, the interstate bridge of the SR520 highway in Washington State, USA adopted the socket connection technology [4]. The states of Minnesota and Louisiana in the United States have standardized the design of socket connections for pile foundations caps [2]. In 2016, one of the ramp bridges of the second north section of the Shanghai Jiamin Elevated Bridge also adopted a socket connection structure. The Huaihe River Bridge and the Zhongpaihe Bridge in Anhui Province, the eastern extension of the Jiangbei Expressway in Hubei Province, and a highway bridge in Shandong Province also use socket structures. In 2019, the socket precast pier connection technology was adopted for the viaduct project of the Beijing–Xiong'An Expressway. There are many other cases of socket connections for precast columns and foundations, which can be found in other countries [2].

At present, many scholars have carried out many research works on socket connection structures [8–12], such as Osanai Y. (1996) [13], Canha R. (2009) [14], Mohebbi A. and Saiidi M. (2018) [15,16], Wang Z.Q. (2019) [17], Cheng Z. (2021) [18,19], and Zhang G. (2021) [20]. Research on socket connections has been focused on whether the mechanical properties of socket bridge piers are similar to those of integral cast-in-place bridge piers, and the reasonable embedded depth of the cap, but there are few studies on the thickness of the bottom plate in the socket connection [5]. Chan T. K. studied the behavior of pre-cast reinforced concrete pile caps and the ultimate load-carrying capacity with three specimens. The research showed that the current design equations for conventionally cast in situ construction can be used to predict the failure loads of the pre-cast units, although the predictions may be conservative in certain cases [21]. Cheng Z. experimentally evaluated the side shear strength of preformed socket connections with various connection parameters. The test results showed that the side shear mechanism in the preformed socket connections could provide significant resistance, facilitating transfer of large vertical loads [22].

The force mechanism of the pile cap is relatively complicated. The thickness of pile cap in socket connections is usually determined by the embedded depth of the column and the thickness of the bottom plate. This paper studied the thickness of the cushion cap through finite element analysis based on the improved socket connection proposed, established the finite element model correctly through experiments, and then discussed the influence of different factors on the vertical bearing capacity of the socket structure. Finally, a formula for calculating the vertical bearing capacity of bearing structure was proposed, which can be available for reference at the time of specification preparation.

2. Analysis of Influencing Factors on Socket Connections

2.1. Structure Detail and Composition

In traditional socket-type structures, the precast pier body is directly inserted into the reserved hole of the foundation, and a certain thickness of mortar is laid at the bottom. There are no steel bars connected between the bridge pier and the foundation. The foundation can be cast-in-place or precast. The advantage of this connection structure is that the construction process is simple and the amount of on-site work is limited [4], as shown in Figure 1.

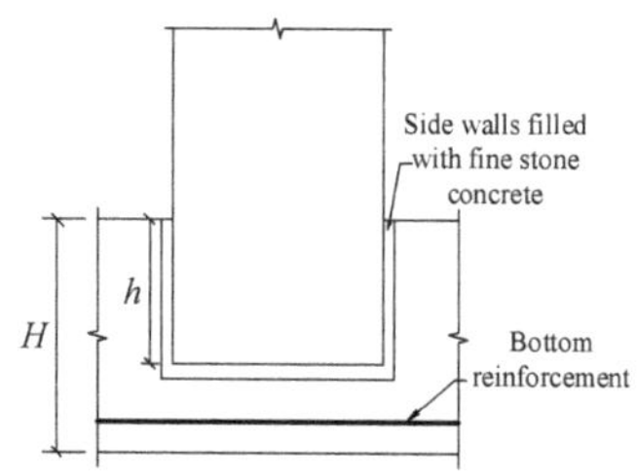

Figure 1. Traditional socket type.

Compared with the cast-in-place pier, the socket-type bridge cap does not meet the equivalent design, while the bridge pier completely meets the equivalent design. Its construction technology requirements are not high, and the social recognition is high, but its scope of application is narrow. It is generally considered that it is only suitable for connections between the column and the pile cap, and not suitable for other occasions. From the perspective of overall force transmission, this method weakens the vertical bearing capacity of the cap, and the damaged part may be transferred from the pier to the cap. Therefore, appropriate measures should be taken to ensure that damage only occurs in the plastic hinge area of the pier. If there are construction defects, this will have little impact on safety, and there is no safety risk in the normal operation stage. However, under extreme loads the safety factor of vertical bearing capacity will be reduced. For various precast assembly technologies, as long as the materials are qualified and the construction and design meet the design standards, it should be considered a suitable assembly solution. The most appropriate assembly scheme for each part is recommended in Table 1 below.

Table 1. Proposed solution of the assembly scheme.

Position	Assembly Method	Reason
Bridge piers and caps	Socket type or grout sleeve	The construction requirements are the highest in the unfavorable stress parts such as compression, bending, shear, and torsion, so the pier and cushion cap must be rigidly connected
Bridge piers and bent cap	Sleeve or Bellows	It has strong adaptability to construction defects, and the defective structure will not collapse because the connection between the pier and the bent cap can be rigid, semi-rigid, or even hinged
Piers and piers	Sleeve	Meet shear and compression requirements
Bent caps and caps	Prestressed	There is prestress in the bent cap, and the prestressed type can make full use of the original advantages

With the improvement of the requirements of modern precast structures, the traditional socket connections cannot meet engineering needs anymore, so an improved socket structure needs to be proposed. To improve the bearing capacity of the cap, the side of the socket in contact with the pier can be used as a breakthrough. By adding trapezoidal shear keys on the side contact surface and using the shear keys to provide shear force, the resisting moment of the contact part can be increased, thereby reducing the socket's depth, as shown in Figure 2. At the same time, an anti-punch hanging bar is added inside the cap to increase the vertical punching-shear bearing capacity of the cap. It should be noted that U-shaped steel bars are reserved at the bottom of the groove of the cap, and after the core-filled concrete is poured in the pipe pier, an effective constraint will be formed between the bottom of the groove and the bottom of the pier. This constraint applies only to the structural connection between hollow piers and socket caps, and this type of structure cannot be used for solid piers. The existence of this structure makes the socket depth of hollow piers shorter than that of solid piers of the same diameter. This structure is related to the flexural capacity of bridge piers under horizontal loads. The details can be found in reference [6].

2.2. Full-Scale Bridge Piers

According to the abovementioned improvement ideas, the new socket design scheme of the centrifugal pipe pier, as shown in Figures 3 and 4, is finally proposed. According to the existing results, the ratio of the socket depth of the pile cap to the diameter of the pier was 0.7. The ratio of socket depth to base plate thickness was 2:1. The height of the cap was 1.5 m, the socket depth was 1 m, and the thickness of the bottom plate was 0.5 m. The insertion end of the pipe pier adopted rough interfaces, and the groove of the cap was made of a large-diameter corrugated pipe to form a groove, which could form a restraint effect on the caulking concrete. Anti-punch hanging bars were set at the bottom of the groove of the cap to enhance the shear resistance. The diameter of the reinforcing bars of the bottom plate was 25 mm, the diameter of the other bars was 16 mm, and the spacing between the bars was 100 mm. The pier was poured with C70 (C70 means the standard compressive strength of the concrete is 70 MPa) concrete and filled with C40 concrete. The longitudinal steel bars were HRB400 (HRB400 means hot-rolled ribbed bars with yield strengths of 400 MPa) steel bars, where 36 steel bars had a diameter of 28 mm in the outer ring and 18 steel bars had a diameter of 16 mm in the inner ring, and HPB300 (HPB300 indicates hot-rolled plain steel bars with yield strengths of 300 MPa.) spiral stirrups had a spacing of 100 mm.

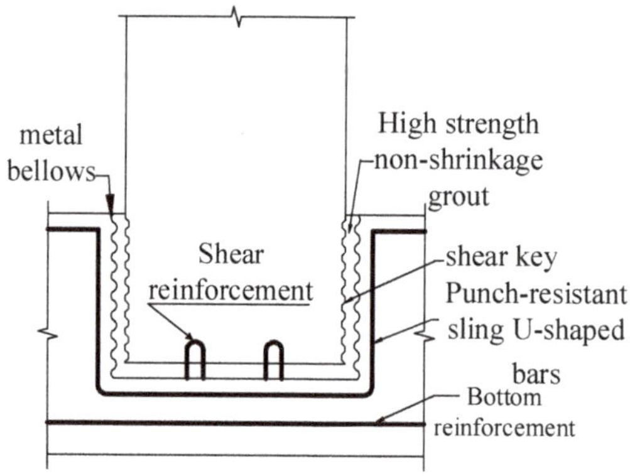

Figure 2. Improved socket type.

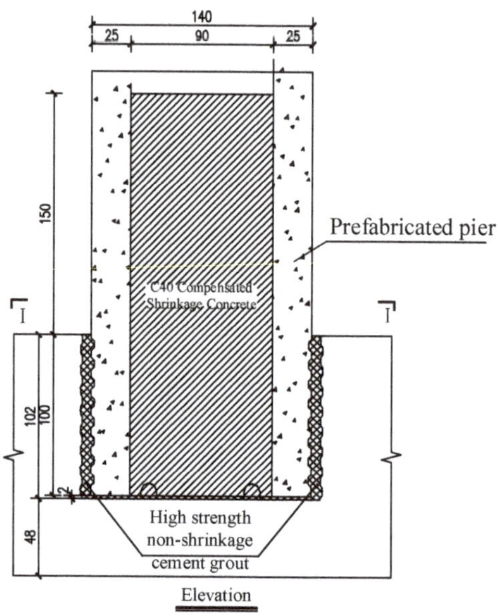

Figure 3. Concrete pier and column structure.

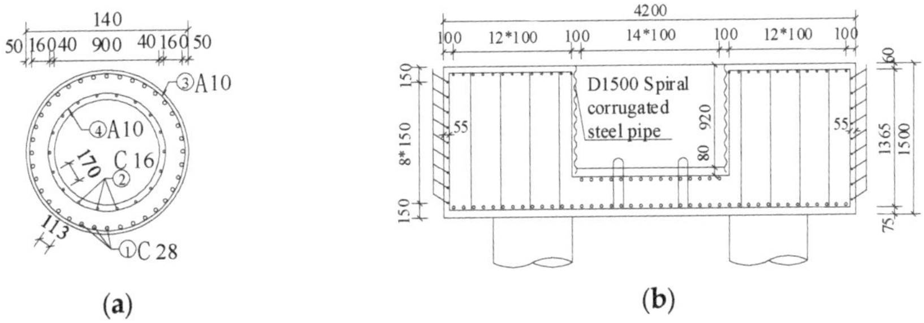

Figure 4. Reinforcement diagram of the full-scale bridge pier. (**a**) Pier section; (**b**) cap reinforcement.

2.3. Model Specimens

During the experimental study, the similarity ratio was 1:2. The diameter of the steel bar of the model cap and the base plate of the cap was 18 mm; the diameter of the other bars was 12 mm, as shown in Figure 5. To meet the requirements of the length of the longitudinal reinforcement, the bottom of the pier was provided with a perforated plug welding of the steel plate with the longitudinal reinforcement of the pier.

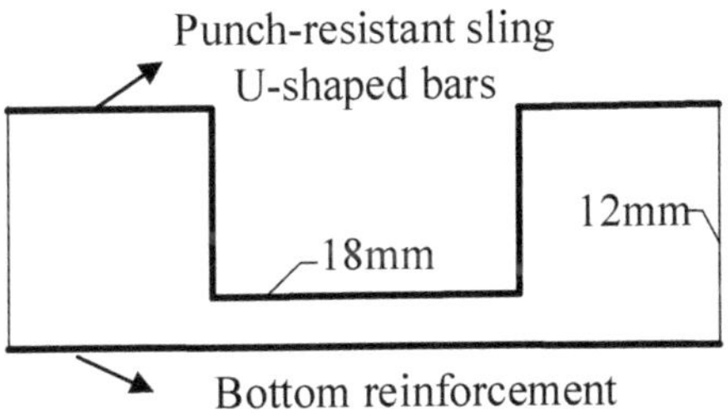

Figure 5. Diameter distribution diagram of the bottom.

The specimens were poured with C70 concrete and filled with C40 concrete. The longitudinal steel bars were HRB400 steel bars, where 18 steel bars had a diameter of 20 mm in the outer ring and 9 steel bars had a diameter of 10 mm in the inner ring, and HPB300 spiral stirrups had a spacing of 100 mm. The main reinforcement of the pier column and the steel end plate adopts perforated plug welding, the size of the steel end plate was the same as the section of the pier column, the thickness was 10 mm, and Q235 (Q235 indicates a steel plate with a yield strength of 235 MPa) steel was used. The structure diagram of the scaled pier column is shown in Figure 6.

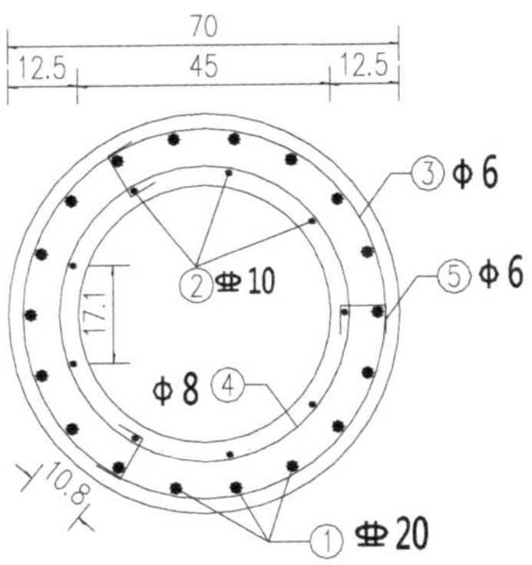

Figure 6. Structure diagram of the scaled pier.

3. Experimental Research

3.1. Specimen Design

In the whole research plan, seven specimens were included, and the research parameters included the depth of bearing insertion, the thickness of the base plate, and the construction method [7]. The focus of the paper was the vertical bearing capacity analysis of precast assembled piers; only specimen S7 was used, as shown in Table 2. Specimen S7 is a precast pier and bearing platform specimen. The caps under the pier column were hollowed out, no bottom plate was provided, temporary supports were set under the caps, and high-strength non-shrinkage cement grouting material was poured. This study focused on the bearing capacity under vertical load, so only the test results of specimen S7 are introduced.

Table 2. Specimen description.

Specimen	Pouring Method	Platform Height/mm	Socket Depth/D	Connection Method	Bottom Plate Thickness/mm
S1	cast-in-place	750		none	
S2	prefabricated	750	0.7	shear key	250
S7	components	750	0.7	shear key	0

Model construction and assembly were divided into three stages, which are the production of steel cages for pipe piers, the centrifugal forming of pipe piers, and the prefabrication and assembly of the caps. Some construction photos are shown in Figure 7.

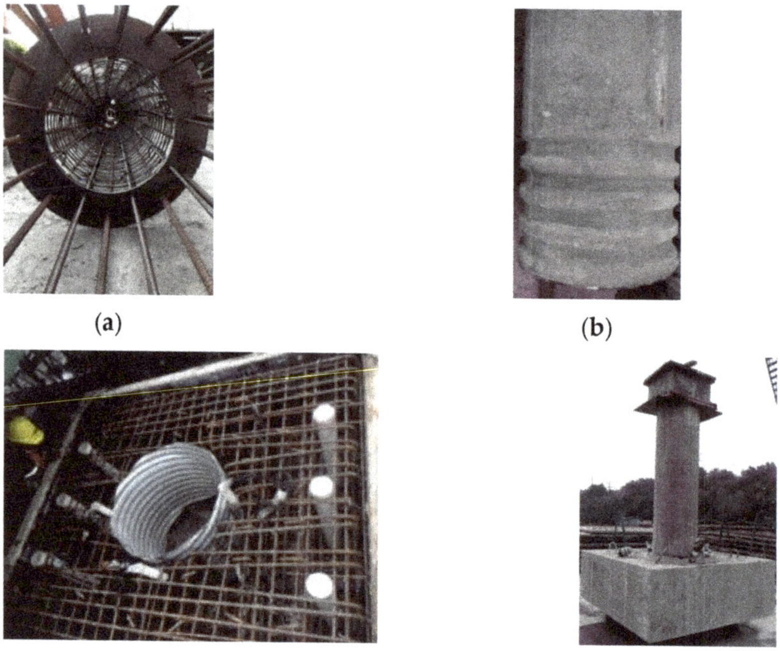

(c) (d)

Figure 7. The model construction process. (**a**) Bridge pier reinforcement cage; (**b**) pier shear key; (**c**) cap steel cage; (**d**) model test piece.

3.2. The Test Loading Device and the Loading Process

Test loading equipment was a 10,000 kN electro-hydraulic servo multifunctional structural test system, as shown in Figure 8. The vertical loading capacity was 10,000 kN in compression and 3000 kN in tension. The maximum stroke was ±300 mm. The horizontal loading capacity was 1500 kN, and the maximum stroke was ±400 mm. The test force measurement range and indication accuracy were 4%–100%, and the displacement resolution was 0.01 mm. Specimen S7 was subjected to axial force by means of the force control and the loading rate was 50 kN/min. The specimen was loaded in axial pressure only and slowly increased until it was damaged. The load increment was about 500 kN, and it was necessary to hold the load and observe the cracks and damage process at each load step. Because of the limitation of the test loading capacity, the maximum loading capacity of the testing machine was 10,000 kN. When the central loading was used, the specimen was not damaged when it reached 10,000 kN, so eccentric loading was applied after unloading in order to obtain the damage pattern of the precast piers. At eccentric distances of 5 cm and 10 cm, the specimens were not damaged. Until the eccentricity of the specimen reached 12.5 cm, the pier was damaged.

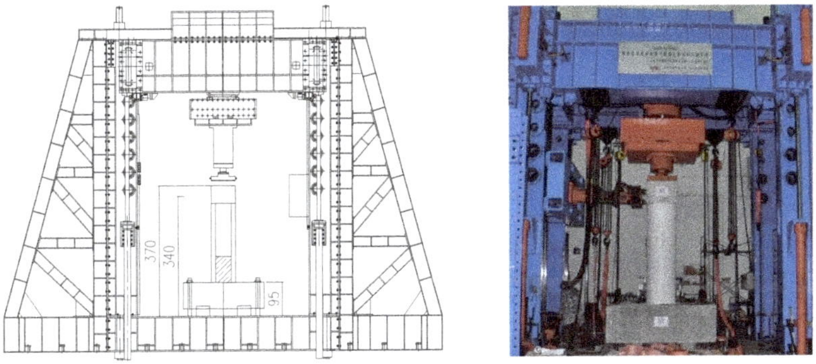

(**a**) (**b**)

Figure 8. Loading device. (**a**) Loading schematic; (**b**) loading photo.

3.3. Test Results

The specimen was subjected to axial compression loading. When the central compression axial pressure was loaded to 4500 kN, a micro-crack was found on the surface of the grouting material on the top surface of the bearing platform, and the crack width was 0.033 mm. When it was loaded to 10,000 kN, the maximum crack width of the grout was 0.08 mm. When the eccentric distance to the south side was changed to 5 cm, the bias loading was carried out. No cracks were found before the loading reached 3000 kN. When the load was 3500 kN, the original crack width of the grouting material was 0.04 mm; when the load was 7500 kN, the width of the first crack was 0.06 mm and the second crack appeared with a width of 0.024 mm. When it was loaded to 10,000 kN, the width of the first crack was 0.07 mm and the width of the second crack was 0.034 mm. Since the limit of the testing machine was reached, the eccentric distance was changed to 10 cm for bias loading. When the loading reached 3500 kN, the width of the first crack was 0.052 mm and the width of the second crack was 0.029 mm. When it was loaded to 10,000 kN, the width of the first crack was 0.072 mm and the width of the second crack was 0.046 mm. Then, the eccentric distance was changed to 12.5 cm for bias loading, and the loading reached 3500 kN; the width of the first crack was 0.06 mm and the width of the second crack was 0.04 mm. Continuing to load, horizontal cracks appeared on the tension side in the south, and vertical cracks appeared on the compression side in the north. When it reached about 9000 kN, there was an explosion, and a large piece of concrete on the compression side at the junction of the north side of the top of the pipe pier and the loading head peeled off, the concrete splashed, and the bearing capacity decreased. One longitudinal bar on the east side buckled, causing three stirrups to break; one longitudinal bar on the east side of the west side buckled, causing two stirrups to break. This shows that under eccentric compression, the unfilled hollow pipe pier becomes the weak link of the precast pipe pier. The damage mode is shown in Figure 9.

Figure 9. Top pier concrete spalling area of specimen S7.

After the experiment, it was found that four cracks appeared in the grouting material, which was approximately evenly distributed along the circumference. A ring-shaped crack appeared between the corrugated pipe and the cap concrete, indicating that the grouting material and the pipe pier formed a whole. Under different loading conditions, the crack width was larger under eccentric loading. There were two to three cracks on the east and west sides and one crack on the north and south sides, developing from bottom to top. The distribution of cracks is shown in Figure 10.

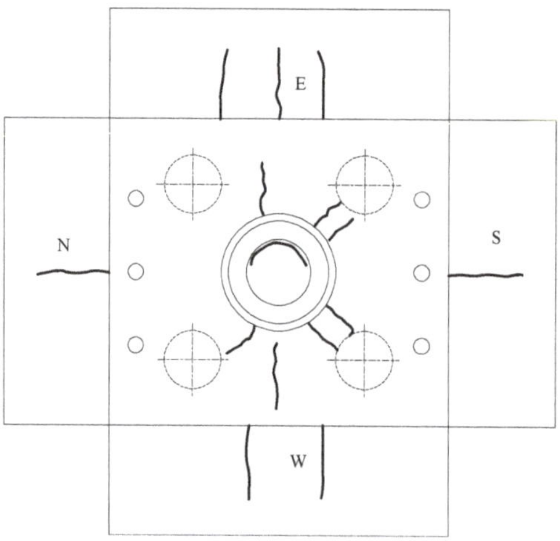

Figure 10. The distribution of cracks on the bottom and side of the cap.

Figure 11 shows the vertical load–displacement relationship of specimen S7 with different eccentricity. It can be seen that as the eccentric distance increased, the stiffness of the load–displacement curve increased at first. At the maximum eccentric distance, the load–displacement curve suddenly dropped, showing the characteristic of brittle failure.

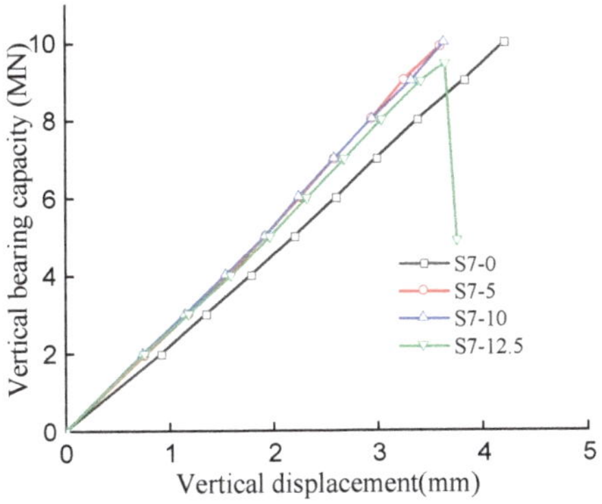

Figure 11. Vertical load–displacement relationship of S7 specimen under different eccentricity.

4. Finite Element Analysis of Specimens

4.1. Establishment and Verification of Model

Taking the 1:2 specimen model as the object, finite element analysis was carried out on the pile cap of different structures and designs, and the influence of different factors on the vertical bearing capacity of the socket connection was analyzed. The concrete material was simulated by the plastic damage constitutive model. Using 3D element modeling, the precast pipe piers were modeled by partitions. The ideal elastic–plastic model was used for the anchoring end and the bottom of the cap, and the damage–plastic model was used for the splicing surface, as shown in Figure 12.

Figure 12. Finite element model of the improved socket connection. (**a**) Precast member; (**b**) sectional view of the model.

The materials used in the model in this paper were mainly concrete and steel bars. The concrete adopted the C3D8R unit and the steel bar adopted the T3D2 unit. The details are shown in Table 3. The number of elements was 10366 and the number of nodes was 14939. The software used was ABAQUS.

The interaction part of the model adopted a Tie (binding) constraint and built-in steel bars. Metal bellows and the concrete surface at the joint were friction-treated. We set the coefficient of friction to 0.6 [23]. The socket part of the pipe pier was set with a width of 70 mm isometric annular shear keys to simulate the helical shear keys on the surface of the metal bellows and fix four piles. The steel bars were built-in and set as truss units. The first analysis step was set as a force-loading step and the second step was set as a displacement-loading step, and the displacement amplitude was added step by step. When the horizontal force was loaded, the vertical axial force was applied to simulate the dead load, and then the horizontal displacement was added until the specimen failed.

Table 3. Material properties of the structural model.

Name	Material Grade	Young's Modulus /GPa	Yield Stress /MPa	Tensile Stress /MPa	Plastic Strain
Concrete	C40	32.5	none		
	C70	39.2			
Steel	HRB300	200	300	450	0.09
	HRB400	200	400	540	0.09

The load–displacement curve obtained by vertical loading is shown in Figure 13. It can be seen that the initial stiffness calculated by the finite element was nearly the same as the test results. The experimental results showed a high capacity with brittle failure, whereas the finite element model showed moderate capacity with ductile behavior, because there was a sudden concrete failure in the top of the pier in the experiment. As the focus of this paper was the ultimate bearing capacity of the pile cap, it can be safely suggested that the vertical bearing capacity calculated by the finite element method is reliable.

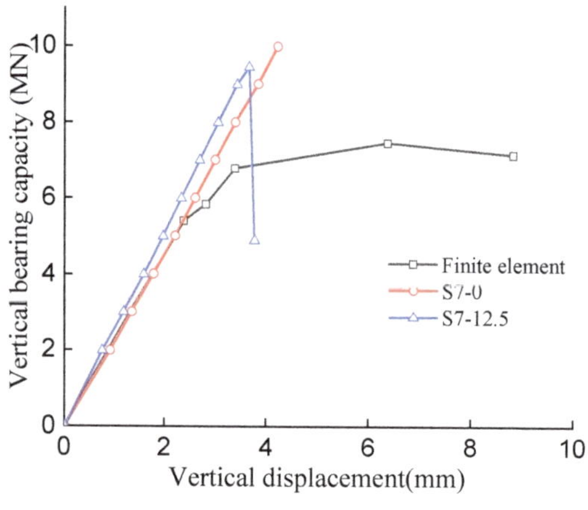

Figure 13. Comparison of the vertical bearing capacity.

It can be seen from the comparison of the above simulation results and test results that the established finite element model has a good simulation effect on the load–displacement curve of the precast socket-type pier. The finite element model can quantitatively simulate the whole process of precast socket-type piers under vertical and horizontal loads. The ultimate vertical bearing capacity of precast socket-type piers is related to the different configuration measures, the reinforcement of the cap, the depth of the socket, the thickness of the base plate of the cap, and the length of the pile, etc. The following parametric analysis of these influencing factors was carried out.

4.2. Parametric Analysis

4.2.1. Different Configuration Measures

To explore the influence of different connection methods of components on the vertical bearing capacity and the contribution of different parts to the overall bearing capacity, a comparison model with only shear keys and only the bottom plate was added. Model diagrams for different connection structures are shown in Figure 14 below. A summary and comparison chart of the vertical bearing capacity of the three comparison models is shown in Figure 15.

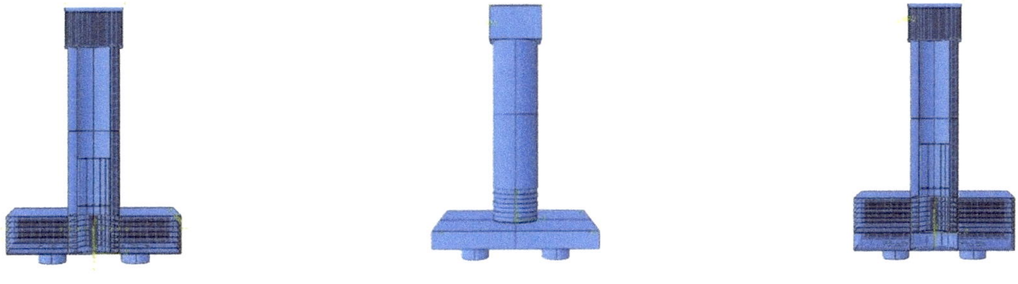

(a) (b) (c)

Figure 14. Diagrams of different structural models. (**a**) Shear key only; (**b**) bottom plate; (**c**) bottom plate + shear key.

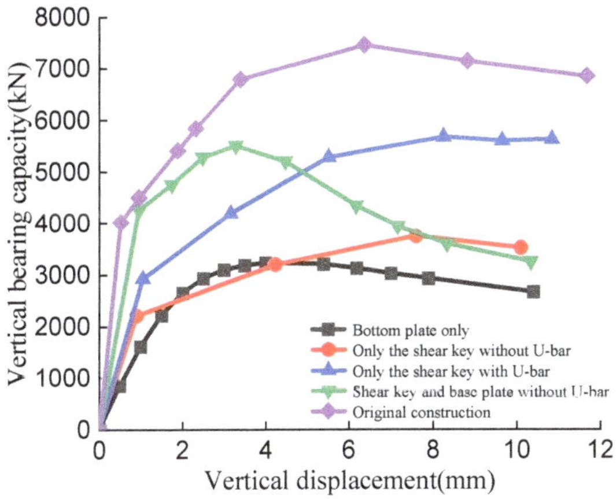

Figure 15. The contribution of different configuration measures.

It can be seen from Figure 15 that the specimen with shear keys, U-shaped steel, and a base plate had the largest vertical bearing capacity, followed by the model with only shear keys, and the ultimate bearing capacity of the base plate-only model was the smallest. Each part of the structure in the whole is listed as shown in Table 4. It can be seen that the vertical bearing capacity of the pile cap was mainly provided by the shear key, the bottom plate, and the U-shaped steel bar, and the contributions of the three parts were 40%, 34%, and 26%, respectively.

Table 4. Contribution rate of different structures.

	U-Shaped Rebar	Shear Key	Bottom Plate	Sum
Peak load/kN	2446	3859	3238	9543
Proportion	26%	40%	34%	100%

4.2.2. The Influence of the Cap Reinforcement

The steel reinforcement inside the cap plays an important role in the vertical bearing capacity of the entire component and is one of the main parts of the vertical bearing capacity contribution. The main factors affecting the vertical bearing capacity of the steel cap are the grade of the steel bar, the reinforcement ratio, and the location. This section mainly studies the effect of reinforcement ratio and U-shaped punching shear reinforcement on the vertical bearing capacity of the cap.

For different reinforcement ratios, there is no need to change the model in the finite element analysis; rather, we can just modify the diameter of the reinforcement material when assigning material properties, and then assign the material to the reinforcement section. The original diameter of the steel bar was 18 mm. For reference analysis, models with steel bar diameters of 8 mm, 16 mm, 20 mm, 25 mm, and no steel bar were created respectively. The finite element analysis results are shown in Figure 16. It can be seen that as the reinforcement ratio increased, the vertical bearing capacity also increased, showing a positive correlation. Specimens with low reinforcement ratios reached the plastic stage earlier than those with high reinforcement ratios. The bearing capacity of the specimens with diameters less than 20 mm appeared to decrease after reaching the maximum bearing capacity. The vertical bearing capacity of the model with a diameter of 8 mm and no steel bar had the largest and most obvious decrease, indicating that the steel bar has a great effect on the vertical bearing capacity. When the longitudinal reinforcement bar of the pile cap was increased from 18 mm to 25 mm, there was a 20% increase in the vertical bearing capacity. It can be seen that increasing the steel bar diameter of the cap is an effective method to improve the vertical bearing capacity of the cap.

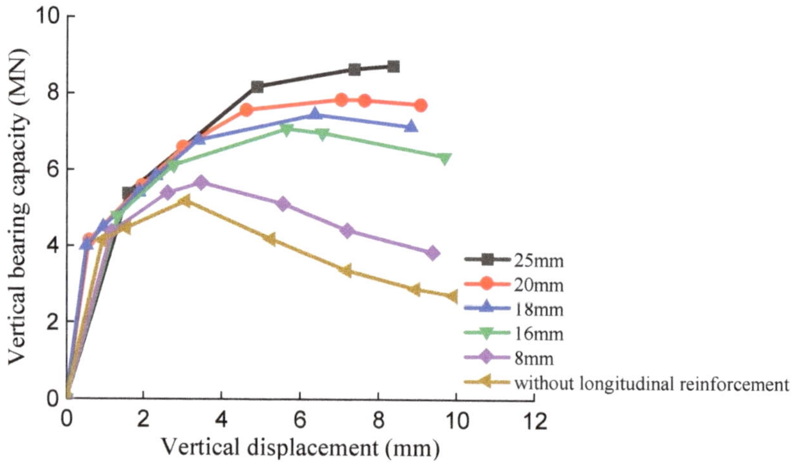

Figure 16. Comparison of different reinforcement ratios.

4.2.3. The Effect of Socket Depth

The socket depth is the most important part of the structure of the socket precast component, and the socket depth affects the horizontal and vertical bearing capacity of the entire structure to a large extent. In the vertical behavior, the socket depth will affect the bottom plate thickness of the bearing plinth. In the case of a certain height of the platform, the greater the socket depth, the smaller the thickness of the reserved bottom plate, and the corresponding vertical bearing capacity will also be greatly reduced. On the horizontal structure, when bearing lateral horizontal force, the smaller the socket depth is, the greater the lateral displacement of the corresponding pier column that is likely to occur. The lateral bearing capacity is poor, and the structure is more prone to damage. The research on socket depth has been focused on the influence on the horizontal bearing capacity, but there is less research on the influence on the vertical bearing capacity. To study the vertical bearing capacity, a set of specimens with different values of socket depth was designed. This group of specimens had the same bottom plate thickness, and the socket depth was the only variable, which was 0.5D, 0.7D, and 1.5D, respectively. The base plate thickness was 0.36D.

The pile cap height was 0.86D, 1.07D, and 1.86D according to the arrangement of socket depth from large to small.

It can be seen from Figure 17 that when the bottom plate thicknesses were the same, the socket depth had a certain influence on the bearing capacity of the component. As the socket depth increased, the vertical bearing capacity of the cap increased. When the socket depth was small, the growth trend of the ultimate bearing capacity was more obvious. When the socket depth was greater than 1.0D, the increase was relatively small. As far as the ultimate bearing capacity is concerned, the bearing depth of 1.5D is twice that of 0.7D, but its limit value was only about 11% larger than that of 0.7D, and the increase was not large. This shows that when the thickness of the pile cap reaches a certain level, the failure mode of the structure will change from the punching failure of the pile cap to the compression failure of the bridge pier, so the final vertical bearing capacity calculated will not increase all the time.

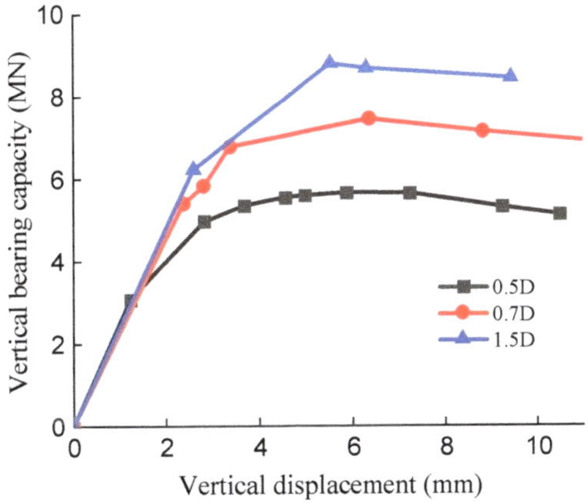

Figure 17. The influence of socket depth.

4.2.4. The Influence of the Thickness of the Base Plate

The precast pier column and the cap were connected by the hollow part in the middle of the cap, and then high-strength grouting material was poured to strengthen the connection wholeness. Under the action of vertical force, in addition to the vertical reaction force generated by the shear key inside the hollow and the U-shaped punching-resistant steel bar inside the cap, there was also a part of the vertical force provided by the base plate at the bottom of the cap which was in contact with the pier column. To explore the influence of different base plate thicknesses on the vertical bearing capacity of components, a total of five finite element models with different plate thicknesses were added, which were 15 cm, 20 cm, 25 cm, 35 cm, and 45 cm. The pile cap heights were 65 cm, 70 cm, 75 cm, 80 cm, and 85 cm in sequence according to the thickness of the bottom plate.

It can be seen from the Figure 18 that as the thickness of the bottom plate increased, the overall vertical bearing capacity increased. As the thickness of the bottom plate increased, the displacement corresponding to the ultimate load also increased. When the thickness of the bottom plate was 45 cm, there was only a slight descending section after reaching the limit. It is not difficult to see that the greater the thickness of the base plate, the greater the punching shear resistance provided, and the vertical bearing capacity also increases. When the plate thickness increased from 25 cm to 45 cm, there was a 21% increase in the vertical bearing capacity.

To explore the thickness of the base plate, the precast components can be compared with the cast-in-place components, as shown in Figure 19. The pier part of the base plate was included in the finite element analysis. If the mechanical properties of the socket-type cap were completely equal to the cast-in-place result, the thickness of the bottom plate needed to be increased to 450 mm. At this time, the steel strain of the base plate of the two caps was nearly the same as the deformation of the center of the cap.

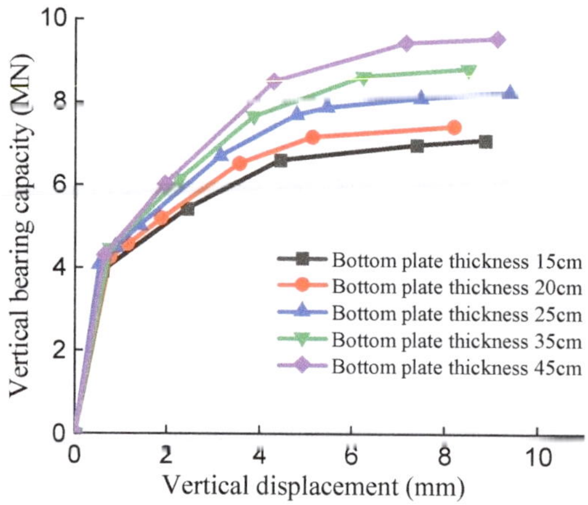

Figure 18. Comparison of different base plate thicknesses.

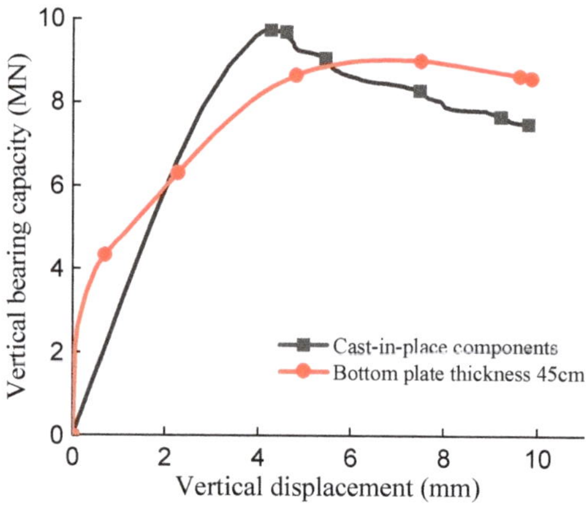

Figure 19. Comparison of the vertical bearing capacity between cast-in-situ and precast bottom plates.

4.2.5. The Effect of Pile Length

The pile cap is an important part of the socket-connected precast pier. As a component directly connected to the ground, the pile is a vertical force-transmitting component that transmits the force on the superstructure to the ground. The length of the pile reflects the restraint at the bottom of the cap. The shorter the pile length is, the more likely the bottom of the cap is to be consolidated at the position of the pile. The longer the pile length is, the closer the cap bottom is to the hinge at the position of the pile. To explore the influence of the pile length on the vertical bearing capacity of the members, on the basis of the S2 specimen, only the pile length of the specimen was changed, and other conditions remained unchanged; pile lengths of 0.2 m, 0.5 m, 1.5 m, and 2.5 m were analyzed by the finite element method. The results are summarized in Figure 20. It can be seen that with the increase of the pile length, the vertical bearing capacity of the member gradually decreased. Regardless of the length of the pile, its bearing capacity curve was relatively flat and had better performance. When considering the flexible deformation of the pile, the vertical bearing capacity of the pile cap decreased. In the case analyzed in this paper, when the pile length increased from 0.2 m to 2.5 m, the vertical bearing capacity decreased by 12%. Next, it is necessary to carry out the vertical bearing capacity analysis considering the combined action of piles and soil in detail.

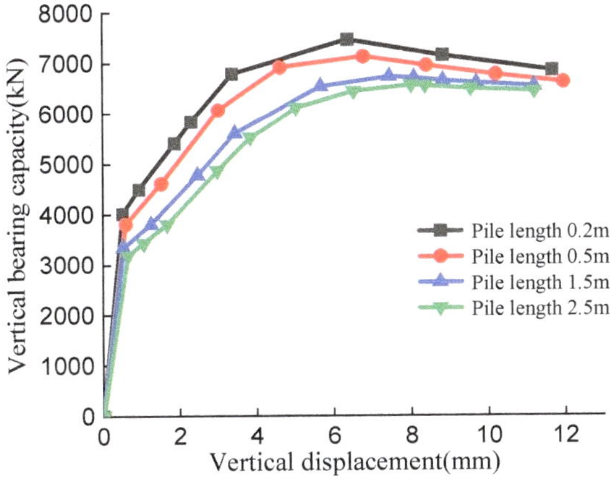

Figure 20. Comparison of different pile lengths.

4.2.6. Influence of Friction Coefficient

The coefficient of friction is a property of the interaction between different materials. Different materials, different grades of concrete, and rebar interactions will have different coefficients of friction. The choice of the basic friction coefficient of 0.6 in this paper is based on the information provided by the previous literature and the consideration of the safety of the vertical bearing capacity. To explore whether different friction coefficients will affect the results of finite element analysis, finite element models with friction coefficients of 0.4, 0.8, 1, and 1.2 were assigned for simulation analysis. Figure 21 shows the vertical displacement–load diagram of each friction coefficient obtained by vertical displacement loading on the basis of specimen S2, only changing the friction coefficient of the material interaction.

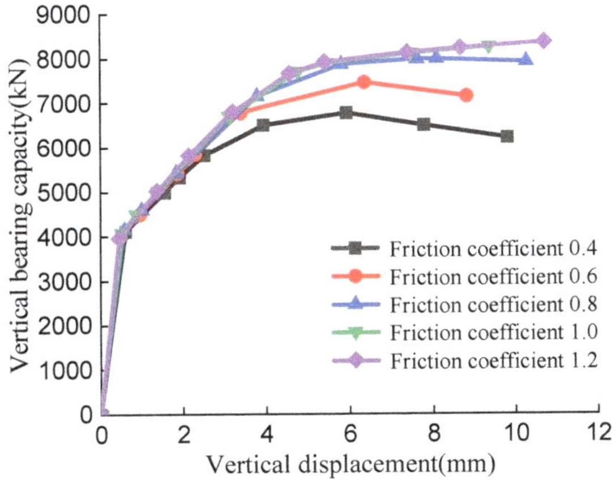

Figure 21. Comparison of different friction coefficients.

It can be seen from the above analysis that the increase of the friction coefficient increased the vertical bearing capacity of the structure and the selection of different friction coefficients had a certain impact on the vertical bearing capacity. The larger the friction coefficient is, the greater the vertical bearing capacity is. With the increase of the friction coefficient, the increase degree of the vertical bearing capacity decreased significantly. The final finite element analysis results of the two friction coefficients of 1.0 and 1.2 are not significantly different. The exact value of friction coefficient needs to be determined by future experimental research.

5. Analysis of Full-Scale Bridge Piers

5.1. Finite Element Modeling and Analysis

To explore the behavior of the original socket connection, a full-scale cast-in-place model and a precast model were established respectively, and vertical displacement loading was performed to obtain the displacement and load curves.

The displacement–load curve of the finite element results is drawn in Figure 22, and the vertical load limit values of full-scale and precast piers are listed in Table 5 below. It can be seen that the vertical bearing capacity of the full-scale member was about 4 times that of the scaled model, which is in line with the theoretical calculation and proved the correctness and accuracy of the model. The results of the finite element analysis can simulate the actual engineering situation well.

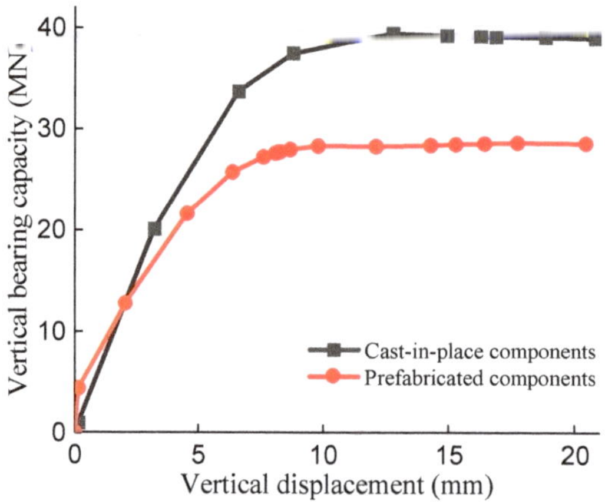

Figure 22. Displacement–load comparison diagram of full-scale socket connection.

Table 5. Summary of the vertical bearing capacity (kN).

Type of Cap	Full-Scale	Reduced Scale1:2
Cast-in-place cap	39,534	9752
Precast cap	28,673	7456

5.2. Bending Moment and Axial Force Correlation Analysis

To further study the factors affecting the mechanical properties of the full-scale model, parameter analysis of axial compression was performed. In practical engineering applications, the pile part of the cushion cap needs to be inserted deep into the ground to ensure the stability of this component, and the pile length is larger than the full-scale model mentioned above. To make the model closer to the actual project, a full-scale model with a pile length of 5 m was added. First, the model was subjected to vertical displacement loading to obtain the maximum vertical load N of the model, and then horizontal displacement loading was applied to the components with 0.1, 0.2, 0.3, 0.4, and 0.5 times the limit of the axial load N to obtain the horizontal bearing load; 0.1 N is 0.1 times the limit axial force N. Taking the length of the bridge pier as the moment lever arm, the ultimate bending moment that the member can bear in the plastic hinge region can be obtained. The full-scale precast model was equally magnified according to the experimental model (Figure 4) with a scale of 1:2, as was the full-scale cast-in-situ model.

In addition, a full-scale model with a bottom plate thickness of 0.8 m was added. The model was established to explore the effect of the bottom plate thickness on the horizontal load.

The maximum horizontal load force was extracted from the horizontal displacement load curve under different axial loads. Taking the height of the bridge pier as the lever arm, M–N correlation diagrams of different models were made. Assuming that the cracking load is the load corresponding to the reinforcement stress of 200 MPa in the pier, the relationship between the horizontal cracking load and the axial force is also shown in Figure 23.

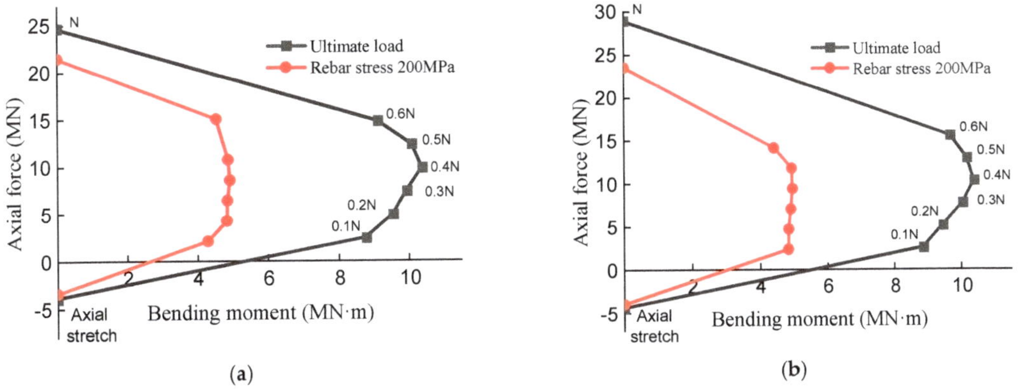

Figure 23. Summary of the M−N diagram. (**a**) Bottom plate thickness is 0.5 m; (**b**) bottom plate thickness is 0.8 m.

It can be seen that when the axial pressure ratio was small, with the increase of the axial pressure, the horizontal bearing capacity of the pier increased, and the corresponding bending moment also increased, showing a positive correlation. When the axial pressure reached about 0.4, there was a situation where the maximum horizontal load occurred. When continuing to increase the axial pressure, the bending moment did not increase but decreased. This is because as the axial pressure continued to increase, the phenomenon of bias pressure appeared under the combined action of the horizontal load, and the corresponding second-order effect was produced on the pier column, which reduced the bearing capacity.

Under the abovementioned conditions, the vertical ultimate bearing capacity was not much different when the bottom plate thickness was 0.5 m and 0.8 m, respectively. This is related to the failure form of the bridge pier mentioned above. Therefore, based on the abovementioned full-scale model, a cap-only loading model was added to carry out by vertical displacement loading. When the thickness of the bottom plate increased, the vertical bearing capacity of the precast cap also increased accordingly. When the thickness of the base plate increased from 0.5 m to 0.8 m, there was a 20% increase in the vertical bearing capacity.

To explore the vertical bearing capacity gap between the prototype cast-in-place pier scheme and the precast pier scheme, the prototype cast-in-place pier with a height of 1.5 m and a pile length of 5 m was vertically loaded, and the results are shown in Figure 24. It can be seen that the vertical bearing capacity of the precast scheme with a base plate thickness of 0.5 m was 72% of that of the prototype cast-in-place scheme. When the thickness of the bottom plate was increased to 0.8 m, that is, the height of the cap was 1.8 m and the bearing capacity of the precast cap was about 87% of that of the cast-in-place cap. The vertical bearing capacity of the two was relatively closer. At the same time, considering the cost factor, it is finally suggested that the thickness of the base plate of the full-scale precast bridge pier scheme should be 0.8 m, that is, the total height of the precast cap is 1.8 m.

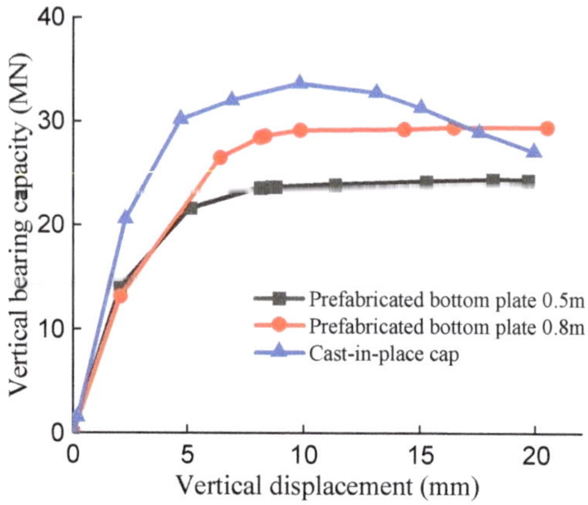

Figure 24. Comparison of different base plate thicknesses.

6. The Derivation of Formula

6.1. Force Model Analysis

It is assumed that the vertical bearing capacity of the cap is mainly composed of three parts: the bending of the bottom plate, the shear key of the side wall, and the suspension of the hanging bars. According to the improved socket structure proposed in this paper, the overall force layout of the components is shown in Figure 2, including side shear keys, hanging bars, and bottom bars. The composition of the vertical bearing capacity of specific components is shown in Figure 25.

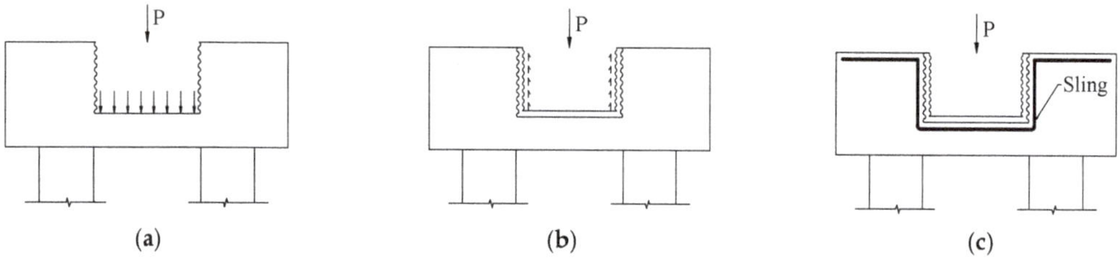

Figure 25. Components of vertical bearing capacity. (**a**) Bottom plate bending; (**b**) side wall shear key; (**c**) hanging bar suspension.

Shear keys provide vertical loads on the one hand and bending resistance on the other. The two are superimposed together to participate in resisting external loads, so when calculating the vertical bearing capacity, only part of the shear keys is considered to participate. In the process of vertical loading, half of all shear keys was assumed in the paper.

6.2. Design Formula

Referring to the existing punching design formula in China, the following design formula is proposed.

$$\gamma_0 F_{ld} \leq 0.35\beta_h f_{td} U_m h_0 + 0.75 f_{sd} A_{su} + V_s \quad (1)$$

In the formula, γ_0 is the safety factor; F_{ld} is the design value of maximum punching force for socket connection; β_h is the dimension effect coefficient for the height of the bottom plate of the socket connection. When $h \leq 300$ mm, β_h is 1.0 h; when $h \geq 800$ mm, β_h is 0.85 h. In between, β_h takes the value by straight line insertion; f_{td} is the design value of tensile strength of concrete axial center of bottom plate of the socket connection; U_m is the perimeter of the cross-sectional area of the broken blank cone at a distance of $h_0/2$ from the pier bottom (top) action surface of the retaining socket member; h_0 is the distance from the bottom of the pier to the main reinforcement of the bottom plate of the socket connection; f_{sd} is the design value of tensile strength of vertical U-shaped steel bar around the socket connection; A_{su} is the total cross-sectional area of the vertical U-shaped reinforcement around the socket connection; V_s is the shear force value provided by the shear key [24,25].

$$V_S = \frac{1}{2}\theta \cdot DX \cdot 0.42 f_{cu}^{2/3} \tag{2}$$

In the formula, f_{cu} is the concrete compressive strength grade; D is the specimen pier diameter; θ is the shear key area angle value and is suggested to take $\pi/4$.

6.3. Formula Validation

The scaled model and full-scale model studied in this paper were calculated using the abovementioned formulas. The concrete of cap was C40, the design value of axial tensile strength $f_t = 1.71$MPa, socket depth X = 0.5 m, the thickness of the bottom plate was 0.15, 0.2, 0.25, 0.35, and 0.45 m, respectively, protective layer thickness c = 30 mm, allowable shear stress of concrete $[\tau_c]$ was 1.35 MPa. The remaining correlation coefficients were taken according to the requirements of the abovementioned formula. We substituted the data into the abovementioned formula for calculation, and summarized all calculation results as shown in Table 6. In practical engineering use, to ensure the safety of use, the cracking load and the development of cracks are usually used as the standard for normal operation, and the crack width of 0.2 mm is usually used as the standard. It was assumed that the maximum stress of the tensile reinforcement shall not exceed 200 MPa. Therefore, the stress–load diagram of the steel bar at the maximum stress of the steel bar in the finite element model was selected, and the vertical load data with a stress of 200 MPa was extracted as the design value, which is summarized in the following table. This result can fully reserve the safe bearing capacity space to achieve the purpose of ensuring the safety of use.

Table 6. Summary of calculation results.

Calculation Formula	Bottom Plate Thickness/mm	Design Formula Calculated Value/kN	Finite Element Simulation Value/kN (Cracking Load)	Safety Multiple
Specification for Design of Concrete Structures of the scaled model	150	678	2860	1.26–4.22
	200	971	2976	
	250	1265	3163	
	350	1855	3054	
	450	2453	3091	
Design formula of scaled model	150	1809	2860	1.33–1.6
	200	1894	2976	
	250	1979	3163	
	350	2149	3054	
	450	2322	3091	
Full-scale model Design formula	500	7026	7962	1.13

From the comparison of the results in the above table, it can be seen that the design load of the model was taken as the cracking value of the finite element load. The calculated values of the existing formula and the design formula proposed in this paper were all smaller than the value of the finite element model, which further shows the reliability of the model and the correctness of the design formula in this paper. The multiples of the safety factor were all greater than 1, indicating the reliability of the calculated value of the formula. By comparing the formula value and safety factor, it shows that the design formula proposed in this paper can be well applied.

7. Conclusions

1. The use of rough interfaces at the bottom of the pier, U-shaped hanging bars in the bottom groove, high-strength grouting material in the seam, and large-diameter metal corrugated pipes are effective measures to enhance the vertical bearing capacity of the socket cap. Compared with the tradition socket type, the improved socket type can improve the vertical bearing capacity of the cap and can be used for occasions with high vertical bearing capacity requirements.
2. In the case of a certain height of the platform, the greater the depth of the socket, the smaller the thickness of the bottom plate. Similarly, the smaller the socket depth, the greater the thickness of the bottom plate. Therefore, there is a negative correlation between the socket depth and the bottom plate thickness, that is, one increases and the other decreases. The purpose of the research is to find a socket depth that can maximize the bearing capacity of the components and reduce the cost under the condition that the height of the cap is fixed. Therefore, it is necessary to select an optimal socket depth according to the force characteristics to meet these two conditions.
3. The vertical bearing capacity of the pile cap is mainly provided by shear keys, the bottom plate, and the U-shaped steel bar. The contribution of the three parts is about 40%, 34%, and 26%, respectively.
4. The vertical bearing capacity of the pile cap increased with the increase of the diameter of the hanging bars and the longitudinal bars of the bottom plate. With the increase of socket depth, the flexural bearing capacity of the pier increased. When the plate thickness was increased from 25 cm to 45 cm, there was a 21% increase in the vertical bearing capacity. When the thickness of the bottom plate increased to a certain extent, the vertical bearing capacity of the pile cap was greater than that of bridge piers. The vertical bearing capacity of the pile cap was not a weak link. When considering the flexible deformation of the pile, the vertical bearing capacity of the pile cap decreased. When the equivalent pile length increased from 0.2 m to 2.5 m, there was a 12% increase in the vertical bearing capacity.
5. When the thickness of the base plate of the cap was 0.8 m and the total height was 1.8 m, the vertical bearing capacity of the precast cap was about 87% of that of the cast-in-place cap. Considering the cost factor, it is finally recommended to select 0.8 m for the thickness of the base plate of the full-scale precast bridge pier scheme in the case.
6. This paper proposed an estimation formula for the vertical bearing capacity of the socket cap, which has a certain safety factor and can be used for design reference. Additional experimental research is needed to verify the accuracy of the finite element analysis results and formula calculation in this paper.

Author Contributions: Conceptualization, J.G. and X.Y.; methodology, J.G. and L.L.; formal analysis, L.L.; funding acquisition, S.L.; validation, S.L. and X.Y.; writing—original draft, J.G. and L.L.; writing—review and editing, S.L. and X.Y. All authors have read and agreed to the published version of the manuscript.

Funding: This research was funded by China Railway 22TH Bureau Group 3RD Engineering Corporation Limited, grant number CR2203-S2-SX308-JS-2021-009; and funded by Shanghai Institute of Technology, grant number ZQ2021-28; and also funded by Shanghai Science and Technology Commission, grant number 2100240100.

Institutional Review Board Statement: Not applicable.

Informed Consent Statement: Not applicable.

Data Availability Statement: Not applicable.

Acknowledgments: Authors would like to acknowledge the finical support provided from China Railway 22TH Bureau Group 3RD Engineering Corporation Limited and Shanghai Urban Construction Design Research Institute (Group) Co., Ltd.

Conflicts of Interest: All authors declare no conflict of interest.

References

1. Zhang, Q.; Alam, M.S. State-of-the-art review of seismic-resistant precast bridge columns. *J. Bridge Eng.* **2020**, *25*, 03120001. [CrossRef]
2. Culmo, M.P.; Marsh, L.; Stanton, J. *NCHRP Web-Only Document 242: Recommended AASHTO Guide Specifications for ABC Design and Construction*; The National Academies Press: Washington, DC, USA, 2018.
3. Wang, Z.; Qu, H.; Li, T.; Wei, H.; Wang, H.; Duan, H.; Jiang, H. Quasi-static cyclic tests of precast bridge columns with different connection details for high seismic zones. *Eng. Struct.* **2018**, *158*, 13–27. [CrossRef]
4. Khaleghi, B.; Schultz, E.; Seguirant, S.; Marsh, L.; Haraldsson, O.; Eberhard, M.; Stanton, J. Accelerated bridge construction in Washington State: From research to practice. *PCI J.* **2012**, *57*, 34–49. [CrossRef]
5. Han, Y.; Liu, Z.H.; Wang, L.L. Test for influence of socket connection structure on dynamic response of prefabricated pier under vehicle collision. *KSCE J. Civ. Eng.* **2022**, *26*, 1188–1202. [CrossRef]
6. Zeng, Z.; Xu, Y.; Wang, Z.G. Experimental study of the axial bearing capacity of prefabricated pier-cap with socket connection. In Proceedings of the 9th Asia-Pacific Young Researchers and Graduates Symposium, Shanghai, China, 22–25 December 2019.
7. Xu, Y.; Zeng, Z.; Wan, Z.G. Experimental studies of embedment length of precast bridge pier with socket connection to pile cap. *Eng. Struct.* **2021**, *233*, 111906. [CrossRef]
8. Sadeghian, P.; Fam, A. Bond-slip analytical formulation toward optimal embedment of concrete-filled circular FRP tubes into concrete footings. *J. Eng. Mech.* **2010**, *136*, 524–533. [CrossRef]
9. Kurama, Y.C.; Sritharan, S.; Fleischman, R.B.; Restrepo, J.I.; Henry, R.S.; Cleland, N.M.; Ghosh, S.K.; Bonelli, P. Seismic-resistant precast concrete structures: State of the art. *J. Struct. Eng.* **2018**, *144*, 03118001. [CrossRef]
10. Azizinamini, A. Accelerated bridge construction. *J. Bridge Eng.* **2020**, *25*, 02020002. [CrossRef]
11. Marsh, M.L.; Wernli, M.; Garrett, B.E. *NCHRP Report 698: Application of Accelerated Bridge Construction Connections in Moderate-to-High Seismic Regions*; The national Academies Press: Washington, DC, USA, 2011.
12. Haraldsson, O.; Janes, T.; Eberhard, M.; Stanton, J. Seismic Resistance of Socket Connection between Footing and Precast Column. *J. Bridge Eng.* **2013**, *18*, 910–919. [CrossRef]
13. Osanai, Y.; Watanabe, F.; Okamoto, S. Stress transfer mechanism of socket base connections with precast concrete columns. *ACI Struct. J.* **1996**, *93*, 266–276.
14. Canha, R.M.F.; Borja Jaguaribe, K., Jr.; de Cresce El Debs, A.L.H.; El Debs, M.K. Analysis of the behavior of transverse walls of socket base connections. *Eng. Struct.* **2009**, *31*, 788–798. [CrossRef]
15. Mohebbi, A.; Saiidi, M.S.; Itani, A.M. Shake table studies and analysis of a precast two-column bent with advanced materials and pocket connections. *J. Bridge Eng.* **2018**, *23*, 04018046. [CrossRef]
16. Mohebbi, A.; Saiidi, M.S.; Itani, A. *Development and Seismic Evaluation of Pier Systems w/Pocket Connections, CFRP Tendons, and ECC/UHPC Columns*; Center for Civil Engineering Earthquake Research, Department of Civil and Environmental Engineering, University of Nevada: Reno, Nevada, 2017.
17. Wang, Z.Q.; Li, T.T.; Qu, H.Y.; Wei, H.; Li, Y. Seismic performance of precast bridge columns with socket and pocket connections based on quasi-static cyclic tests: Experimental and numerical study. *J. Bridge Eng.* **2019**, *24*, 04019105. [CrossRef]
18. Cheng, Z.; Liu, D.; Li, S.; Wang, J.; Zhang, J. Performance characterization and design recommendations of socket connections for precast columns. *Eng. Struct.* **2021**, *242*, 112537. [CrossRef]
19. Cheng, Z.; Sritharan, S. Outdoor test of a prefabricated column-pile cap-pile system under combined vertical and lateral loads. *J. Bridge Eng.* **2020**, *25*, 04020052. [CrossRef]
20. Zhang, G.D.; Han, Q.; Xu, K.; Du, X.; He, W. Experimental investigation of seismic behavior of UHPC-filled socket precast bridge column-foundation connection with shear keys. *Eng. Struct.* **2021**, *228*, 111527. [CrossRef]
21. Chan, T.K.; Poh, C.K. Behaviour of precast reinforced concrete pile caps. *Constr. Build. Mater.* **2000**, *14*, 73–78. [CrossRef]
22. Cheng, Z.; Sritharan, S. Side shear strength of preformed socket connections suitable for vertical precast members. *J. Bridge Eng.* **2019**, *24*, 04019025. [CrossRef]
23. AASHTO GSCB INTERIM-2003. *Guide Specifications for Design and Construction of Segmental Concrete Bridges (Revision 2)*; American Association of State Highway Transportation Officials: Washington, DC, USA, 1999.
24. Lai, L. *Analysis on Mechanicl Charteristics of Improved Socket Type Bridge Pier and Caps*; Shanghai Institute of Technology, Academic Degree of Master of Engineering: Shanghai, China, 2022.
25. Li, J. *Shear Behavior and Continuous Rigid-Frame System of Precast Segmental Bridges in the Fourth Nanjing Yangtze River Bridge*; Southeast University, Academic Degree of Master of Engineering: Nanjin, China, 2010.

buildings

MDPI

Article

Feasibility Evaluation of Novel High-Damping Rubbers as Energy-Dissipation Material under Axial Dynamic Load for Damper Devices

Tzyy Wooi Teh, Chee Ghuan Tan * and Mohd Zamin Jumaat

Department of Civil Engineering, Faculty of Engineering, Universiti Malaya, Kuala Lumpur 50603, Malaysia
* Correspondence: tancg@um.edu.my

Abstract: High-damping rubber (HDR) material has been widely used in bearings for seismic-isolation devices in structures. Nevertheless, HDR has not yet been developed in dampers to reduce the response of structures to earthquake excitations by dissipating the energy applied to the structures under direct axial load. The purpose of this paper was to evaluate the feasibility of using novel hyperelastic composite material (HECM), which is an HDR material, in experimental investigations to determine its damping ratio, compressibility, and elasticity behavior under axial dynamic load for the development of novel dampers in the future. First, a series of tests on HECM was conducted using the double-shear method to determine the most suitable sample for a purely dynamic compression test. Subsequently, the HECM was used in a device working as a scaled-down damper under both direct tension and compression dynamic load conditions, and pure direct compression dynamic load conditions were tested. Various thicknesses of the HECM (6, 8, and 10 mm) used in the testing damper were examined under a constant force with various frequencies of 0.01, 0.1, 0.25, and 0.5 Hz. The results show that the 10 mm thick HECM can provide a high damping ratio of 10% to 13% under axial conditions. Hence, this study is important for evaluating HECM, which has the potential for use in developing a full-scaled rubber damper system to resist axial force in the future. The damper is a novel rubber damper with high damping capability to dissipate energy under axial load. Furthermore, the damper can serve as an alternative choice that is more durable and overcomes the current weaknesses of passive dampers.

Keywords: high-damping rubber; hyperelastic composite material; axial dynamic load; energy dissipation; seismic; damper

Citation: Teh, T.W.; Tan, C.G.; Jumaat, M.Z. Feasibility Evaluation of Novel High-Damping Rubbers as Energy-Dissipation Material under Axial Dynamic Load for Damper Devices. *Buildings* **2022**, *12*, 1917. https://doi.org/10.3390/buildings12111917

Academic Editor: Fulvio Parisi

Received: 25 September 2022
Accepted: 3 November 2022
Published: 7 November 2022

Copyright: © 2022 by the authors. Licensee MDPI, Basel, Switzerland. This article is an open access article distributed under the terms and conditions of the Creative Commons Attribution (CC BY) license (https://creativecommons.org/licenses/by/4.0/).

1. Introduction

High-damping rubber (HDR) is a passive vibration-control material used in seismic-isolation or base-isolation systems. The common types of base isolators that utilize HDR are high-damping natural rubber bearing (HDRB) and lead rubber bearing (LRB) [1]. HDRB is an elastomeric bearing that consists of thin layers of high-damping rubber and steel plates with an alternate layer. The low shear modulus of the elastomer controls the horizontal stiffness of the bearings, whereas steel plates provide high vertical stiffness and prevent the bulging of the rubber [2]. HDR can also enhance the seismic resilience of structures such as school buildings and tunnel-lining structures by acting as isolators in structural frames or in the joints between structures [3,4] and will be a new approach to consider in seismic resilience. HDRB provides a horizontal damping ratio of 10–20% [5,6] and can be improved by incorporating additional fine carbon blocks, resins, oils, and proprietary fillers. Natural rubber facilitates flexibility through its ability to move and return to its normal position [2,7]. According to our literature review, HDR is mainly used in bearing or base isolation that uses the lateral behavior of HDR in shear [8] but not in the axial force behavior of HDR. The restoring force characteristics, the effect of loading frequency, and the effect of large displacement are among the important properties that need to be designed, and it was stated that the restoring force characteristics of HDR are similar to those of a linear viscous damper [9]. Restoring forces are expressed by stiffness K and linear damping coefficient C. The dynamic properties and strength of HDR show the possibility of the benefits of seismic dampers because (1) they permit energy dissipation, even for small horizontal deformations or micro-vibrations [10]; (2) there is a lower permanent deformation after seismic activities; and (3) the increase in temperature is small, which does not affect the dynamic performance. Typically, HDR can function well from −30 °C to +50 °C [11,12], and the damping ratio seems to be not correlated with temperature [13].

Dampers currently available in the industry can be utilized according to structural requirements, such as friction dampers, metallic dampers, buckling restrained braces, tuned liquid/mass dampers, and viscoelastic and viscous-fluid dampers [14,15]. Dampers work under direct axial conditions; viscous-fluid velocity-dependent devices in the form of cylinder type and viscoelastic (VE) dampers are popularly used for energy dissipation under seismic and wind-induced dynamics due to their excellent performance and reduction effect on the relative displacement of structures [16,17]. However, when the end of the damper moves with respect to the others, viscoelastic materials or fluids undergo shearing behavior. The heat produced from the shearing action dissipates energy. The mechanical or dynamic properties that require attention are the response of the storage modulus, loss modulus, damping coefficient, and loss factor to the frequency of motion, strain amplitude, and temperature [18]. As the number of cycles increases, the increase in temperature reduces the storage and loss moduli, resulting in a reduction in stiffness and energy dissipation capacity [19–22]. The performance of the viscous fluid is significantly reduced once the liquid leaks [23], which leads to frequent maintenance of the damper, thereby increasing the overall life cost in the design life period of the structure. Viscous-fluid dampers can operate at ambient temperatures ranging from −40 °C to 70 °C, but variations in temperature in the range of 0 °C to 50 °C result in variations in the damping ratio of +44% to −25% [21]. However, other dampers, such as displacement-dependent devices, do not carry vertical loads [24,25], and their behavior mainly depends on displacement rather than velocity. Both devices are well-defined in EN 15129 [26]. Viscous dampers generally provide a range of damping ratios of 10% to 30% [27], with an average of approximately 15%. In other words, if any device achieves more than a 10% damping ratio, it has a function similar to that of a viscous damper when using a direct axial under a dynamic load.

Hence, HECM will overcome all the weaknesses of the dampers above if it is successfully used. The proposed HECM was designed using displacement-dependent behavior under a vertical load similar to that of a velocity device. EN 15129 does not cover the direct testing requirements of HECM under axial force; hence, the testing procedures in this study referred to the fundamental testing requirements specified in EN 15129 under a high-damping elastomer with modification using direct axial load. Once the novel hyperelastic composite material is developed, a novel type of damper can be developed in the future to best fit it due to its advantages over other dampers, as described above. The successful development of this HECM, on the other hand, will create the opportunity for a replacement material for those currently used in passive dampers, such as friction dampers, viscoelastic dampers, and metallic dampers, and will potentially have an edge due to the HECM providing higher damping, being durable in nature, and exhibiting good performance under shearing.

2. Material Testing Method

2.1. Hyperelastic Composite Material

The damping testing of the material for the HECM was conducted before the axial dynamic loading. Dynamic tests were conducted to determine the damping ratio according to Clause 8.2.4.2.5 in EN 15129:2009. Although the actual behavior of the HECM in the product is directly under an axial load, the testing of the HECM under Clause 8.2.4.2.5 in EN 15129:2009 is the direct shear method. There is no provision under EN 15129 for material testing directly under axial load for elastomer materials. Hence, the materials were tested using the provision in EN 15129. The test method in accordance with EN 15129 is straightforward and determines which HECM composition has the highest damping ratio under shear, which has a direct correlation in the axial direction. Finally, the most suitable HECM was chosen after a number of different samples were obtained.

2.2. Test Procedures

The double-shear test piece was used to characterize the HECM during material development, particularly to determine the mechanical properties of the material. Under normal circumstances, a double-shear test piece was used to determine the damping property of the elastomer. The configuration and dimensions of the double-shear test piece are illustrated in Figure 1. Dynamic tests were performed using an INSTRON servo-hydraulic testing machine, as shown in Figure 2. The load cell capacity of the testing machine was 25 kN with a stroke displacement of ±75 mm.

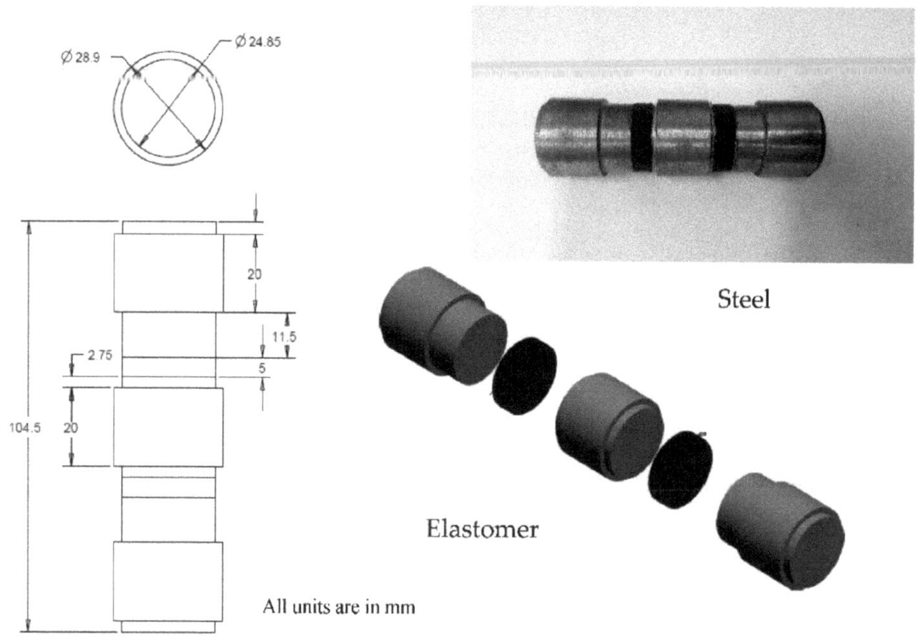

Figure 1. Configuration of double-shear test sample.

A wave matrix module integrated with an INSTRON servo-hydraulic testing machine was used to program the dynamic tests. All the double-shear test pieces were dynamically tested by varying the strain amplitudes from 5% to 10%, 20%, 50%, 70%, 100%, and 150% at a frequency of 0.5 Hz. These testing conditions were applied according to EN15129:2009 clause 8.2.2.1.3.2: effect of the strain amplitude. Five consecutive cycles were programmed for each strain amplitude to record the force and displacement data. The tests were performed in ascending order of the strain amplitude. The dynamic horizontal force–displacement data were analyzed cycle by cycle using Equation (1), as stated in BS EN 15129:2009 Annex G.5. The method of analysis expresses stiffness as

$$K_b = \frac{F^+ - F^-}{d^+ - d^-} \tag{1}$$

where F and d are the horizontal force and the displacement, respectively. d^+ and d^- are the maximum and minimum displacement values in the cycle, respectively, and F^+ and F^- are the force values at those displacements.

Figure 2. Double-shear testing using INSTRON servo-hydraulic testing machine.

The dynamic test measures the damping or hysteresis energy loss within an elastomer subjected to cyclic deformation. The damping ratio ξ was calculated using Equation (2), as stated in EN15129 Annex G.5. The equivalent viscous damping ratio ξ is expressed as follows:

$$\xi = \frac{2H}{\pi K_b (d^+ - d^-)^2} \tag{2}$$

where H is the area of the hysteretic loop, and K_b is the effective stiffness.

All the force and displacement data were used to calculate the damping ratio. The calculations were executed using the data generated by the INSTRON machine and plotted in the Excel graph shown in Figure 3. The results of the damping ratio according to the BS EN 15129:2009 document for rubber samples are listed in Table 1. G is the shear modulus of rubber. According to BS EN 15129:2009, the damping ratio is the third loading cycle of 15.25%. This HECM was chosen because it had the highest damping ratio among others and, therefore, was used as a sample in the dynamic axial load model.

Table 1. Result data for force–displacement.

Cycle	G	γ	d^+	d^-	F^+	F^-	H	K_b	ξ	f
-	(MPa)	(%)	(mm)	(mm)	(kN)	(kN)	(kN·mm)	(kN/mm)	(%)	(Hz)
1	0.4288	50.6	2.7806	−2.7813	0.10424	−0.10863	0.30854	0.03827	16.59	0.03
2	0.4343	50.6	2.7806	−2.7813	0.10169	−0.1139	0.2941	0.03876	15.61	0.03
3	0.4401	50.6	2.7804	−2.7812	0.10211	−0.11635	0.29107	0.03928	15.25	0.03
4	0.4441	50.6	2.7806	−2.7813	0.10261	−0.11788	0.28953	0.03964	15.03	0.03
5	0.4048	75.8	4.171	−4.1716	0.14277	−0.15861	0.6063	0.03613	15.35	0.03

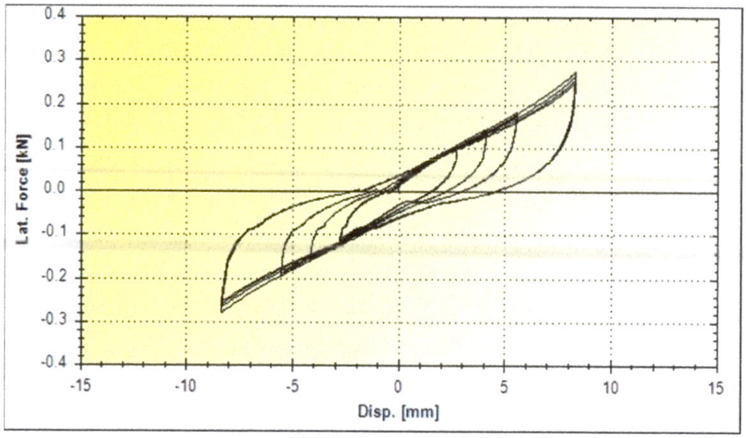

Figure 3. Force–displacement test curve graph.

3. Development of Hyperelastic Composite Material under Dynamic Axial Loads

The above HECM was placed in a scaled-down damper, which was subjected to a direct axial load. Figure 4 shows the isometric view and dimensions of the proposed hyperelastic composite material used for testing. It consists of two discrete pistons mounted on a steel plate, which act in compression alternately with each high-damping rubber when axial pushing and pulling forces are exerted. The pushing and pulling movements simulate the rotation in the beam–column joint of the structural framing under dynamic loading. The HECM has energy-dissipation capability to reduce the amplitude, velocity, and acceleration of vibration with minimal reaction force. This mechanism is contained within a cylindrical steel tube casing, which confines the boundary condition of the mechanical system.

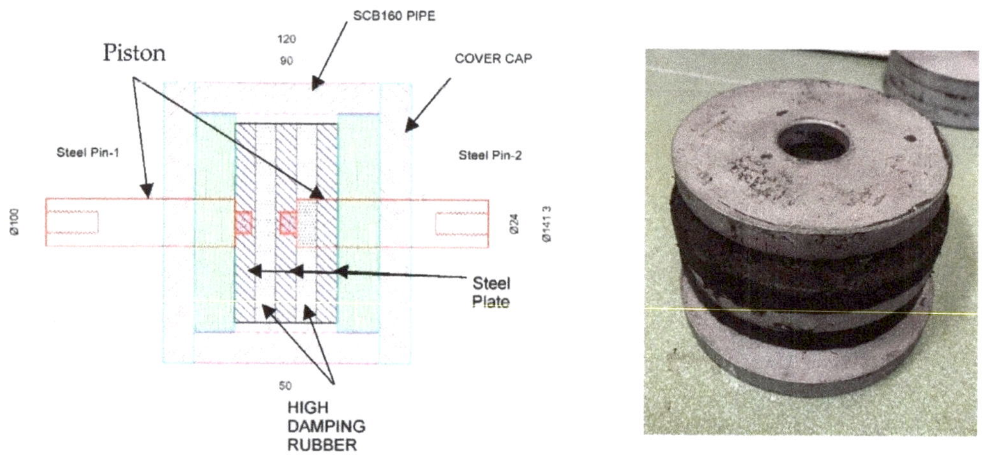

Figure 4. Typical isometric view and dimension of the proposed scaled-down hyperelastic composite damper.

The steel parts are all corrosion-protected according to international standards, depending on the location, environmental conditions, and required degree of protection. Upon request, a special arrangement can cater to the wider needs of the environmental conditions. The rubber thickness was controlled by the allowable design pressure stated in EN 1337-3 [28]. For this sample test, a maximum thickness of 10 mm was used. There were three types of samples with thicknesses of 6 mm, 8 mm, and 10 mm. According to EN 15129 clause 8.2.4.1.5.2, the cyclic frequency should be 0.5 Hz and at least 0.01 Hz for a bearing isolator in a shear damping test since there is no provision for testing under compression conditions. Thus, these frequency values were used as a basis for testing. The interval frequencies 0.1 Hz and 0.25 Hz were used to obtain the damping behavior and correlation between 0.01 Hz and 0.5 Hz. Thus, each type of sample was tested at different frequencies of 0.01 Hz, 0.1 Hz, 0.25 Hz, and 0.5 Hz. The test plan is described in subsequent sections. The effects of the rubber thickness and loading frequency on the device's damping ratio value were investigated.

4. HECM under Dynamic Axial Loads

4.1. Test

The testing of the proposed HECM was conducted, and the HECM was tested using the double-shear method as per the BS EN 15129 requirements described in Section 2. The actual product, however, was tested using the axial compression load method. The HECM damping behavior in the axial direction was expected to be lower than in the shear direction. During the dynamic cyclic loading test, both rubber layers alternately experienced tension and compression. Generally, high-damping rubber has a better performance in compression conditions, but the HECM performance is reduced under tension. As a result, two samples of dampers with 10 mm thick HECM were produced to test the testing mechanism. The first sample was designed to ensure that the two rubber layers experienced tension and compression alternately during the dynamic cyclic loading test. The second sample was designed to ensure that the rubbers were only subjected to pure compression for both layers. The force, F applied in Equation (1), is the axial force.

4.2. Testing on the Device Sample with Tension–Compression Behavior

Testing was performed using a Dynacell machine, as shown in Figure 5. The sample was fixed at the soffit, and the top piston induced a 10 kN compression and tension force at a cyclic frequency of 0.01 Hz. When the tension force was applied, the steel plate at the center, which was mounted on the top piston, exerted a compression force on the top-layer rubber, while the bottom layer was in tension. Similarly, when a compression force was applied, the steel plate exerted a compression force on the bottom rubber layer, while the top layer was in tension. Table 2 and Figure 6 show the data and force–displacement graph for the first sample, respectively. The results showed that the sample produced a symmetrical hysteretic loop graph with a damping ratio of 8.48% using Equation (2), as shown in Table 2 and Figure 6. This is almost half the value obtained using the shear method. This indicates that the damping ratio of the damper may be significantly reduced by the tension behavior. The reduction factor due to the tension condition can be verified by comparing the results under purely compression conditions. Hence, it is vital to ensure that the rubber is under compression. If the tension behavior reduced the damping, then to produce maximum damping under natural behavior, the second sample purely under compression was proposed to test the subsequent performance.

Table 2. Result data for force–displacement under tension and compression.

Cycle	d^+	d^-	F^+	F^-	H	K_b	ξ	f
-	(mm)	(mm)	(kN)	(kN)	(kN·mm)	(kN/mm)	(%)	(Hz)
1	0.22448	−0.22199	9.8674	−10.198	1.1466	44.942	8.15	0.00995
2	0.22425	−0.22182	9.7901	−10.044	1.1687	44.463	8.41	0.01002
3	0.22486	−0.22202	9.7799	−9.8674	1.1700	43.966	8.48	0.00999

Figure 5. Testing on the product sample using Dynacell machine.

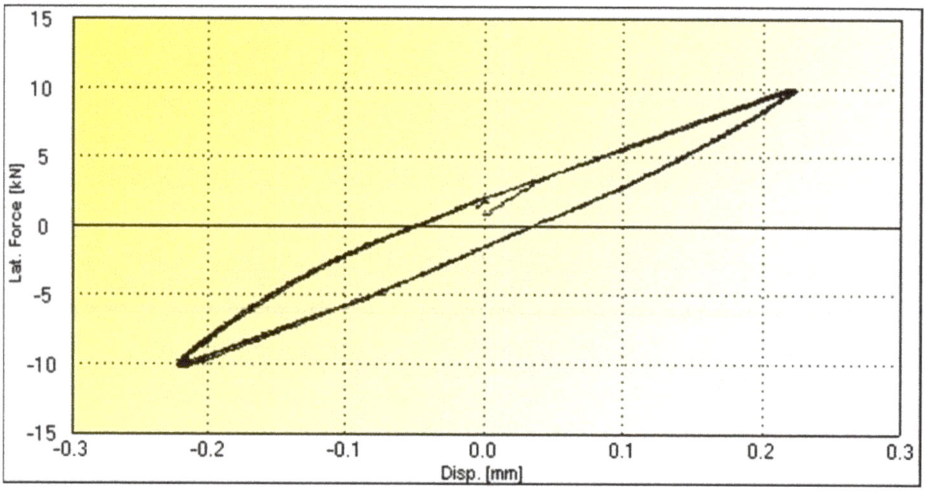

Figure 6. Force–displacement graph for sample under compression and tension.

4.3. Testing of the Device Sample with Pure Compression Behavior

To examine the rubber behavior under pure compression, the product sample was modified with no adhesive applied at the surface of the center steel plate, as shown in Figure 4. The sample was enclosed to confine the rubber during the cyclic loading. This is similar to the smaller scale of an actual damper. According to the results presented in Table 3 and Figure 7, the profile shape of the hysteretic graph is similar to that shown in Figure 6, except for the portion where the force was close to zero. This result illustrates the actual condition of the product, where the piston was released from one side of the rubber before engaging the opposite side of the rubber. There was a very short period during which the piston was in the reset zone. This is represented in the graph by the fact that the line does not continue straight and instantly drops when near zero force before it increases in force again, as highlighted in the graph. This slightly reduced the damping of the product. Nevertheless, the sample under pure compression exhibited a 13.7% damping ratio using Equation 2 under the third cycle, as shown in Figure 7. This shows that the damping ratio increased by more than 50%, compared with the previous test. As the result under pure compression produced better results, in subsequent testing, we used a sample under compression for different thicknesses of high-damping rubber and frequencies.

Table 3. Result data for force–displacement under pure compression.

Cycle	d^+	d^-	F^+	F^-	H	K_b	ξ	f
-	(mm)	(mm)	(kN)	(kN)	(kN·mm)	(kN/mm)	(%)	(Hz)
1	0.5585	−0.5629	8.6552	−11.307	4.8635	17.801	13.83	0.01
2	0.5584	−0.563	8.3737	−11.055	4.7244	17.326	13.8	0.01
3	0.5584	−0.5629	8.2284	−10.883	4.6131	17.044	13.7	0.01

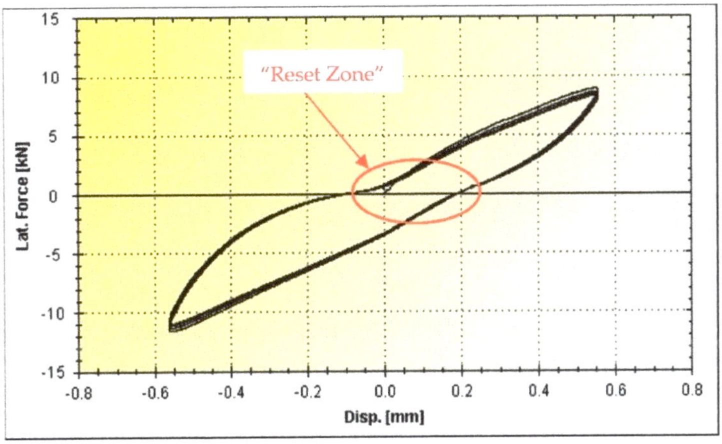

Figure 7. Product testing, showing force–displacement test and results for product under cyclical compression.

4.4. The Effects of Rubber Thickness and Load Frequency under Pure Compression

To further investigate the effects of rubber thickness and different frequencies on the damping characteristics, three more sample dampers of 6, 8, and 10 mm were tested using four frequencies of 0.01, 0.1, 0.25 Hz, and 0.5 Hz. The damping ratio result from the testing was obtained using the third cycle as per EN 15129:2009. All the data from the above tests are presented in Figure 8 and Table 4 for dampers that are 6 mm thick, Figure 9 and Table 5 for dampers 8 mm thick, and Figure 10 and Table 6 for dampers that are 10 mm thick. The results are summarized in Figure 11. According to the results, when using thinner rubber, the damping ratio percentage significantly decreased, especially from 10 mm to 8 mm. It was further reduced from 8 mm to 6 mm but at a slightly lower rate. When the rubber thickness is reduced, the allowable design pressure on rubber increases [28], and stiffness and Keff increase. However, the axial displacement under the same load was reduced. This significantly reduced the damping behavior that mitigated the force. The applied frequency affected the damping behavior of high-damping rubber. This can easily be explained by the fact that, at a higher frequency, the rubber stiffness is not fully utilized and thus does not provide more displacement. In other words, the applied force is not fully mobilized into the rubber material. This is illustrated in the above graph, according to which the damping ratio was reduced after the applied frequency was increased. Fortunately, the rate of the decrease in the damping ratio became slower when the frequency continued to increase. For instance, when using 10 mm thick rubber, the damping ratio was reduced from 13.7 Hz to 12.3 Hz when the frequency increased 10 times from 0.01 Hz to 0.1 Hz. However, the damping ratio rate was reduced from 13.7 Hz to 10.05 Hz, even though the frequency increased 50 times, from 0.01 Hz to 0.5 Hz.

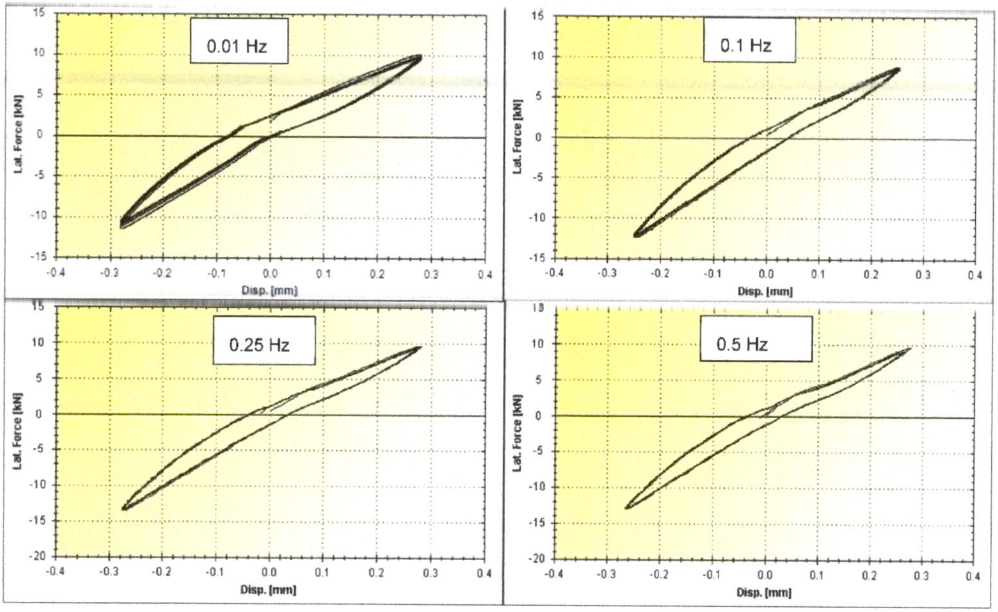

Figure 8. Plot of device testing, showing force–displacement for 6 mm thick HDR material at different frequencies.

Table 4. Result data for 6 mm thick HDR material at different frequencies.

6 mm Thick Rubber Cycle	d^+	d^-	F^+	F^-	H	K_b	ξ	f
-	(mm)	(mm)	(kN)	(kN)	(kN·mm)	(kN/mm)	(%)	(Hz)
1	0.2798	−0.2804	9.9608	−11.344	1.45	38.031	7.73	0.01
2	0.2799	−0.2806	9.7562	−11.047	1.3644	37.115	7.45	0.01
3	0.2798	−0.2804	9.5962	−10.794	1.3179	36.397	7.35	0.01
1	0.25449	−0.2496	8.72	−12.377	1.072	41.852	6.42	0.099
2	0.25465	−0.2499	8.5428	−12.125	1.0607	40.964	6.48	0.1001
3	0.25407	−0.2499	8.53	−12.069	1.0481	40.873	6.43	0.0999
1	0.28144	−0.2759	9.5915	−13.538	1.1903	41.5	5.88	0.2447
2	0.27696	−0.2754	9.321	−13.344	1.1562	41.033	5.88	0.2503
3	0.27609	−0.2750	9.3281	−13.278	1.1472	41.019	5.86	0.2498
1	0.27821	−0.2642	9.5691	−12.954	1.0325	41.526	5.38	0.4871
2	0.2656	−0.2656	9.0762	−12.873	1.0025	41.318	5.47	0.4995
3	0.26711	−0.2646	9.0518	−12.9	1.0051	41.286	5.48	0.4995

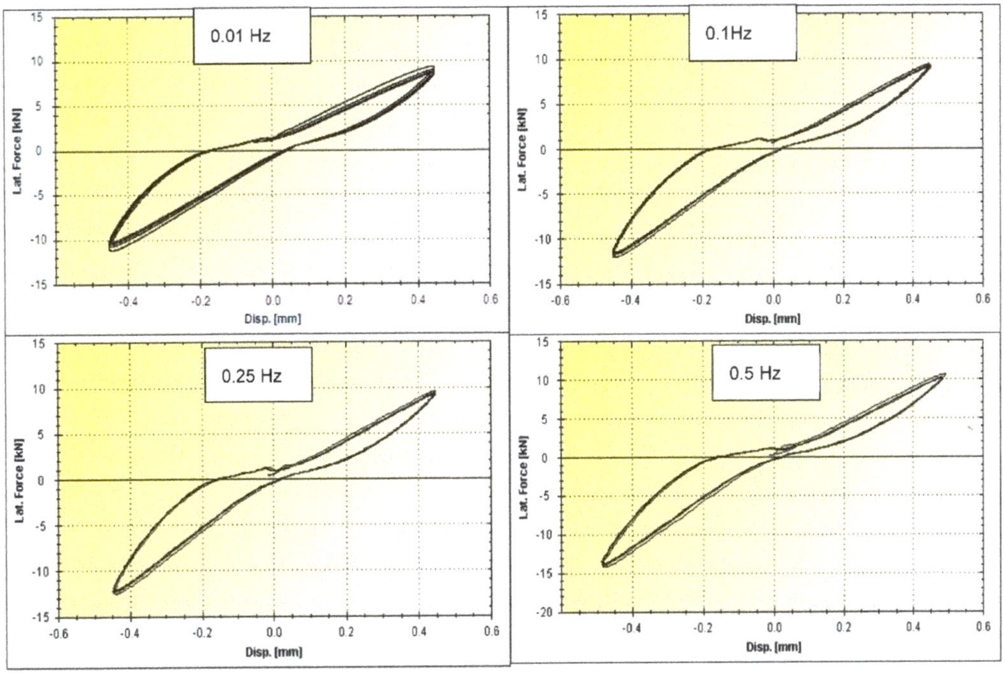

Figure 9. Plot of device testing, showing force–displacement for 8 mm thick HDR material at different frequencies.

Table 5. Result data for 8 mm thick HDR material at different frequencies.

8 mm Thick Rubber Cycle	d^+	d^-	F^+	F^-	H	K_b	ξ	f
-	(mm)	(mm)	(kN)	(kN)	(kN·mm)	(kN/mm)	(%)	(Hz)
1	0.4484	−0.4524	9.101	−10.93	2.9893	22.236	10.55	0.01
2	0.4484	−0.4524	8.7585	−10.528	2.7383	21.411	10.03	0.01
3	0.4483	−0.4524	8.582	−10.31	2.6171	20.975	9.79	0.01
1	0.45331	−0.4527	9.143	−11.712	2.5299	23.019	8.52	0.1
2	0.45295	−0.4520	9.0148	−11.612	2.4025	22.793	8.19	0.1
3	0.45308	−0.4524	8.9376	−11.29	2.3557	22.338	8.19	0.1
1	0.44954	−0.4478	9.4756	−12.375	2.3989	24.35	7.79	0.248
2	0.44757	−0.4481	9.2494	−12.255	2.2409	24.009	7.41	0.2494
3	0.44712	−0.4471	9.199	−12.017	2.2025	23.726	7.39	0.25
1	0.49519	−0.4836	10.535	−14.111	2.761	25.179	7.29	0.4924
2	0.48599	−0.4849	10.171	−13.889	2.5212	24.781	6.87	0.4995
3	0.48632	−0.4849	9.9704	−13.596	2.473	24.264	6.88	0.5

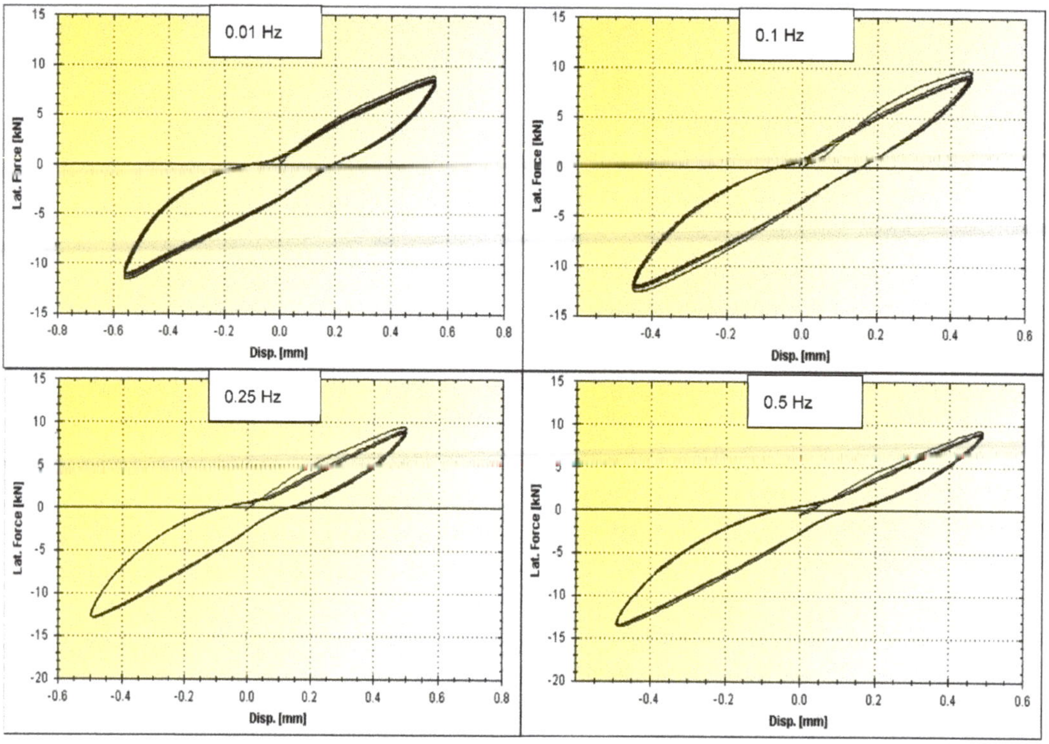

Figure 10. Plot of device testing, showing force–displacement for 10 mm thick HDR material at different frequencies.

Table 6. Result data for 10 mm thick HDR material at different frequencies.

10 mm Thick Rubber Cycle	d^+	d^-	F^+	F^-	H	K_b	ξ	f
-	(mm)	(mm)	(kN)	(kN)	(kN·mm)	(kN/mm)	(%)	(Hz)
1	0.5585	−0.5629	8.6552	−11.307	4.8635	17.801	13.83	0.01
2	0.5584	−0.563	8.3737	−11.055	4.7244	17.326	13.8	0.01
3	0.5584	−0.5629	8.2284	−10.883	4.6131	17.044	13.7	0.01
1	0.4556	−0.4514	9.5276	−12.266	3.9322	24.029	12.67	0.1002
2	0.4556	−0.4507	9.0664	−11.973	3.7401	23.216	12.49	0.1
3	0.4551	−0.4505	8.9405	−11.754	3.6215	22.853	12.3	0.1
1	0.5028	−0.4989	9.2421	−12.672	3.9203	21.878	11.37	0.2466
2	0.5000	−0.4995	8.7878	−12.607	3.6054	21.404	10.73	0.2504
3	0.4998	−0.4983	8.6859	−12.739	3.5238	21.466	10.49	0.2503
1	0.4893	−0.4914	9.0509	−13.369	3.5421	22.861	10.26	0.4916
2	0.4898	−0.4896	8.9326	−13.444	3.3499	22.846	10.18	0.5003
3	0.4894	−0.4895	8.8238	−13.193	3.2781	22.492	10.05	0.5

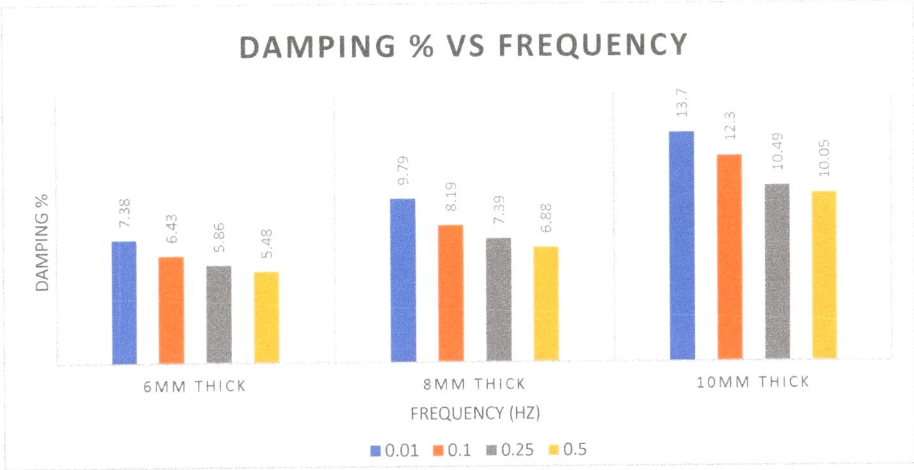

Figure 11. Plot of damping result with different thicknesses of rubber and frequencies.

5. Finite Element Analysis of Hyperelastic Composite Damper

5.1. FE Modeling

To corroborate the proposed hyperelastic composite material properties used for future studies, FE analyses were conducted. HECM properties in the modeling were simulated based on the laboratory results in Section 4 since the 10 mm thick HECM only provided more than a 10% damping ratio, as shown in Section 4. Thus, a detailed finite element analysis was performed using 10 mm thick damping at the loading frequency of 0.01 Hz and 0.5 Hz, respectively, where these two frequencies are covered in EN 15129. ABAQUS software was used at this stage due to its ability to consider the inelastic behavior of rubber elements.

Three-dimensional solid elements were used to define the steel plates and viscoelastic layers. The viscoelastic layers were fully bonded to the steel plates; therefore, tie constraints were used to simulate the interactions. Element types of C3D8R (8-node linear brick, reduced integration with hourglass control) and C3D20H (20-node quadratic brick, hybrid with linear pressure) were used for steel and viscoelastic layers, respectively. For boundary or restraint, one side of the damper was fully restrained against all translation and rotation movements, while the other side was free to move. Figure 12 shows the finite element model of the conventional viscoelastic damper in ABAQUS software.

The density of the viscoelastic material obtained was 1200 kg/m3. Both hyperelasticity and viscoelasticity properties were used to define the behavior of the composite material. ABAQUS uses the strain energy potential, rather than Young's modulus and Poisson's ratio, to relate stresses to strains for hyperelastic materials. The strain energy potential defines the strain energy stored in the material per unit of reference volume (volume in the initial configuration) as a function of the strain at that point in the material. Several forms of strain energy potentials are available in ABAQUS to model most commercial elastomers. Generally, for the hyperelastic material models available in ABAQUS, a strain energy potential can be defined directly by specifying material coefficients, or ABAQUS automatically determines approximate values for the coefficients through provided experimental test data. In this paper, the neo-Hookean model was used to define hyperelasticity through experimental tensile test data obtained for the composite material. The neo-Hookean model is recognized as the best method for predicting the nonlinear stress–strain behavior of materials undergoing large deformations with satisfactory approximation when the parameters are based on only one test. The meshing of the FE model is shown in Figure 13.

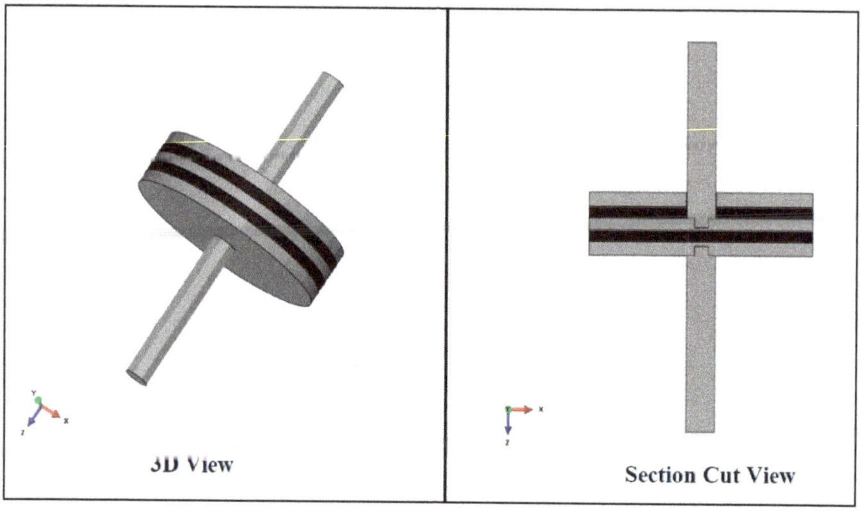

Figure 12. The 3D model and section view of the FE model.

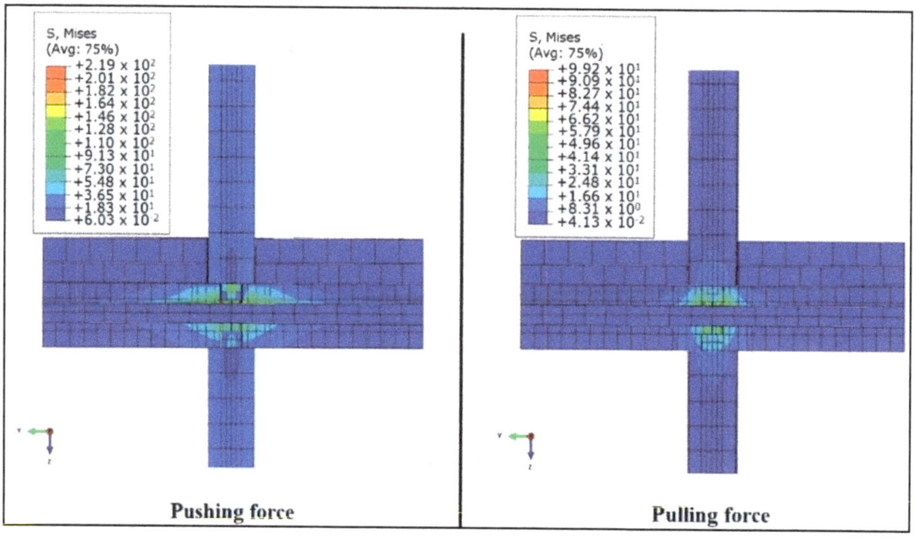

Figure 13. Meshing and stress contour of the FE model.

Viscoelasticity properties in ABAQUS are defined in four ways: the direct specification of the Prony series parameters, the inclusion of creep test data, the inclusion of relaxation test data, or the inclusion of frequency-dependent data obtained from sinusoidal oscillation experiments. To model the viscoelastic material in this research, viscoelasticity was implemented through the direct specification of the Prony series parameters. The Prony series represents the mechanical analogy of viscoelastic material behavior more rationally, where the linear elastic properties are represented by springs, and the time-dependent viscous properties are represented by the dashpots.

5.2. Validation of FE Model

For the validation of the FE model of the scaled-down hyperelastic composite damper, the performance of the FE model under dynamic loads was investigated. The aim was to check whether the dynamic responses of the simulated viscoelastic damper were in good agreement with the experimental results. For the validation of the dynamic performance of the viscoelastic damper model, displacement-controlled analysis was applied to the model under the frequency of 0.01 Hz and 0.5 Hz up to +0.43 mm. The load–displacement results obtained from the FE analysis were compared with the experimental results and are shown in Figure 14. It is obvious from the figures that the results of the experimental tests and FE model became stable after a few cycles. The shape of the hysteresis loops obtained from the FE model and experiment were in agreement.

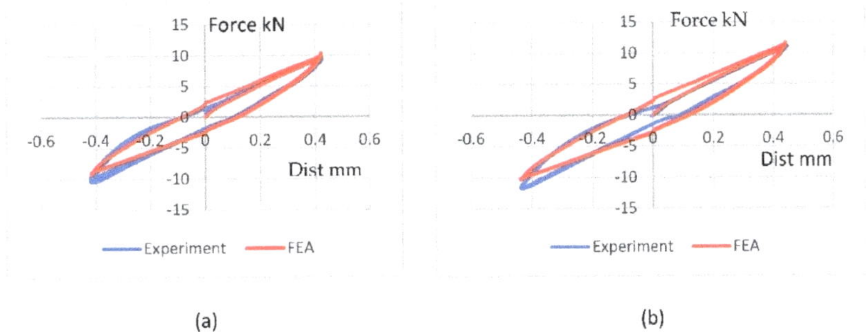

Figure 14. Comparison of load–displacement curves under (**a**) 0.01 Hz and (**b**) 0.5 Hz.

The Prony series coefficients were derived based on the mechanical properties of the composite material used in the scaled-down damper at various frequencies and strains. Table 7 presents the Prony series coefficients used in HECM to generate the graphs in Figure 14, where g is the dimensionless shear relaxation modulus associated with relaxation times, respectively.

Table 7. Prony series coefficients of HECM.

g	τ
0.01279	5.00×10^{-4}
0.27523	0.005
0.28005	0.0159
0.20402	0.05
0.10778	0.159
0.05928	0.5
0.02867	1.59
0.01549	5
0.00753	15.9
0.00399	50
0.00195	159
0.00106	500
0.000472524	1590
0.000313733	5000
7.94×10^{-5}	15,900
0.000121258	50,000
3.59×10^{-5}	500,000
1.14×10^{-5}	5,000,000

6. Conclusions

The results above demonstrated that the HECM, a type of HDR, is not limited to the shear behavior but also exhibits good performance in direct axial compression. The HECM can provide a damping ratio of approximately 10% to 13% under axial conditions instead of purely under shear conditions and, therefore, can be developed into a novel system for rubber dampers resisting axial forces in the near future. This damper is expected to provide similar performance and competitiveness as a viscous damper as long as the damping ratio is greater than 10% [27]. According to EN 15129:2009, the testing frequency used should be 0.5 Hz for the damping testing. The above result showed that the damping ratio was higher than 10% when using a 10 mm thick sample at 0.5 Hz. On the other hand, HECM can provide more durable and consistent behavior, which provides new advantages for the damper industry. This HECM can be used as a damper material under direct axial force conditions, such as those experienced by current viscous dampers. As a result, HECM can be used to develop a new damper for the industry in the near future.

Dampers should have larger deformation and higher axial capacity to accommodate the requirements of dynamic events on structures. Pure HECM is not sufficient to meet these requirements. Hence, a combination of HECM and layers of steel plates should be developed for further study.

Author Contributions: Conceptualization, C.G.T. and T.W.T.; methodology, C.G.T. and T.W.T.; software, T.W.T.; validation, C.G.T., and M.Z.J.; formal analysis, C.G.T. and T.W.T.; investigation, T.W.T.; data curation, T.W.T.; writing—original draft preparation, T.W.T.; writing—review and editing, C.G.T. and T.W.T.; supervision, M.Z.J.; funding acquisition, C.G.T. All authors have read and agreed to the published version of the manuscript.

Funding: Universiti Malaya Impact-Oriented Interdisciplinary Research Grant Programme (IIRG004B-2021IISS).

Data Availability Statement: The data presented in this study are available on request from the corresponding author. The data are not publicly available due to the confidentiality requirements of the lab.

Acknowledgments: This research was financially supported by the Universiti Malaya Impact-Oriented Interdisciplinary Research Grant Programme (IIRG004B-2021IISS). The testing facilities provided by Kossan Rubber Industries Bhd are gratefully acknowledged.

Conflicts of Interest: The authors declare that they have no affiliations with or involvement in any organization or entity with any financial interest in the subject matter or materials discussed in this manuscript.

List of Symbols

G	Shear modulus
γ	Shear strain
ξ	Viscous damping ratio
K_b	Effective stiffness
H	Area of the hysteretic loop
d^+	The maximum displacement values in the cycle
d^-	The minimum displacement values in the cycle
F^+	The force values at maximum displacement
F^-	The force values at minimum displacement
f	Frequency

References

1. Gu, H.; Itoh, Y. Ageing Behaviour of Natural Rubber and High Damping Rubber Materials Used in Bridge Rubber Bearings. *Adv. Struct. Eng.* **2010**, *13*, 1105–1113. [CrossRef]
2. Kerileng, K.; Dundu, M. Base Isolation Systems in Multi-Storey Structures. In Proceedings of the International Conference on Structural Engineering Research, Sydney, Australia, 20–22 November 2017; pp. 203–209.

3. Samadian, D.; Ghafory-Ashtiany, M.; Naderpour, H.; Eghbali, M. Seismic resilience evaluation based on vulnerability curves for existing and retrofitted typical RC school buildings. *Soil Dyn. Earthq. Eng.* **2019**, *127*, 105844. [CrossRef]
4. Huang, Z.; Zhang, D.; Pitilakis, K.; Tsinidis, G.; Huang, H.; Zhang, D.; Argyroudis, S. Resilience assessment of tunnels: Framework and application for tunnels in alluvial deposits exposed to seismic hazard. *Soil Dyn. Earthq. Eng.* **2022**, *162*, 107456. [CrossRef]
5. Desai, M.; John, R. Seismic performance of base isolated multi-storey building. *Int. J. Sci. Eng. Res.* **2015**, *6*, 84–89.
6. Park, K.H.; Fujiwara, Y.; Mazda, T.; Kajita, Y. Evaluation of mechanical properties considering hysteresis characteristic of high damping rubber bearing. *J. Phys. Conf. Ser.* **2020**, *1687*, 012019. [CrossRef]
7. Atam, E. Friction Damper-Based Passive Vibration Control Assessment for Seismically-Excited Buildings Through Comparison With Active Control: A Case Study. *IEEE Access* **2018**, *7*, 4664–4675. [CrossRef]
8. Yan, Y. Influence of High Damping Rubber Bearing on Seismic Performance of The Bridge. *IOP Conf. Ser. Earth Environ. Sci.* **2020**, *510*, 052058. [CrossRef]
9. Fujita, S.; Fujita, T.; Kasahara, Y.; Suizu, Y.; Furuya, O.; Teramoto, T.; Kitamura, H. *Energy Absorption Characteristics of High Damping Rubber Damper for Vibration Control of High-Rise Buildings*; SMiRT-12; IASMiRT: Anaheim, CA, USA, 1993; pp. 243–248.
10. Teramoto, T.; Kitamura, H.; Ozaki, H.; Furuya, O.; Morikawa, S.; Suzuki, S. Practical Application of high-damping rubber dampers to slender buildings. In Proceedings of the 11th World Conference on Earthquake Engineering WCEE, Acapulco, Mexico, 23–28 June 1996.
11. Gan, Z.P.; Hayashikawa, T.; Matsumoto, T.; He, X.W. Seismic response analysis of base-isolated bridge subjected to long duration earthquake in low temperature environment. *J. Struct. Eng. JSCE* **2015**, *61A*, 335–343.
12. Fuller, K.; Ahmadi, H.; Goodchild, I.R.; Magonette, G.; Taucer, F.; Dumoulin, C. Rubber-Based Energy Dissipators for Earthquake Protection of Structures. In Proceedings of the 12th WCEE, Auckland, New Zealand, 30 January–4 February 2000.
13. Itoh, Y.; Gu, H.; Satoh, K.; Yamamoto, Y. Long-Term Deterioration of High Damping Rubber Bridge Bearing. *Doboku Gakkai Ronbunshuu A* **2006**, *62*, 595–607. [CrossRef]
14. Castaldo, P.; Tubaldi, E.; Selvi, F.; Gioiella, L. Seismic performance of an existing RC structure retrofitted with buckling restrained braces. *J. Build. Eng.* **2020**, *33*, 101688. [CrossRef]
15. Min, K.-W.; Kim, J.; Kim, Y.-W. Design and test of tuned liquid mass dampers for attenuation of the wind responses of a full scale building. *Smart Mater. Struct.* **2014**, *23*, 045020. [CrossRef]
16. Zhu, R.; Guo, T.; Mwangilwa, F. Development and test of a self-centering fluidic viscous damper. *Adv. Struct. Eng.* **2020**, *23*, 2835–2849. [CrossRef]
17. Dolati, S.S.K.; Mehrabi, A.; Dolati, S.S.K. Application of Viscous Damper and Laminated Rubber Bearing Pads for Bridges in Seismic Regions. *Metals* **2021**, *11*, 1666. [CrossRef]
18. Sarwar, W. Viscoelastic Material as Energy Dissipater Viscoelastic Damper for Building Structures to Mitigate the Seismic Vibration. *Civ. Environ. Eng. Rep.* **2019**, *29*, 41–49. [CrossRef]
19. Kanitkar, R.; Harms, M.; Crosby, P.; Lai, M.L. Seismic retrofit of a steel moment frame structure using vis-coe-lastic dampers. *J. Earthq. Technol.* **1998**, *35*, 207–219.
20. Xu, Y.; Xu, Z.; Guo, Y.; Huang, X.; Dong, Y.; Li, Q. Dynamic Properties and Energy Dissipation Study of Sandwich Viscoelastic Damper Considering Temperature Influence. *Buildings* **2021**, *11*, 470. [CrossRef]
21. Constantinou, M.C.; Symans, M.D. *Experimental & Analytical Investigation of Seismic Response of Structures with Supplemental Fluid Viscous Dampers*; NCEER-92-0032; National Center for Earthquake Engineering Research: Buffalo, NY, USA, 1992.
22. Chang, K.C.; Tsai, M.H.; Chang, Y.H.; Lai, M.L. Temperature Rise Effect of Viscoelastically Damped Structures Under Strong Earthquake Ground Motions. *J. Mech.* **1998**, *14*, 125–135. [CrossRef]
23. Konstantinidis, D.; Makris, N.; Kelly, J.M. In-situ condition assessment of seismic fluid dampers: Experimental studies and challenges. *Meccanica* **2014**, *50*, 323–340. [CrossRef]
24. Alexandru, C. Variety of displacement Dependent Devices Used for Bridges and Viaducts Certified By ICECON SERT SRL. *Acta Uiversitatis Cibniensis-Tech. Ser.* **2017**, *69*, 121–129. [CrossRef]
25. Sadek, F.; Mohraz, B.; Taylor, A.W.; Chung, R.M. Passive Energy Dissipating Devices for Sesismic Applications. In *Building and Fire Research Laboratory*; Report NISTIR 5923; National Institute of Standarts and Technology: Gaithersburg, MD, USA, 1996.
26. *EN 15129*; Anti-Seismic Devices. Comité Européen de Normalisation (CEN): Brussels, Belgium, 2009.
27. Reza, A.M. A study on the damping ratio of the viscous fluid dampers in the braced frames. *Eur. Online J. Nat. Soc. Sci.* **2014**, *3*, 1223–1235.
28. *EN 1337-3:2005*; Structural Bearings—Part 3: Elastomeric Bearings. Technical Committee CEN/TC 167. BSI: London, UK, 2009.

buildings

MDPI

Article

Aftershock Fragility Assessment of Continuous RC Girder Bridges Using a Modified Damage Index

Zhengnan Wang [1], Xiaowei Deng [2], Xiheng Luo [2], Xinzhi Dang [1] and Junjun Guo [1,*]

1 State Key Laboratory of Disaster Reduction in Civil Engineering, Tongji University, 1239 Siping Road, Shanghai 200092, China
2 Tongji Architectural Design (Group) Co., Ltd., Shanghai 200092, China
* Correspondence: guojj@tongji.edu.cn

Abstract: Aftershock fragility is usually calculated conditioned on a range of potential post-mainshock damage states. The post-mainshock damage can be identified using damage indices, the latter being frequently associated with displacement-based parameters such as the maximum drift ratio or the residual displacement. However, when the reliable simulation of a structural system in a specific post-mainshock damage state is the objective of a numerical study, using such damage indicators may not assure the structure experiencing a homogeneous level of damage due to different mainshocks characteristics, which induce the aftershock fragility results unreliable. Along these lines, the current study presents a damage evaluation methodology mainly used for aftershock fragility assessment. It aims to reduce the variation of damage levels derived by using different mainshock seismic motions. The methodology presented herein includes: (i) the introduction of a damage index defined by comparing the monotonic pushover curve of the intact and post-earthquake damaged structure; (ii) the description of a finite element (FE)-based scheme that enables to quantify of the proposed damage index; and (iii) a deterioration-related modeling technique that can capture both strength and stiffness degrading performance of structural systems exposed to earthquake-induced excitations. The latter is essential to support the FE-based quantification scheme for the damage index. This methodology evaluation methodology can be primarily used for calculating the aftershock fragility assessment for a multi-span RC continuous girder bridge. The back-to-back incremental dynamic analysis framework uses a larger number of mainshock-aftershock artificial sequences to generate the aftershock fragility curves. The AS fragility results obtained via MBDI are compared with that via maximum drift ratio in terms of the ability to reduce the variation of residual capacities obtained using different mainshocks to induce a specific damage state but collapse by the same aftershock. The comparison shows a more robust relationship of MBDI with the residual capacity. It is found that MBDI, as well as its quantification approach proposed in this study, is a more effective damage predictor than the widely used displacement-based indices for AS fragility assessment.

Keywords: aftershock fragility assessment; damage index; RC columns; incremental dynamic analysis

Citation: Wang, Z.; Deng, X.; Luo, X.; Dang, X.; Guo, J. Aftershock Fragility Assessment of Continuous RC Girder Bridges Using a Modified Damage Index. *Buildings* **2022**, *12*, 1675. https://doi.org/10.3390/buildings12101675

Academic Editor: Fabrizio Greco

Received: 19 August 2022
Accepted: 8 October 2022
Published: 12 October 2022

Copyright: © 2022 by the authors. Licensee MDPI, Basel, Switzerland. This article is an open access article distributed under the terms and conditions of the Creative Commons Attribution (CC BY) license (https://creativecommons.org/licenses/by/4.0/).

1. Introduction

During major earthquake events, it is common to observe strong aftershocks within a short time window after the mainshock [1–3]. For example, the 2011 Tohuku earthquake (Japan) of moment magnitude M_W equal to 9 was followed by numerous aftershocks of M_W higher than 5, while the shortest time interval between two successive events was no longer than a few days [4]. Such a short time window observed for the occurrence of successive aftershock events renders it nearly impossible to retrofit the damaged structures promptly. Hence, a mainshock-affected structure is usually associated with a higher vulnerability to additional structural failures when exposed to subsequent aftershock(s), even when the aftershock's intensity is lower than the mainshock. For example, the mainshock of the Kocaeli earthquake (Turkey, 17.08.1999, M_W = 7.4) damaged various structural systems.

However, many of those structures eventually collapsed by an aftershock of lower magnitude (M_W = 5.9) that occurred almost a month after the mainshock [5]. Therefore, from a public safety perspective, it is essential to reliably assess the increased vulnerability of a mainshock-damaged structure subjected to potential aftershock(s) and hence, support the relevant post-earthquake decision-making.

The aftershock (AS) fragility, defined as the probability of exceeding a particular damage state given the intensity of an aftershock and conditioned on a specific damage level induced by a mainshock, is a valuable tool to assess the seismic performance and quantify the vulnerability of structures to damages from earthquakes [6–21] and have been widely used for different types of structures, such as infilled RC structures [22], containment structures [23], and cast-in-place RC industrial structures [24]. Under successive ground motions, the structures may suffer from an increased number of hysteresis loops compared to a single ground motion event. Therefore, it is relevant for AS fragility to consider a damage index that captures the cumulative damage and the different deterioration mechanisms (e.g., stiffness and strength degradation) that govern the cyclic performance of structural systems subjected to seismic motions.

Most of the existing studies on AS fragility has adopted displacement-based damage indices, such as the maximum drift ratio or the residual displacement, due to their calculation-related simplicity (e.g., [10]). However, several studies have found that even if the maximum or residual displacement of a structure under different ground motions is identical, the degree of structural damage could be quite different [25,26]. Therefore, a relatively deficient representation of the structural damage can be seen by using those damage indices, and the latter can adversely affect the reliability of AS fragility since the simulation of a specific post-mainshock damage state using such damage indices may fail to ensure a homogeneous damage level due to different mainshocks.

Along these lines, the Park and Ang (PA) damage index [27,28], widely used in the literature, evaluates seismic damage by linearly combining the large deformation induced by the earthquake excitation and the corresponding energy dissipation. This integrated (i.e., hybrid) damage index is comprehensively calibrated against a large amount of observed seismic damages, including shear and bond failures. However, several difficulties have arisen with using the proposed equation for the PA index. A critical problem is related to the weighting factor, which has to be assigned to the energy dissipation term and has to be assigned arbitrarily because the factor varies with structural configuration and section properties [29]. Furthermore, the linear relationship between the two terms of the PA damage index has been frequently found inappropriate for capturing structural damage efficiently [29].

An alternative, still hybrid, damage index was presented by Bracci et al. (the Bracci damage index, BDI) [30]. BDI is defined in terms of the loss ratio of the monotonic load-deformation curve area due to the extremal loading. The loss area of a load-deformation curve is caused by both the strength degradation and the irrecoverable deformations. Bracci et al. [30] suggested a quantification procedure for BDI, in which the simplified bilinear moment-curvature relationship is employed to represent the actual physical property of the structure, and the yield curvature, being necessary for the BDI calculation, is modified after each cycle accounting for both the plastic deformation and the stiffness degradation. The formulation of the BDI in terms of the moment-curvature relationship renders this damage index sufficiently correlated with the seismic performance [29]; hence, it can favor the reliable simulation of post-mainshock damage, which is of primary interest for the AS fragility. However, some application-related restrictions are caused by the proposed quantification procedure for the BDI. First, the BDI analytical expression includes a weighting factor related to the degrading bilinear moment–curvature curve and is defined arbitrarily [29]. The latter undermines the reliability of the estimated damage that is also disfavored by the relatively simple bilinear moment-curvature curve currently used [31]. Most importantly, since the quantification procedure has to modify the moment-curvature curve after each cycle, the BDI is associated with a demanding calculation that restricts its application in engineering practice.

In order to solve the problems mentioned above, the central objective of the current study is the development of a damage evaluation framework oriented to favor the AS fragility assessment and lead, eventually, to fragility results of increased reliability compared to existing schemes. The integrated and novel methodology presented herein consists of (i) an advanced damage index, (ii) a finite element (FE)-based quantification procedure for this damage index, and (iii) a deterioration modeling technique that supports the quantification procedure. Significantly, the damage index is developed after the appropriate modification of the existing BDI and accounts not only for the peak and residual displacement but also for the stiffness and strength deterioration, the latter being relevant for the seismic response of structures. The proposed quantification procedure of the modified BDI damage index is based on FE analysis with the use of a deterioration modeling technique that effectively captures strength and stiffness degrading performance. The proposed damage evaluation framework is applied herein for the AS fragility assessment of a multi-span reinforced concrete (RC) continuous girder bridge. At the same time, a large number of mainshock-aftershock artificial sequences are used within the B2B-IDA (back-to-back incremental dynamic analysis) scheme to generate AS fragility curves. The performance of the current damage evaluation procedure is comparatively assessed with existing damage indices regarding the variation of collapse capacity of the damaged structures induced by different mainshocks. The major contribution of the proposed methodology is the refinement of an effective predictor for evaluating cumulative damage for AS fragility assessment, which can provide powerful support for the performance-based earthquake design of infrastructures considering aftershocks.

2. Modified Bracci Damage Index and FE-Based Quantification

This section presents the modified BDI damage index, currently proposed for the AS fragility assessment of structures subjected to earthquake hazards. The necessary quantification procedure of the modified BDI is also described herein based on FE model analysis.

2.1. Definition of the Modified Bracci Damage Index

Identical to the original definition of the BDI, the modified one herein (MBDI) employs the philosophy of assessing the seismic-induced structural damage in terms of the loss ratio of the monotonic load-deformation curve area compared to the one corresponding to the intact structure. Especially in the MBDI model, D_r, as shown in Figure 1, is defined as the total area under the post-earthquake reloading pushover curve. This area initiates from the residual displacement point and, via the reloading curve, merges into the degraded post-earthquake monotonic curve. D_f is defined as the area under the unloading curve that starts from the ultimate point of the intact monotonic backbone curve and ends by the zero-force point. Finally, D_0 is the total area under the intact monotonic load-deformation curve of the structure under investigation.

Based on the above, the area loss of the monotonic load-deformation curve during the earthquake can be calculated as follows:

$$D_{loss} = D_0 - D_r \quad (1)$$

Furthermore, the maximum loss of area is:

$$D_p = D_0 - D_f \quad (2)$$

Hence, MBDI can be calculated as follows:

$$MBDI = \frac{D_{loss}}{D_p} \quad (3)$$

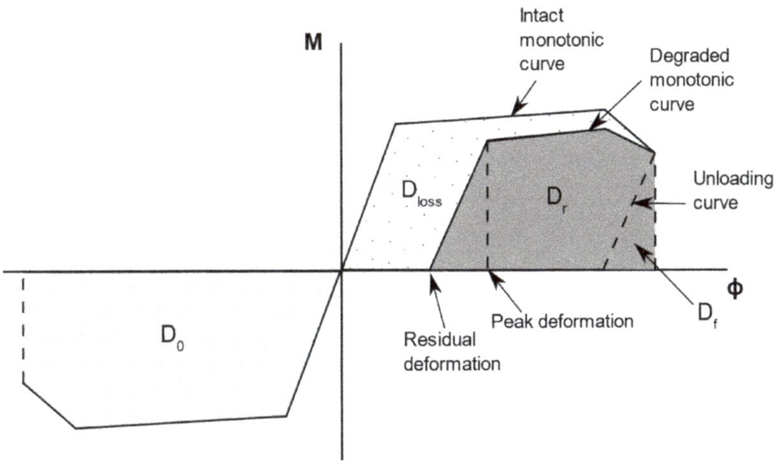

Figure 1. Schematic illustration of the MBDI [29].

2.2. FE-Based Quantification of MBDI

Following the MBDI definition, the main challenge for its rigorous quantification is determining the post-earthquake monotonic reloading pushover curve of the structure under study. To this end, an FE-based quantification procedure consists of five steps, as illustrated in Figure 2.

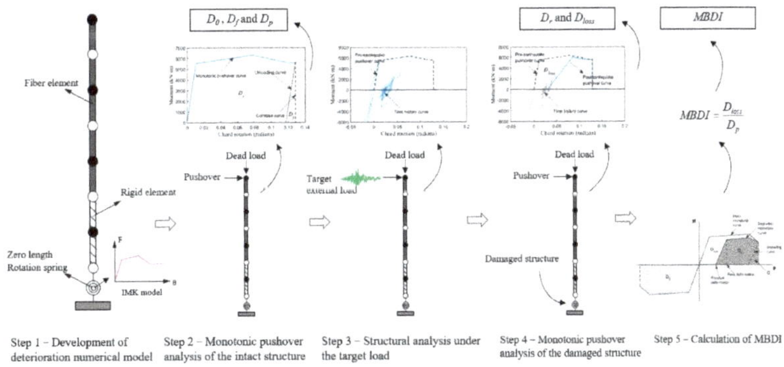

Figure 2. Flow chart of the proposed procedure for the FE−based quantification of MBDI.

Step 1—Development of deterioration numerical model: The structure under study needs to be numerically simulated with an appropriate deterioration constitutive model that can capture the structural degrading performance of structural elements accounting for the earthquake-induced cumulative damage, such as the strength deterioration in the backbone curve as well as the cyclic deterioration of strength and stiffness under seismic loading. The deterioration model constitutes a crucial part of the MDBI quantification procedure; hence, mode details are provided in the following section.

Step 2—Monotonic pushover analysis of the intact structure: A monotonic pushover analysis needs to be performed by using the finite element model of the intact structure. This inelastic static analysis should lead the structure to respond just before its final performance phase, in which the collapse mechanism can be triggered (see the blue line in Figure 3a). By executing this step, both D_0 and D_f can be calculated. Eventually, the maximum possible area loss, D_p, can be estimated according to Equation (2).

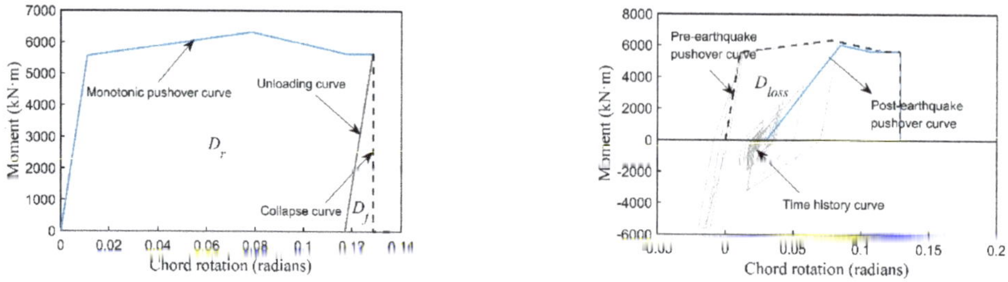

(a) (b)

Figure 3. Schematic illustration of the MDBI quantification procedure: (**a**) Monotonic pushover curve of the intact structure; (**b**) Monotonic post–earthquake pushover curve of the damaged structure.

Step 3—Structural analysis under the target load: The finite element model of the structure under study will be used for its structural analysis when exposed to the target external loading. Either quasi-static or dynamic loads can be considered. At the same time, various outcomes from the structural analysis (e.g., the moment–rotation relationship presented by the dark gray line in Figure 3b) can be used to simulate the cyclic degradation behavior under the target load according to the cumulative deterioration model developed in Step 1.

Step 4—Monotonic pushover analysis of the damaged structure: The analysis of the structure under the target external loading should be followed by the performance of a back-to-back monotonic pushover analysis until the collapse in order to determine the residual pushover curve of the damaged structure (i.e., the blue line in Figure 3b). By doing so, D_r and D_{loss} can be determined according to Equation (1).

Step 5—Calculation of MBDI: The damage index proposed herein can be calculated according to Equation (3).

3. Deterioration Modeling Technique

The deterioration modeling is essential herein to support the FE-based quantification procedure for the MDBI. Along these lines, lumped plasticity (LP) models have been widely seen to provide quite reliable results compared with fiber element models performing, especially at lower levels of deformation that drive the losses. Additionally, both strength and stiffness degradation, usually occurring as the structure approaches collapse, have been sufficiently captured by LP models [32–35], the latter being challenged in real applications due to the rather cumbersome calculation of the essential backbone curve parameters. Therefore, the deterioration modeling technique proposed herein exploits the advantages of the lumped plasticity models while developing a straightforward method to determine the relevant backbone curve parameters. Next, the modeling technique is demonstrated via its application for bridge columns.

3.1. Lumped Plasticity Model of Bridge Columns

The modeling method described by the current study consists of three different types of elements, including a zero-length rotation spring for the bottom of the column and rigid and fiber-based elements (Figure 4c). Among them, the zero-length rotation spring and the rigid element are combined to act as the plastic hinge zone of the column. The length of the rigid element, as shown in Figure 4a, is estimated according to Equation (4), recommended by the Caltrans Seismic Design Criteria (Caltrans 2010), which considers the effects of strain localization and softening.

$$L_p = 0.08L + 0.22f_{ye}d_{bl} \geq 0.044f_{ye}d_{bl} \tag{4}$$

where L is the column length, f_{ye} is the expected yield strength of longitudinal reinforcement, and d_{bl} is the nominal bar diameter of longitudinal reinforcement.

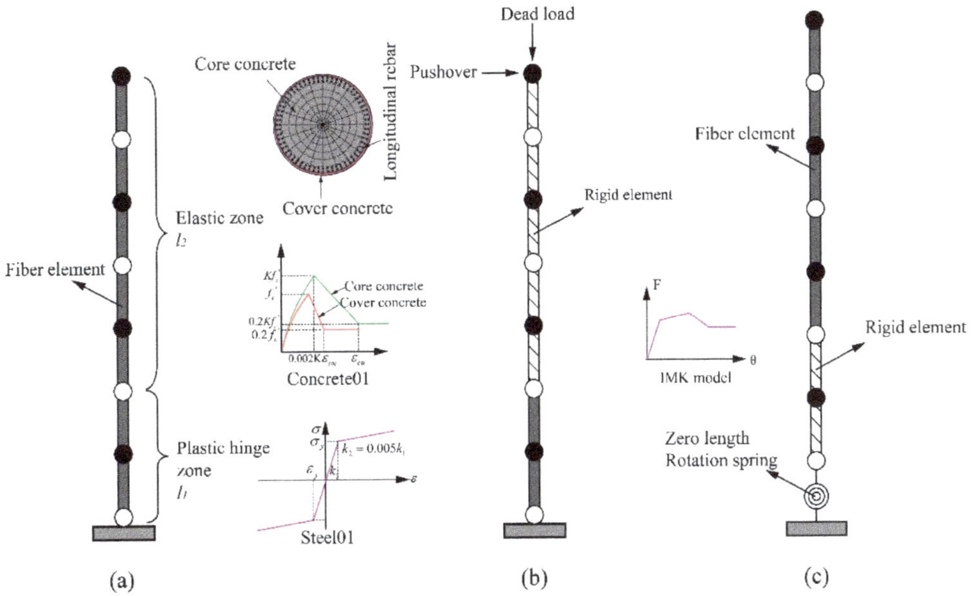

Figure 4. Procedure for determining the backbone curve parameters: (**a**) the fiber-based column model, (**b**) conducting monotonic pushover analysis to determine the parameters for the zero-length rotation spring, and (**c**) the final lumped plasticity column model.

The zero-length rotation spring employs the modified Ibarra–Medina–Krawinkler deterioration hysteretic model (IMK model) proposed by Lignos and Krawinkler [36,37]. This model is defined by a moment–rotation backbone curve and associated hysteretic rules. It can capture the strength and stiffness deterioration of RC columns up to the onset of structural collapse. The peak-orientated model is selected here as the associated hysteretic rules in this study.

3.2. *Definition of Lumped Plasticity Model*

The lumped plasticity model needs the definition of the backbone curve parameters and the cyclic deterioration parameters.

3.2.1. Definition of Backbone Curve Parameters

The parameters of the IMK model can be classified into two categories: the parameters that define a quad-linear backbone curve and energy-based cyclic deterioration parameters that define four cyclic strength and stiffness degradation modes. Even though Lignos and Krawinkler [36,37] developed the median parameters for RC components using semi-empirical relationships, such empirical equations would introduce uncertainty into the backbone curve, considerably influencing the simulation accuracy. Therefore, a procedure to determine backbone curve parameters is proposed here. The basic process is summarized as follows:

(1) Model the column using fiber-based elements.

The primary advantage of the fiber-based model is that it is convenient to simulate the monotonic behavior of concrete and reinforce steel materials employed in the fiber section. The disadvantage is that it is challenging to capture cyclic deterioration behavior. Since the cyclic deterioration behavior does not influence determining the backbone curve of the column, it is reasonable to use the fiber-based model as a reference to determine the backbone curve parameters for the IMK model.

Columns are modeled with displacement-based beam-column elements using the OpenSees platform [38]. The one-dimensional stress-strain response of concrete and reinforcing steel are simulated using the Concrete01 and Steel01 material models, respectively. The plastic hinge length (l_1) is defined according to Equation (4), as illustrated in Figure 4a.

(2) Replace the elements outside the plastic zone (l_2) with rigid beam elements and conduct a monotonic pushover analysis.

The monotonic pushover result will determine the backbone curve parameters for the IMK model.

The elements outside the plastic zone are replaced with rigid beam elements to secure the backbone curve of the lumped mode that can represent the monotonic behavior with sufficient accuracy, as shown in Figure 5. Otherwise, the elastic performance of these elements will be elastic, introducing error in the reference backbone curve.

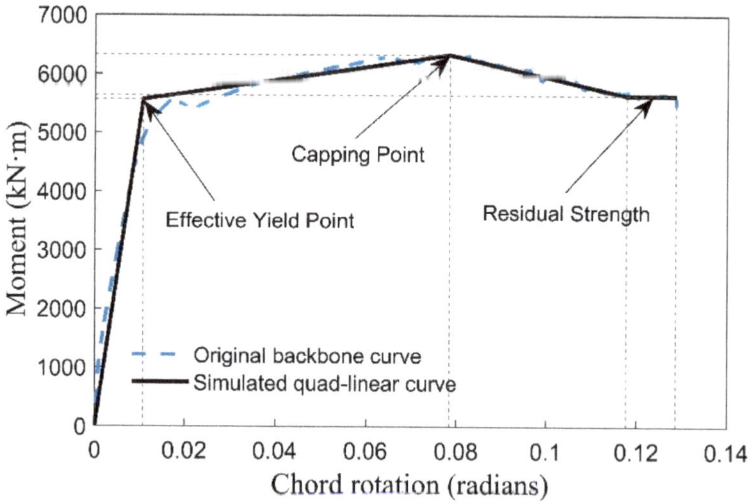

Figure 5. Schematic illustration of simulating the backbone curve with the quad-linear model.

(3) Determine the backbone curve parameters for the IMK model according to the reference pushover curve.

As shown in Figure 5, using the quad-linear model to simulate the backbone curve and determine the backbone curve parameters for the IMK model according to the values of the four turning points in the multi-linear model. It is noted that the (equal) integration area method is used to define the multi-linear model. The comparison between the monotonic pushover curves of the original fiber-based model and the finally obtained lumped-based model illustrates that the lumped-based model can reliably represent the monotonic force–deformation relationships of the column.

3.2.2. Determine Cyclic Deterioration Parameters

The uncertainty from the empirical equations for cyclic deterioration parameters has been reported not to cause significant errors [31]. Therefore, the empirical equation obtained from [32] is employed as follows,

$$\Lambda = 170 \times 0.27^{v} \times 0.1^{s/d} \times \theta_p \quad (5)$$

where Λ is the normalized energy dissipation capacity, v is the axial load ratio, s is the stirrup spacing, d is the column depth, and θ_p is the pre-capping plastic rotation for monotonic loading (difference between yield rotation and rotation at the maximum moment).

3.3. Validation of the Deterioration FE Modeling Technique

An available quasi-static test of an RC column under reversed cyclic loading is studied to verify the proposed modeling technique. Figure 6 compares the simulated and tested hysteretic behavior of the RC column. It is observed that the numerical analysis curve shows a good agreement with the experimental results. In addition, the comparison between the test results (blue dotted line) with the monotonic backbone curve (black dotted line) in Figure 6 also illustrates that the test specimen exhibits a significantly high level of nonlinearity and deterioration. Thus, it is concluded that even when the column experiences a high level of nonlinearity and deterioration under cyclic loadings, the numerical model can still reasonably capture the hysteretic behavior with high accuracy.

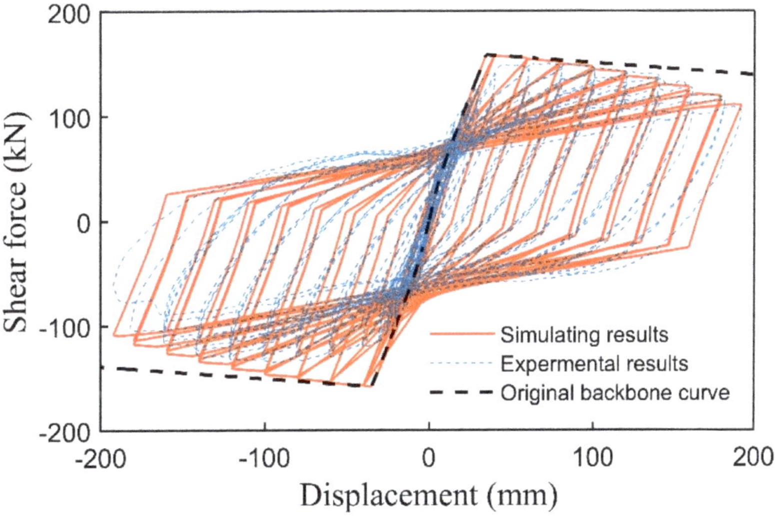

Figure 6. Validation with quasi-static test results.

To summarize, the effectiveness of the proposed deterioration modeling technique in representing the hysteresis behavior strongly supports the quantification of MBDI. It takes both the advantage of fiber-based models in the simplicity to predict the monotonic pushover curve and the lumped-based models in the simulation of cyclic deterioration behaviors. With the integration of MBDI, the FE-based quantification procedure, and the deterioration modeling method, the methodology provides a novel perspective and a feasible way to evaluate the damage. Its advantages over the method suggested by Bracci are summarized as follows:

(1) It is applicable to quantify damage under static cyclic loading and dynamic excitation.
(2) The process is simple and convenient to implement because it avoids tedious update works of the backbone curve.
(3) With sophisticated hysteretic models being used, it can consider all the critical deterioration behaviors of RC components compared to the bi-linear model employed by Bracci.
(4) The residual monotonic pushover curve area after external loading is influenced by many factors, such as stiffness and strength deterioration, peak displacement, and residual displacement, meaning that the proposed methodology can reflect that all of these factors account for cumulative damage.
(5) Considering such multiple factors, the Modified BDI has a stronger relationship with the residual seismic capacity, which facilitates assessing the AS fragility.

4. Case Study: Numerical Model, Earthquake Actions, and the Damage States

The current study focuses on the AS fragility assessment of an RC continuous girder bridge, which is numerically modeled following the deterioration technique presented in Section 3. The detailed description of the bridge numerical model is as below, along with strong ground motions, the latter being necessary to represent the seismic excitation of the bridge structure. The damage states of the bridge model are also defined since they constitute an inherent part of any fragility calculation.

4.1. Description of Case Study Bridge

A multi-span RC continuous girder bridge, a common type of highway bridge in China, is selected as the testbed structure for the AS fragility analysis. The bridge has three units, each consisting of three spans, and has a deck width of 9.8 m. The spans in the first and last units of the bridge were 30 m long each, and the length of the three middle spans in the central unit was chosen to be 30 m, 50 m, and 30 m, resulting in an overall bridge length of 290 m. Each bridge column has a circular cross-section of 1524 mm in diameter along 12.10 m in height. Especially, 28 longitudinal steel bars of 32.8 mm diameter and transverse (spiral) reinforcing bars of 12 mm diameter and 89 mm spacing were used to reinforce the cross-section of the bridge columns. The compressive strength of concrete for columns is 33.0 Mpa, while the yield strength of reinforcing bars is 465 Mpa. The column aspect ratio was calculated well above 2.5, indicating the dominant flexural behavior of the column that, in turn, may lead to flexural failure due to the formation of a plastic hinge. The single-column bents and the box girder were linked with sliding rubber bearings. In each unit, one column can be found along the longitudinal direction designed to be the critical component; hence, fixed bearings have been placed at the top of those elements to decrease their vulnerability to structural failures. Figure 7 shows the thorough review of plans for the middle unit of the bridge, which is numerically modeled in this study.

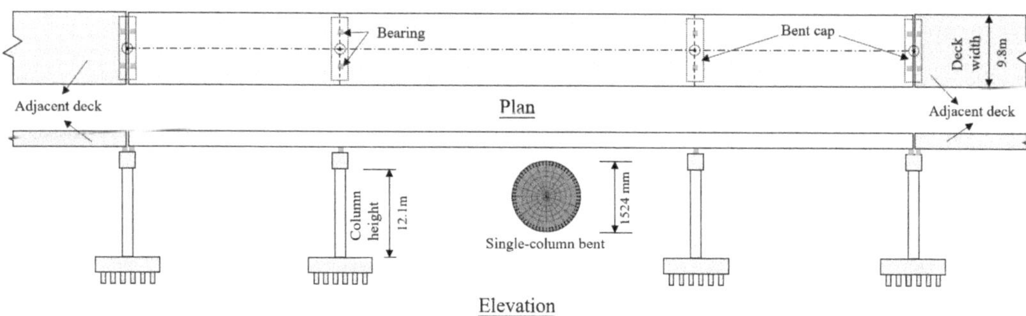

Figure 7. Schematic illustration of the bridge layout.

The bridge was modeled using OpenSees Finite element code [38], as shown in Figure 8. It is noted that only the seismic performance of the middle unit of the bridge is investigated here, with the other two units being the boundary conditions. The critical column, i.e., the third column from the left, was modeled using the previous deterioration modeling technique. Additionally, the soil-structure interaction effects were considered. The first two natural vibration periods in the longitudinal direction of the FE model of the bridge were 0.99 s and 0.17 s, reflecting the bridge's dominant first mode of vibration.

4.2. Strong Ground Motions

The calculation of the AS fragility requires using the bridge finite element model to perform response history analysis. To this end, mainshock-aftershock earthquake sequences were generated using a set of 30 earthquake ground motions already selected and used by Vamvatsikos and Cornell [39]. These earthquake motions, used herein both as mainshock and aftershock ground excitations, have been recorded at firm soil sites during past seismic events occurring in the western USA (California), with moment magnitude and closest distance to fault rupture ranging between 6.5 and 6.9 and 15 and 33 km, respectively. Additionally, the intensity of the selected earthquake motions, in terms of peak ground acceleration (PGA), varies between 0.04 and 0.63 g. It should be noted, though, that the measure used in the current study to quantify the intensity of earthquake motions is the first mode of spectral acceleration, $Sa(T_1)$, calculated at a damping ratio of 5%. The response spectra of the 30 ground motions are illustrated in Figure 9. The use of the 30 seismic motions, for which more details can be found elsewhere [39], eventually allowed the generation of 900 artificial mainshock-aftershock sequences since each strong ground motion was used both as mainshock and aftershock excitation. Such an artificial generation of mainshock–aftershock sequences has been widely adopted for relevant studies due to the difficulty of finding a sufficient number of recorded sequence events.

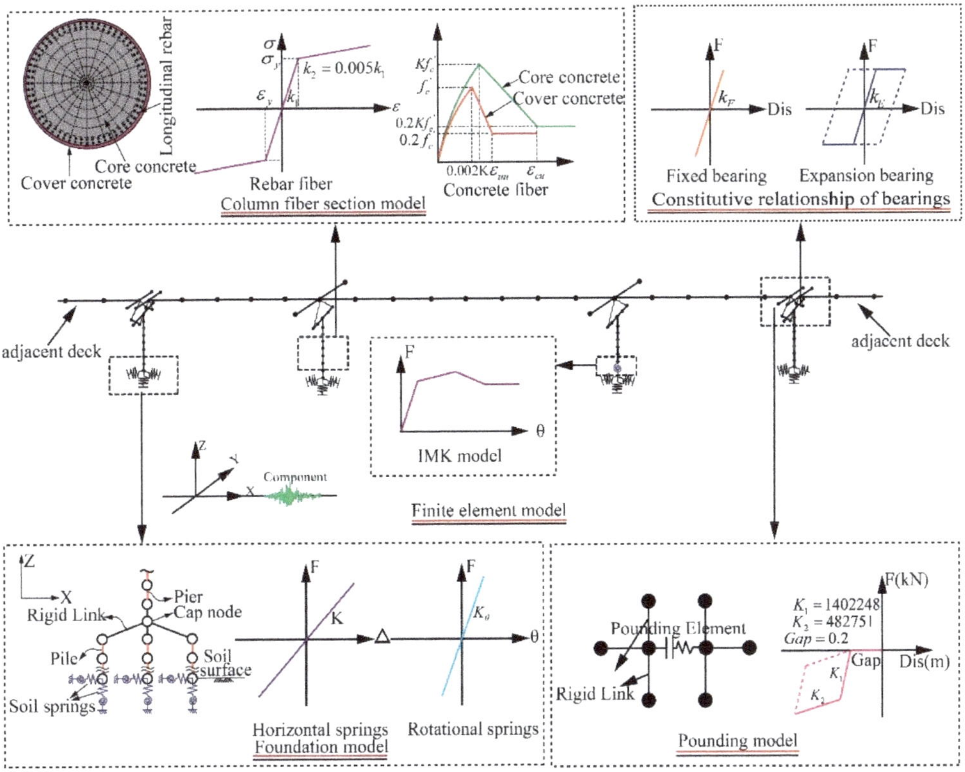

Figure 8. Numerical model of the bridge and its various components.

4.3. Definition of Damage States

The MBDI is employed herein as the seismic demand (also called engineering demand parameter, EDP) to quantify the structural damage induced to the multi-span RC continuous girder bridge when subjected to the seismic hazard. Along these lines, the correlation between physical damage and the currently proposed MBDI for various damage states needs to be established to facilitate the fragility calculation, especially to enable the numerical realization of different levels of damaged structures. This study categorizes the damage as minor, moderate, and extensive, while the last damage level adopted herein corresponds to the collapse.

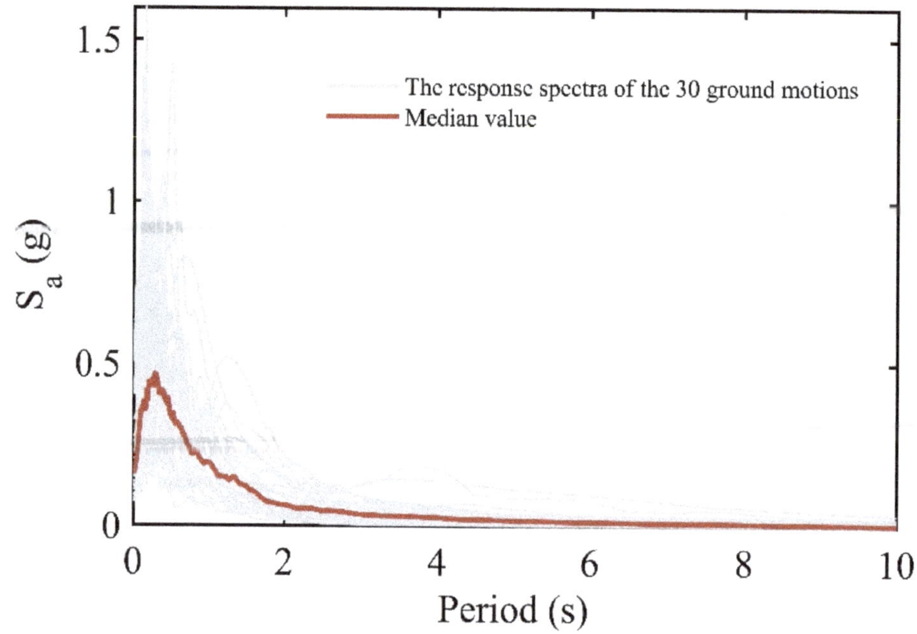

Figure 9. The response spectra of the 30 ground motions.

The correspondence between the physical damage and the damage states, the latter defined in terms of MBDI, are shown in Table 1. The maximum drift ratio for each damage state can be obtained first through monotonic pushover analysis on the intact numerical model. Step 1 and Step 2 are introduced in detail in Section 2.2. The pushover result is illustrated in Figure 10a. The maximum drift ratio corresponding to each damage state can be determined numerically when it reaches the physical damage. According to the pushover result, the maximum drift ratio for each damage state is 0.0446, 0.0784, 0.1035, and 0.1286, respectively, as listed in Table 1. Based on the above, Step 3, Step 4, and Step 5, introduced in detail in Section 2.2, are carried out to obtain the values of MDBI corresponding to each limit state, as shown in Figure 10b. In Step 3, a monotonic pushover is conducted as the target load until it reaches the maximum drift ratio corresponding to a specific damage state, then unloaded until zero force. In Step 4, conduct a back-to-back monotonic pushover analysis in the same direction as the structural model until collapse. In Step 5, the values of MDBI corresponding to each limit state are calculated according to Equation (3). Finally, the threshold of the MBDI corresponding to the damage states are 0.02, 0.21, 0.57, and 1 for minor damage, moderate damage, extensive damage, and collapse, respectively.

Table 1. Damage states defined in the current study.

DS	Limit State	Physical Damage	Maximum Drift Ratio	MBDI
DS1	Minor	Spalling of cover concrete	0.33	0.02
DS2	Moderate	Core crushing	0.57	0.21
DS3	Extensive	Bar buckling	0.76	0.57
DS4	Collapse	Large drift	0.94	1.00

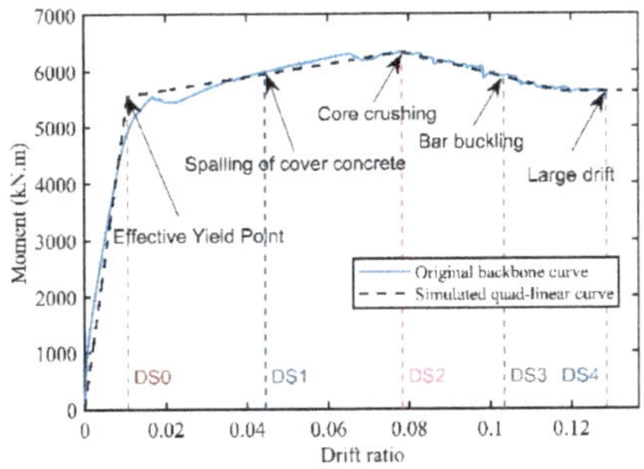

(a)

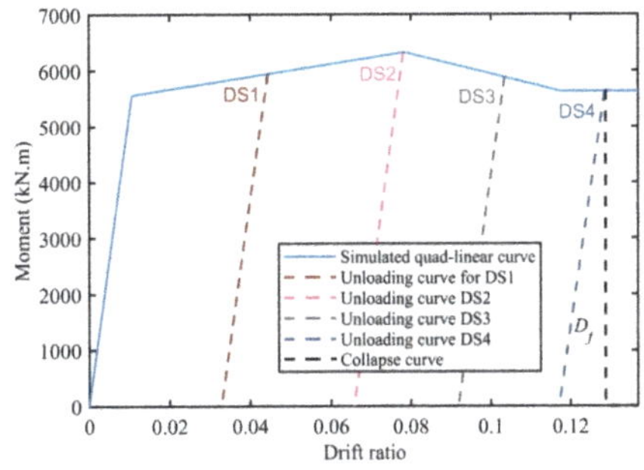

(b)

Figure 10. (**a**) The pushover curves for the study case, and (**b**) the definition of damage states in terms of Maximum Drift ratio and MBDI.

5. Aftershock Fragility Assessment

5.1. The B2B-IDA Methodology for Aftershock Fragility Assessment

The incremental dynamic analysis (IDA) approach is widely used to generate fragility curves. An IDA involves a series of nonlinear time-history analyses of the structure subjected to incrementally increasing ground excitations. The intensity of each ground motion record in the ensemble is scaled to multiple levels to force the structure to undergo the entire range of behavior, from elastic to inelastic and finally to collapse.

The engineering community has developed different numerical schemes to assess AS fragility [6–18]. The latter can be mainly classified according to how the mainshock damage and the assessment of the post-mainshock capacity are simulated. Among those schemes, the so-called back-to-back incremental dynamic analysis (B2B-IDA) framework is widely used to calculate AS fragility of structural systems of varying size, geometry, material, importance, and complexity [10]. According to the B2B-IDA framework, post-mainshock damaged structure simulation is facilitated by incrementally scaling each mainshock strong ground motion until a specific level of damage is induced. Next, the AS fragility curves can be generated by performing IDA on the mainshock-damaged structure under a set of selected aftershock excitations.

This paper employs the B2B-IDA methodology to generate AS fragility curves. The process involves the following steps:

(1) First, scale each mainshock incrementally to determine the intensity of the mainshock to induce the structure to reach exactly each specified damage state.
(2) Second, perform a nonlinear time history analysis of the structure in each post-mainshock damage state subjected to mainshock-aftershock sequences (so-called "back-to-back" dynamic analyses). The mainshock serves as a means to induce the structure to sustain a certain level of post-mainshock damage, while the aftershock is scaled incrementally to calculate the AS fragility according to the EDP during the aftershock.
(3) Last, determine the probability distribution of aftershock intensity values associated with each post-aftershock damage state conditioned on each post-mainshock damage state over all the 900 pairs of specific mainshock-aftershock sequences.

5.2. Aftershock Fragility Results

Scaling an earthquake record to obtain a realization of the building in each post-mainshock damage state is an iterative process, and the MBDI is used as EDP in the process. The intensities of the mainshocks to induce each damage state are firstly listed in Table 2.

Table 2. Seismic intensities of the mainshocks to induce each damaged state.

Ground Motion	Seismic Intensity (Sa(T1),g)			
	DS1	DS2	DS3	DS4
1	1.30	2.24	2.96	3.23
2	1.70	3.42	3.88	3.95
3	1.56	2.60	2.90	3.06
4	1.88	2.98	3.71	4.24
5	1.77	2.63	2.86	2.91
6	1.13	2.65	2.85	2.93
7	1.49	5.45	[illegible]	6.50
8	1.33	2.03	2.16	2.24
9	0.98	1.28	1.60	1.69
10	1.50	1.64	1.68	1.70
11	1.11	1.58	1.89	2.18
12	1.05	2.22	3.77	4.96
13	1.47	2.42	2.67	2.80
14	0.75	1.14	1.36	1.54
15	1.25	1.67	1.89	2.02
16	0.74	1.11	1.28	1.50
17	1.51	2.30	2.91	3.29
18	1.04	1.26	1.36	1.44
19	1.40	2.16	2.80	3.22
20	1.16	1.58	1.86	2.03
21	1.12	2.21	4.15	4.84
22	1.32	2.10	2.63	3.52
23	1.45	3.73	6.81	7.71
24	3.38	4.60	5.01	5.25
25	2.26	4.42	4.83	5.00
26	1.44	2.28	2.79	3.22
27	1.06	1.47	1.73	1.94
28	0.86	1.10	1.20	1.26
29	1.00	1.65	2.03	2.24
30	0.81	1.05	1.24	1.35

Based on these, the back-to-back nonlinear time history analysis (NTHA) of the structure subjected to mainshock-aftershock sequences is performed. The AS fragility curves for each post-mainshock damage state are generated through the probability distribution of the EDPs obtained from 900 artificial mainshock-aftershock sequences. The IDA curves of the structure subjected to aftershocks conditioned on each post-mainshock damage state are shown in Figure 11. In the IDA method, the probability distribution of the EDPs for each damage state over all the 900 seismic sequences is assumed to be lognormal distribution. It must be validated through the hypothesis-testing method, e.g., the Kolmogorov–Smirnov test. The AS fragility curves are determined as the cumulative distribution function (CDF) of the EDPs in terms of the intensity of the aftershock. The IDA curves of the structure subjected to aftershocks conditioned on each post-mainshock damage state are shown in Figure 11. It is found that the aftershock IDA curves have a distinct upward tendency around the damage state corresponding to the mainshock damaged state. This phenomenon is even more evident for the higher structural damage sustained during the mainshock. Moreover, by comparing the IDA curves in the three subfigures, it is also found that the more severe damage caused by the mainshock, the lower the seismic intensity of the aftershock required to cause a certain degree of damage.

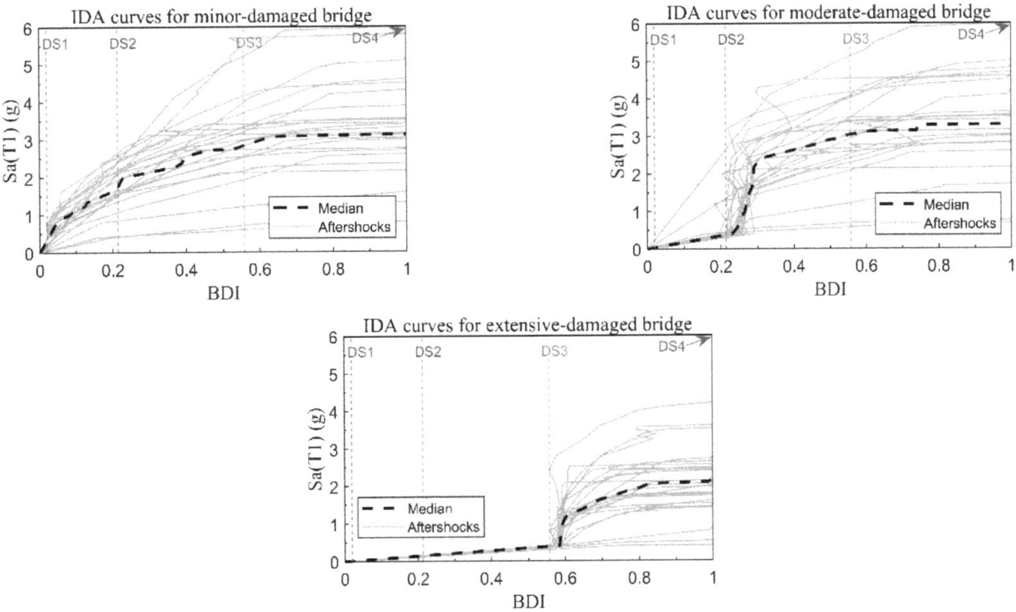

Figure 11. IDA curves of aftershocks for the bridge in minor, moderate, and extensive damage states.

Figure 12 compares the AS fragility conditioned on different mainshock-damage states. As shown in Figure 12a, only the minor-damage aftershock fragility for the intact mainshock-damage state varies with the AS intensity, and the fragility for other mainshock-damage states turns out to be 1.0 constantly. This is because the structure already experienced a higher level than the minor damage due to the mainshock and can only sustain a greater extent of damage after an aftershock. Along this line, it is also observed that the moderate aftershock fragility for the moderate and extensive mainshock-damaged structure in Figure 12b and the extensive aftershock fragility for the extensive mainshock-damaged structure in Figure 12c are constantly 1.0. For others, the aftershock fragility curves increase with the AS intensity. In addition, it is also found that the higher the damage sustained during the mainshock, the greater the probability that the bridge will exhibit a more severe damage state when subjected to an aftershock of a given intensity. For example, Figure 12d compares the collapse fragility conditioned on the different mainshock-damage states. Specifically, when an aftershock with an intensity of 3.0 g occurs, the probability for the structure in a minor damage state (due to the mainshock) to collapse is 0.2. If the structure sustains extensive damage from the mainshock, the probability of collapse will sharply rise to 0.78. Therefore, the post-mainshock damage level will significantly influence the vulnerability of the structure to withstand an aftershock. In other words, when a structure sustains a higher damage level from a mainshock, it will be much more fragile. A similar trend can also be found in Figure 12a–c.

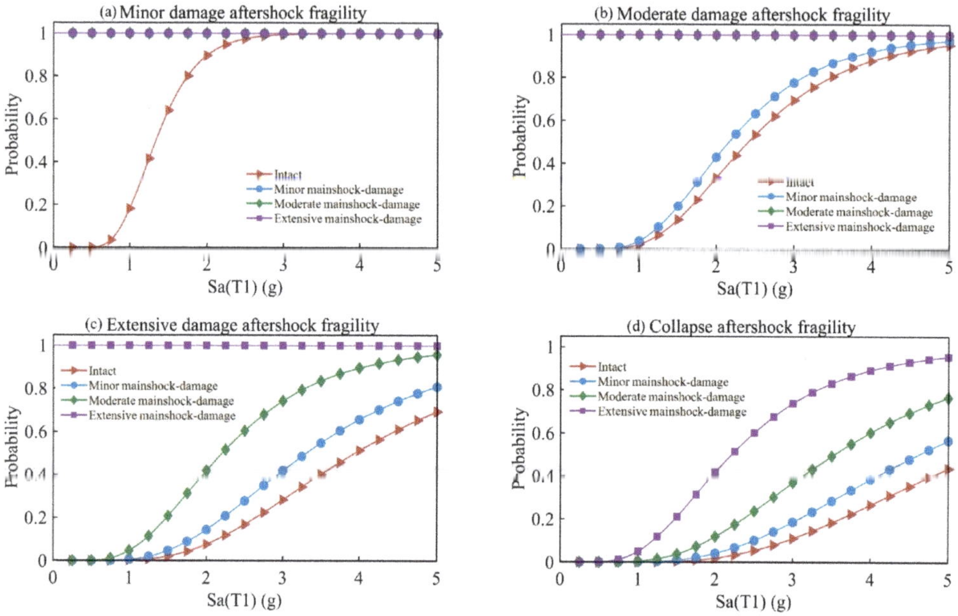

Figure 12. Aftershock fragility curves for different damage states (**a**) minor damage, (**b**) moderate damage, (**c**) extensive damage, and (**d**) collapse damage conditioned on different mainshock-damage states.

6. Validation of MBDI as EDP for Aftershock Fragility

In order to validate the effectiveness of the MBDI as a predictor of AS fragility, the assessment results are compared with those obtained from using the maximum drift ratio as EDP, one of the most widely used displacement-based damage indices [10].

The effectiveness of damage indices as predictors of AS fragility is evaluated in terms of their ability to reduce the deviation of collapse capacity of the structure when simulating a specific damage state using different mainshocks. The collapse capacity under an earthquake record herein is defined as the smallest ground motion spectral acceleration (at T_1 and a damping ratio of 5%) of the record that would induce localized or complete structural collapse in an aftershock. Along these lines, the collapse capacity of a structure can be defined as the mean value of the collapse capacities under a number of ground motions. The deviation of collapse capacity concerned above is quantified via the coefficient of variation, COV (i.e., COV = σ/μ, where σ and μ are the standard deviation and the arithmetic mean value). σ and μ are calculated statistically based on the collapse capacities derived from using different mainshocks to induce a specific damage state. The same set of 30 aftershocks, as mentioned in Section 4, is used in this study to determine the collapse capacities.

Figure 13 illustrates the outcome for the mainshock-damage structure in minor, moderate, and extensive damage states (DS0 = 1, 2, 3) obtained via MBDI and the maximum drift ratio, with each column representing the COV of the residual capacities for the set of 30 mainshocks that induce damage and collapse by each aftershock. The comparison shows that the COVs for MBDI are generally lower than using the maximum drift ratio. In addition, the difference between the COVs via the two damage indices becomes more significant as the mainshock-damage level increases. The lower trend of the COVs for MBDI, when the structure sustains extensive damage from the mainshocks, shows a stronger relationship between MBDI and the residual capacity and therefore demonstrates the advantage of MDBI over maximum drift ratio in simulating severe damage. Based on the above discussion, it is concluded that MBDI, in this case, is a better damage index predictor of AS fragility.

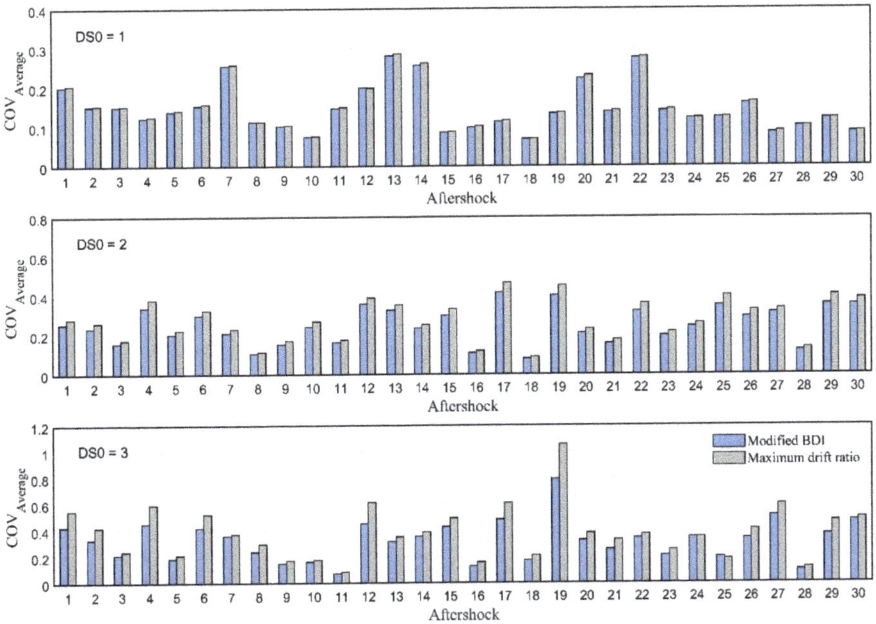

Figure 13. Variability of collapse capacity derived from different mainshocks that induce initial damage states via different damage indices.

7. Summary and Conclusions

This study presents an FE-based damage evaluation methodology for continuous girder bridges used to develop AS fragility curves based on post-mainshock damage conditions. It integrates a combined damage index that accounts for multiple factors associated with cumulative damage, an FE-based quantification procedure, and the supporting deterioration modeling technique. The proposed combined damage index (MBDI) is a modified version based on the damage model developed by Bracci. The FE-based quantification procedure is developed, in which back-to-back monotonic pushover analysis is required. For the deterioration modeling technique, the lumped plastic numerical model of bridge columns is established, consisting of three elements: a zero-length rotation spring at the bottom of the column, a rigid element, and fiber-based elements. The IMK hysteretic model captures the deterioration behaviors under MS-AS sequences. The process of determining the parameters for IMK material (including both the backbone curve and the cyclic deterioration parameters) is introduced. The proposed damage measure methodology can consider nearly all the critical factors related to cumulative damage, such as stiffness and strength deterioration, peak displacement, and residual displacement. A multi-span RC continuous girder bridge is studied and modeled to demonstrate the proposed method using OpenSees. The B2B-IDA framework is applied using 900 MS-AS synthetic sequences generated from 30 earthquake records to calculate the AS fragility conditioned on a different level of post-mainshock damage. The influence of the post-mainshock damage states on AS fragility is investigated. The following broad conclusions emerge.

(1) The aftershock fragility results show that the higher the damage sustained during the mainshock, the greater the probability that the bridge will exhibit a more severe damage state when subjected to an aftershock of a given intensity.

(2) The proposed damage index's effectiveness is investigated in the second part of the study. The AS fragility results obtained via MBDI are compared with those via maximum drift ratio in terms of the ability to reduce the variation of residual capacities obtained using different mainshocks to induce a specific damage state but collapse by the same aftershock. The comparison shows a more robust relationship of MBDI with the residual capacity and therefore demonstrates that MBDI is a better damage index predictor of AS fragility.

(3) The major contribution of the proposed damage measure methodology is the refinement of an effective predictor for evaluating cumulative damage for AS fragility assessment. It enables quantitative evaluation of the increased vulnerability of damaged RC bridges while significantly reducing the variation derived from using different mainshocks to simulate structural damage.

(4) It is noted that the deterioration modeling technique proposed in this study fits well with the structures in which the plastic hinges generate only in finite and fixed positions under external loading. For those structures with uncertain plastic hinge zone, developing other applicable deterioration modeling methods is necessary. Whether the quantification procedure for the MBDI is still suitable needs to be discussed in future studies.

(5) The proposed damage quantification methodology can be extended to other AS fragility analysis frameworks, such as the IDA-Cloud framework. In future studies, the methodology should be adapted to explore its effectiveness in developing AS fragility using the cloud method.

Author Contributions: Conceptualization, Z.W., J.G. and X.D. (Xinzhi Dang); methodology, Z.W.; investigation, Z.W. and J.G.; case study design copyright, X.D. (Xiaowei Deng), X.L. and X.D. (Xinzhi Dang); writing—original draft preparation, Z.W.; writing—review and editing, Z.W. and J.G.; formal analysis, Z.W.; validation, Z.W.; discussion, Z.W., J.G., X.D. (Xiaowei Deng), X.L. and X.D. (Xinzhi Dang). All authors have read and agreed to the published version of the manuscript.

Funding: This research was funded by the National Key Research and Development Program of China (2019YFE0112300), the Shanghai Post-doctoral Excellent Program (2021333), the Post doctoral Innovation Practice Base Program of Shanghai Yangpu District, the Shanghai Rising-Star Program (21QB1406000), and the National Natural Science Foundation of China (51978512).

Data Availability Statement: Not applicable.

Conflicts of Interest: The authors declare no conflict of interest.

References

1. Goda, K. Record selection for aftershock incremental dynamic analysis. *Earthq. Eng. Struct. Dyn.* **2015**, *44*, 157–162. [CrossRef]
2. Goda, K.; Taylor, C.A. Effects of aftershocks on peak ductility demand due to strong ground motion records from shallow crustal earthquakes. *Earthq. Eng. Struct. Dyn.* **2012**, *41*, 2311–2330. [CrossRef]
3. Knopoff, L.; Gardner, J. Higher seismic activity during local night on the raw worldwide earthquake catalogue. *Geophys. J. Int.* **1972**, *28*, 311–313. [CrossRef]
4. Hosseinpour, F.; Abdelnaby, A.E. Fragility curves for RC frames under multiple earthquakes. *Soil Dyn. Earthq. Eng.* **2017**, *98*, 222–234. [CrossRef]
5. Villaverde, R. Methods to Assess the Seismic Collapse Capacity of Building Structures: State of the Art. *J. Struct. Eng.* **2007**, *133*, 57–66. [CrossRef]
6. Bazzurro, P.; Cornell, C.A.; Menun, C.; Motahari, M. Guidelines for seismic assessment of damaged buildings. In Proceedings of the 13th World Conference on Earthquake Engineering, Vancouver, BC, Canada, 1–6 August 2004; p. 1708.
7. Maffei, J.; Telleen, K.; Nakayama, Y. Probability-Based Seismic Assessment of Buildings, Considering Post-Earthquake Safety. *Earthq. Spectra* **2008**, *24*, 667–699. [CrossRef]
8. Polese, M.; Di Ludovico, M.; Prota, A.; Manfredi, G. Damage-dependent vulnerability curves for existing buildings. *Earthq. Eng. Struct. Dyn.* **2013**, *42*, 853–870. [CrossRef]
9. Luco, N.; Bazzurro, P.; Cornell, B.Y.C.A. Dynamic Versus Static Computation Of The Residual Capacity Of A Mainshock-damaged Building To Withstand An Aftershock. In Proceedings of the 13th World Conference on Earthquake Engineering, Vancouver, BC, Canada, 1–6 August 2004.

10. Raghunandan, M.; Liel, A.B.; Luco, N. Aftershock collapse vulnerability assessment of reinforced concrete frame structures. *Earthq. Eng. Struct. Dyn.* **2015**, *44*, 419–439. [CrossRef]
11. Li, Q. Mathematical Formulation of Tools for Assessment of Fragility and Vulnerability of Damaged Buildings. Ph.D. Thesis, Tsinghua University, Beijing, China, 2006.
12. Jeon, J.-S.; DesRoches, R.; Lowes, L.N.; Brilakis, I. Framework of aftershock fragility assessment-case studies: Older California reinforced concrete building frames. *Earthq. Eng. Struct. Dyn.* **2015**, *44*, 2617–2636. [CrossRef]
13. Gaetani d'Aragona, M.; Polese, M.; Elwood, K.J.; Baradaran Shoraka, M.; Prota, A. Aftershock collapse fragility curves for non-ductile RC buildings: A scenario-based assessment. *Earthq. Eng. Struct. Dyn.* **2017**, *46*, 2083–2102. [CrossRef]
14. Wen, W.; Zhai, C.; Ji, D.; Li, S.; Xie, L. Framework for the vulnerability assessment of structure under mainshock-aftershock sequences. *Soil Dyn. Earthq. Eng.* **2017**, *101*, 41–52. [CrossRef]
15. Ebrahimian, H.; Jalayer, F.; Asprone, D.; Lombardi, A.M.; Marzocchi, W.; Prota, A.; Manfredi, G. A performance-based framework for adaptive seismic aftershock risk assessment. *Earthq. Eng. Struct. Dyn.* **2014**, *43*, 2179–2197. [CrossRef]
16. Jalayer, F.; Ebrahimian, H. Seismic risk assessment considering cumulative damage due to aftershocks. *Earthq. Eng. Struct. Dyn.* **2017**, *46*, 369–389. [CrossRef]
17. Shokrabadi, M.; Burton, H.V. Risk-based assessment of aftershock and mainshock-aftershock seismic performance of reinforced concrete frames. *Struct. Saf.* **2018**, *73*, 64–74. [CrossRef]
18. Trevlopoulos, K.; Guéguen, P. Period elongation-based framework for operative assessment of the variation of seismic vulnerability of reinforced concrete buildings during aftershock sequences. *Soil Dyn. Earthq. Eng.* **2016**, *84*, 224–237. [CrossRef]
19. Shinozuka, M.; Kim, S.-H.; Kushiyama, S.; Yi, J.-H. Fragility curves of concrete bridges retrofitted by column jacketing. *Earthq. Eng. Eng. Vib.* **2002**, *1*, 195–205. [CrossRef]
20. Sung, Y.-C.; Su, C.-K. Time-dependent seismic fragility curves on optimal retrofitting of neutralised reinforced concrete bridges. *Struct. Infrastruct. Eng.* **2011**, *7*, 797–805. [CrossRef]
21. Tolentino, D.; Márquez-Domínguez, S.; Gaxiola-Camacho, J.R. Fragility assessment of bridges considering cumulative damage caused by seismic loading. *KSCE J. Civ. Eng.* **2020**, *24*, 551–560. [CrossRef]
22. Furtado, A.; Rodrigues, H.; Varum, H.; Arêde, A. Mainshock-aftershock damage assessment of infilled RC structures. *Eng. Struct.* **2018**, *175*, 645–660. [CrossRef]
23. Zhai, C.-H.; Bao, X.; Zheng, Z.; Wang, X.-Y. Impact of aftershocks on a post-mainshock damaged containment structure considering duration. *Soil Dyn. Earthq. Eng.* **2018**, *115*, 129–141. [CrossRef]
24. Poiani, M.; Gazzani, V.; Clementi, F.; Lenci, S. Aftershock fragility assessment of Italian cast–in–place RC industrial structures with precast vaults. *J. Build. Eng.* **2020**, *29*, 101206. [CrossRef]
25. Li, Y.; Song, R.Q.; Van de Lindt, J.W. Collapse Fragility of Steel Structures Subjected to Earthquake Mainshock-Aftershock Sequences. *J. Struct. Eng.* **2014**, *140*, 1–10. [CrossRef]
26. Baker, J.W.; Cornell, C.A. *Vector-Valued Ground Motion Intensity Measures for Probabilistic Seismic Demand Analysis*; Pacific Earthquake Engineering Research Center, College of Engineering, University of California: Berkeley, CA, USA, 2006.
27. Park, Y.-J.; Ang, A.H.-S.; Wen, Y.K. Seismic damage analysis of reinforced concrete buildings. *J. Struct. Eng.* **1985**, *111*, 740–757. [CrossRef]
28. Park, Y.-J.; Ang, A.H.-S. Mechanistic seismic damage model for reinforced concrete. *J. Struct. Eng.* **1985**, *111*, 722–739. [CrossRef]
29. Williams, M.S.; Sexsmith, R.G. Seismic damage indices for concrete structures: A state-of-the-art review. *Earthq. Spectra.* **1995**, *11*, 319–349. [CrossRef]
30. Bracci, J.; Reinhorn, A.; Mander, J.; Kunnath, S. *Deterministic Model for Seismic Damage Evaluation of RC Structures*; National Academies: Washington, DC, USA, 1989; NCEER-89-0033.
31. Ibarra, L.F.; Medina, R.A.; Krawinkler, H. Hysteretic models that incorporate strength and stiffness deterioration. *Earthq. Eng. Struct. Dyn.* **2005**, *34*, 1489–1511. [CrossRef]
32. Haselton, C.B.; Deierlein, G.G. Assessing Seismic Collapse Safety of Modern Reinforced Concrete Moment-Frame Buildings. *Civ. Eng.* **2008**, *137*, 481–491. [CrossRef]
33. Abdelnaby, A.E. Fragility Curves for RC Frames Subjected to Tohoku Mainshock-Aftershocks Sequences. *J. Earthq. Eng.* **2018**, *22*, 902–920. [CrossRef]
34. Berry, M.P.; Lehman, D.E.; Lowes, L.N. Lumped-plasticity models for performance simulation of bridge columns. *ACI Struct. J.* **2008**, *105*, 270–279.
35. Hosseinpour, F.; Abdelnaby, A.E. Effect of different aspects of multiple earthquakes on the nonlinear behavior of RC structures. *Soil Dyn. Earthq. Eng.* **2017**, *92*, 706–725. [CrossRef]
36. Lignos, D.; Krawinkler, H. *Sidesway Collapse of Deteriorating Structural Systems Under Seismic Excitations*; Report No. TB 172; John, A., Ed.; Blume Earthquake Engineering Research Center, Department of Civil and Environmental Engineering, Stanford University: Stanford, CA, USA, 2009.
37. Lignos, D.G.; Krawinkler, H. Deterioration modeling of steel components in support of collapse prediction of steel moment frames under earthquake loading. *J. Struct. Eng.* **2010**, *137*, 1291–1302. [CrossRef]
38. OpenSees. Open System for Earthquake Engineering Simulation. Available online: http://opensees.berkeley.edu/ (accessed on 7 October 2022).
39. Vamvatsikos, D.; Cornell, C.A. Incremental dynamic analysis. *Earthq. Eng. Struct. Dyn.* **2002**, *31*, 491–514. [CrossRef]

buildings

Article

Optimization of the Seismic Performance of a Steel-Concrete Wind Turbine Tower with the Tuned Mass Damper

Yanchao Yue, Changxin Li *, Kai Jia, Yuhang Zhang and Jingjing Tian

School of Human Settlements and Architecture and Engineering, Xi'an Jiaotong University, Xi'an 710049, China
* Correspondence: lcx1670372195@stu.xjtu.edu.cn

Abstract: To optimize the seismic performance of a new type of steel-concrete tower, a 120 m steel-concrete composite tower model with a tuned mass damper (TMD) was constructed in ABAQUS for simulation analysis. Firstly, a time history analysis was conducted to study the towers with and without a TMD to determine the difference in their accelerations, velocities, and displacements. Then, a frequency spectrum analysis was performed to determine the tower vibration reduction effect of TMDs with different mass ratios. Five different cases were considered to explore the impact of different layouts on the dynamic performance of the tower. The results showed that the TMD had a significant vibration reduction effect on the tower accelerations, velocities, and displacements. The acceleration was reduced the most, while the vibration reduction effect in the middle of the tower was more significant than that at the top of the tower. For the steel-concrete tower studied in this paper, the optimal mass ratio of TMD was found to be 0.01. Placing one TMD at the top and another in the middle of the tower was found to be the optimal TMD arrangement for tower vibration reduction.

Keywords: steel-concrete wind turbine tower; tuned mass damper; mass ratio; layout

Citation: Yue, Y.; Li, C.; Jia, K.; Zhang, Y.; Tian, J. Optimization of the Seismic Performance of a Steel-Concrete Wind Turbine Tower with the Tuned Mass Damper. *Buildings* **2022**, *12*, 1474. https://doi.org/10.3390/buildings12091474

Academic Editors: Bo Wang, Bo Fu and Xinxin Wei

Received: 7 August 2022
Accepted: 12 September 2022
Published: 17 September 2022

Copyright: © 2022 by the authors. Licensee MDPI, Basel, Switzerland. This article is an open access article distributed under the terms and conditions of the Creative Commons Attribution (CC BY) license (https://creativecommons.org/licenses/by/4.0/).

1. Introduction

Using wind energy is one of the most competitive large-scale ways to utilize clean and renewable energy to help solve the major contradiction that affects the sustainable development of humankind: the rapid economic development on the one hand, and the aggravation of environmental pollution, the increase in energy demand, and the depletion of fossil fuels on the other. Wind energy has attracted increased attention all over the world [1], and numerous countries have adopted wind power generation as an important source of renewable energy and developed a large number of wind power generation systems.

The tower is the main structural component of a wind turbine. As the main load-bearing component of a wind turbine, its structural form and materials adopted are the major factors in improving the wind turbine capacity and reducing the project costs. Modern large-scale wind power systems usually adopt tapered towers [2,3]. Experience with large-scale towers reveals that there are still some challenges in using the steel. For example, the cost of the steel tower increases exponentially with the height. Steel corrodes easily and needs regular inspections and maintenance. As the tower height increases, its diameter must also increase, but the steel towers of a very large size can no longer be transported by road [4–6]. Therefore, many different structural forms have been proposed [7–13] to avoid the transportation and manufacturing challenges of all steel towers. The steel-concrete tower, whose lower portion is made of concrete and upper portion is made of steel, can not only make full use of the advantages of convenient installation of steel towers and low maintenance cost of concrete towers, but also solve the problem of excessive bottom tower diameter prohibiting efficient transport. Thus, it is one of the future development directions of large-scale wind turbines.

For large-scale wind turbine towers, although the economic design encourages increasing the outer diameter, the transportation restrictions permit only increasing the wall thickness, thus increasing the cost [14]. Kang et al. [15] pointed out that concrete towers have obvious transportation advantages and larger economic benefits compared to steel towers, especially for taller towers. Because of the higher structural requirements of large wind turbine towers, Kaldellis et al. [16] suggested that the use of steel-concrete towers was economical and effective because of the full utilization of the advantages of the convenient installation of steel towers and low maintenance cost of concrete towers. Pons et al. [17] compared the advantages and disadvantages of traditional steel towers, concrete towers, and steel-concrete towers and concluded that the advantages of steel-concrete towers were more pronounced in large-scale wind turbines.

Moreover, numerous researchers have studied the structural connections of steel-concrete towers. The concrete part is generally prefabricated and then installed at the construction site. To facilitate transportation, it is divided into several sections along the tower height, and each section is further divided into several segments along the longitudinal direction. Kang et al. [15] analyzed the behavior of concrete tower joints under the action of torque by combining existing theory and finite element simulations, pointed out that the reliability of horizontal joints would affect the mechanical characteristics of the prefabricated concrete towers, and demonstrated the inaccuracy of the current theory by studying the torsional strength of a segmented tower structure in detail. Song and Cong [18,19] used ANSYS to analyze the mechanical performance of the horizontal and longitudinal joints of a prefabricated concrete tower. The results showed that adequate design of the horizontal and vertical joints can ensure a good mechanical performance of the tower. Kim et al. [20] proposed a new design method for the steel-concrete connection, and carried out a fatigue load test and a static load test on connection specimens. The experimental results verified that the connection section had not only good fatigue resistance, but also high static strength.

The steel-concrete composite tower is a cantilever structure, and its seismic response characteristics have a significant influence on the structural performance. However, most of the existing seismic performance analyses consider the steel tower structures [21–24], and there are only limited studies on the seismic performance of the steel-concrete towers. On the other hand, many studies have shown that a tuned mass damper (TMD) may significantly reduce the dynamic response of high-rise structures [25–32]. Therefore, this study analyzes the effects of TMDs on the seismic response of the new steel-concrete tower. Furthermore, this study varies the TMD mass ratio and layouts, including multiple TMDs, to examine the resultant changes in the tower displacement, velocity, and acceleration seismic response spectra, as well as optimizes the seismic performance of the new tower using TMDs to provide suggestions for the application of this kind of tower in practical projects.

The reminder of the paper is organized into four sections: Section 2 presents a simplified model of the steel-concrete tower with a TMD created in ABAQUS; Section 3 investigates the TMD impact on the tower dynamic characteristics using the time history analysis method; Section 4 optimizes the TMD structural parameters and layout for the improvement of the dynamic characteristics of the steel-concrete tower using spectral analysis. The final section provides conclusions from the analyses presented in the paper. The results show that the TMD could significantly reduce the accelerations, velocities, and displacements of the tower. The best TMD mass ratio and optimal arrangement for the vibration reduction of the steel-concrete tower were also identified.

2. Establishment and Simplification of Finite Element Model

2.1. Steel-Concrete Composite Tower

2.1.1. Model

As Figure 1a shows, the wind turbine tower established in this paper was composed of a steel tower, steel-concrete connection section, concrete tower, and independent foundation. The design of the steel tower was a circular tower, identical to a traditional steel tower [33]. The steel-concrete connection section was used to connect the steel tower and the concrete tower, as shown in Figure 1b. The concrete tower was divided into 20 sections, and each section was divided into 4 pieces. There were longitudinal and transversal joints between each piece, as depicted in Figure 1c. The 20 concrete sections were connected by 31 prestressed anchor cables, and pieces of the same tower section were connected by the cast-in-place concrete.

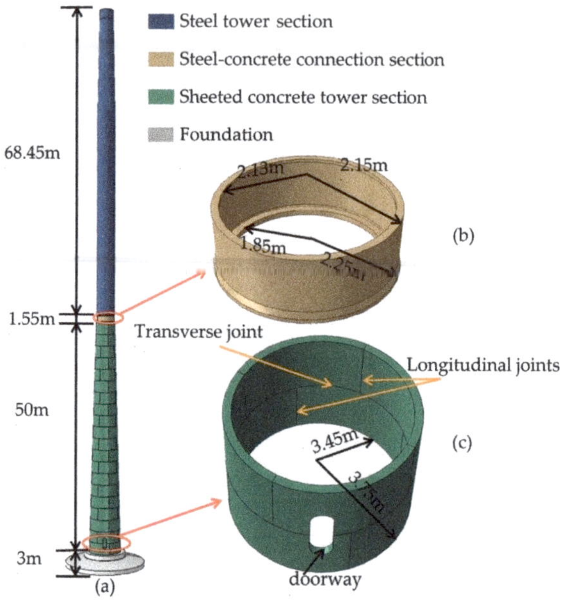

Figure 1. Model of the tower's geometric parameters. (**a**) Model of steel-concrete wind turbine tower; (**b**) Steel-concrete connection section; (**c**) Concrete pieces of 1st and 2nd sections.

2.1.2. Materials

Steel Tower and Conversion Section

The steel tower and steel-concrete connection section were made of Q355 steel, with a Poisson's ratio of 0.3, a yield stress of 355 MPa, and an elastic modulus of 206 GPa. The constitutive model was based on the bilinear kinematic hardening model, as shown in Figure 2. The elastic modulus of Q355 is represented by the slope of the first straight line in the figure. The slope of the second straight line is 1% of the slope of the first straight line [34], indicating that the steel entered the strengthening stage.

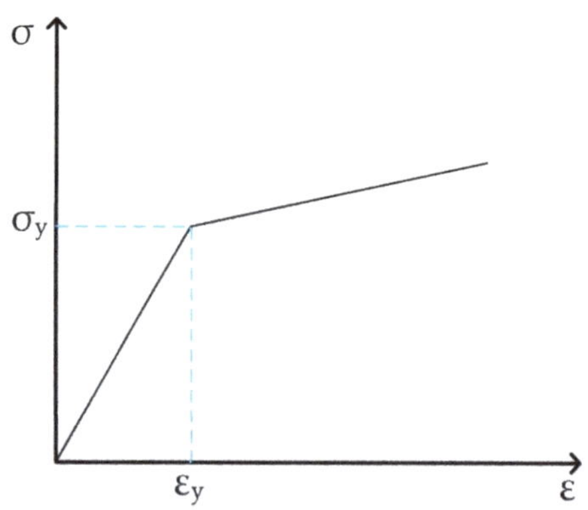

Figure 2. Bilinear kinematic hardening model of Q355.

Concrete Tower

The concrete damaged plasticity (CDP) model was used to simulate the tensile and compressive behavior of concrete under seismic conditions, where the dilation angle was 38°, eccentricity is 0.1, f_{b0}/f_{c0} ratio was 1.16, K was 0.6667, and viscosity parameter was 0.005. In the concrete section, C50 concrete was used, and the corresponding stress–strain curve was obtained in accordance with GB50010-2010 [35], as shown in Figure 3.

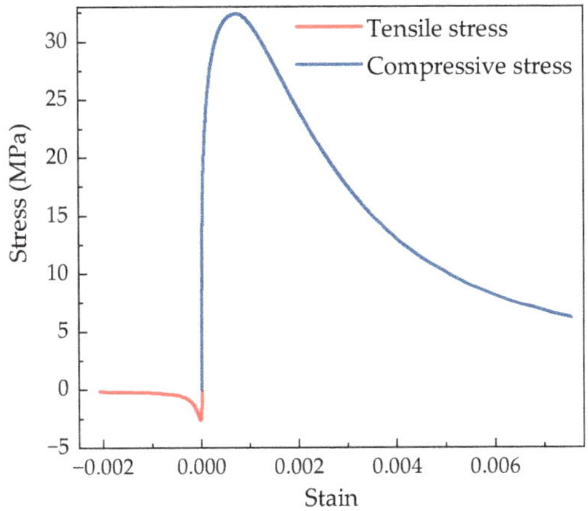

Figure 3. Plastic damage model of C50.

Prestressed Anchor Cables

For the prestressed anchor cables, the linear elastic model was used, and the expansion coefficient was set to 0.0001.

The Foundation

The foundation was not the primary research topic. To simplify the calculation, the C50 concrete linear elastic model was used.

2.1.3. Interaction between Components of Wind Turbine Tower

Between Concrete Pieces

As mentioned above, the 20 concrete sections were connected by 31 prestressed anchor cables, and pieces of the same tower section were connected by the cast-in-place concrete. Thus, surface-to-surface contact friction was utilized in transversal joints, and the model allowed the joint surface to develop a small elastic deformation to simulate relative movement between concrete parts. The friction model of penalty function was adopted [36], and the friction coefficient was taken as 0.4. Hard contact was defined in the normal direction. The contact surfaces would come under pressure if they remained in contact. When there was a gap between contact surfaces, the contact pressure would vanish and no interaction would occur. To simplify the calculations, the binding constraints were used in this study to simulate the cast-in-place concrete connection at the longitudinal joints of the concrete pieces

Prestressed Anchor Cables

There was no bond between the prestressed anchor cables and the concrete sections. As a result, as shown in Figure 4, the MPC multipoint constraint was used to simulate the interaction among the anchor cables, the upper conversion section, and the lower foundation.

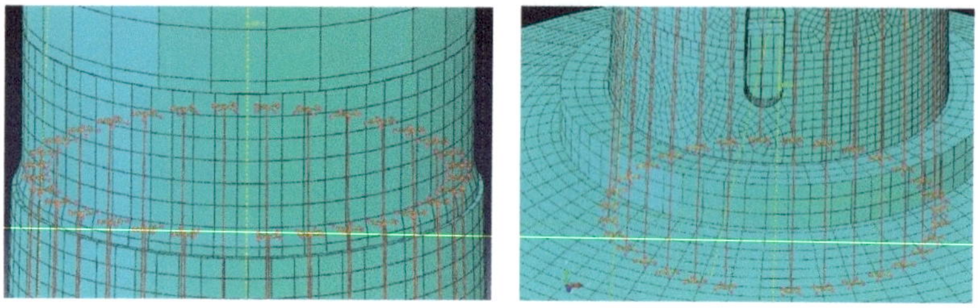

(**a**) (**b**)

Figure 4. Interaction of prestressed anchor cables. (**a**) MPC constraint of conversion section and prestressed anchor cables; (**b**) MPC constraint of foundation and prestressed anchor cables.

The Steel Conversion Section

Flanges connected the steel-concrete connection section to the concrete tower and the steel tower. Tie constraints were used instead of flange connections to simplify the model [37].

Concrete Tower and Foundation

Because the concrete tower and foundation were connected by cast-in-place concrete, the tie constraint was used to simulate the connection.

2.2. *Tuned Mass Damper (TMD)*

2.2.1. Model Parameters

The mechanical model of a TMD is shown in Figure 5. A TMD model must specify the three most important parameters: mass, stiffness, and damping. The mass of a TMD changes with the mass of the main structure. Considering the ease of installation, the TMD mass should not be excessive. The typical range of mass ratios between TMD and the main structure is usually 0.25–2% [38] and generally not more than 5% [39]. In this study, the mass TMD ratios remained within this range.

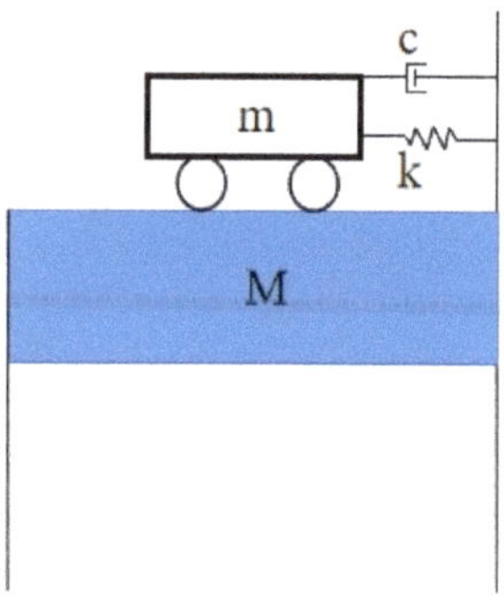

Figure 5. TMD structure diagram.

There are numerous optimization principles for determining the optimal value of TMD stiffness and damping. The three most common are the maximum effective damping, the minimum displacement, and the minimum acceleration [40–43]. The maximum effective damping optimization principle seeks to maximize the equivalent damping of the TMD and main structure, the minimum displacement optimization principle seeks to minimize the main structure displacements, and the minimum acceleration optimization principle seeks to minimize the main structure accelerations. The most widely used optimization principle of maximum effective damping was adopted in this study.

The following procedure was used to calculate the optimal TMD stiffness and damping values [38]: first, the TMD and primary system mass ratio $m = m_d/m$ was determined, where m_d is the mass of the TMD, and m is the mass of the main structure. The empirical formula for the TMD optimal frequency ratio is as follows:

$$\lambda_d = \frac{1}{\sqrt{1+1.5\mu}}. \tag{1}$$

The TMD vibration frequency is calculated using the following formula:

$$\omega_d = \lambda_d \omega_0, \tag{2}$$

where λ_d is the TMD to primary system optimal frequency ratio, ω_d is the TMD vibration frequency, and ω_0 represents the vibration frequency of the main structure. Then, the optimal TMD spring stiffness can be determined as follows:

$$k_d = m_d {\omega_d}^2. \tag{3}$$

The empirical formula for the TMD optimal damping ratio is as follows:

$$\xi_d = \sqrt{\frac{\mu(1-0.75\mu)}{4}}, \tag{4}$$

and the optimal TMD damping is as follows:

$$c_d = 2m_d\omega_d\xi_d. \tag{5}$$

2.2.2. TMD Simplified Model

The TMD was made up primarily of a mass block, a spring, and a damper. These three components could be simplified when modeling the TMD. The mass block could be broken down into lump mass points, each assigned a mass. One end of a spring or damper was linked to the lump mass point, while the other was linked to the tower [44]. To avoid excessive deformation at the point (117 m coupling point) where the TMD connected with the tower, an additional point was defined and coupled with the entire section of the tower using rigid beam elements, causing the TMD to act on the entire tower section rather than a single point. Figure 6 shows the model.

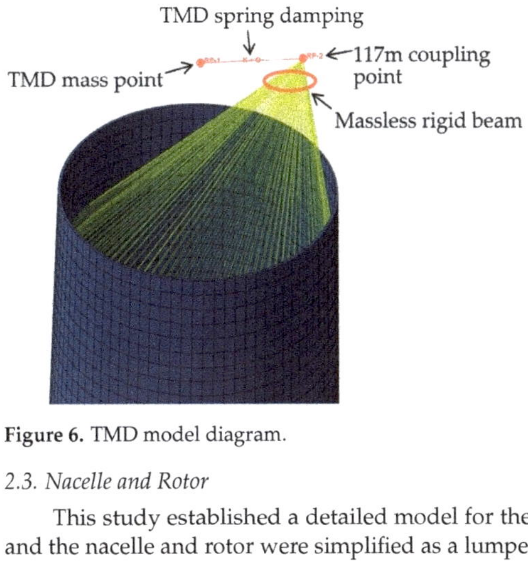

Figure 6. TMD model diagram.

2.3. Nacelle and Rotor

This study established a detailed model for the tower only to simplify the analysis, and the nacelle and rotor were simplified as a lumped mass added to the top of the tower.

The nacelle and rotor effects on the tower included their gravity, as well as the bending moment, torque, and shear force generated during operation, as shown in Table 1.

Table 1. Effect of the wind turbine on the tower.

Load Application Position (m)	M_z (N·m)	M_y (N·m)	F_x (N)	F_y (N)
118.196	14,049,710	919,540	214,520	2,361,460

3. Influence of TMD on Dynamic Characteristics of the Tower

3.1. Analysis Method

The first six modes of the tower without a TMD are shown in Figure 7. As can be seen, the top of the tower and its middle section developed the largest lateral displacements. Hence, TMDs were assumed to be installed at the top of the tower (at a height of 117 m) and at a height of 75 m, to investigate their impact on the dynamic properties of the tower structure. The mass ratio of a TMD to the tower structure was assumed as 0.01, and the mass of the main structure was assumed as 1,068,023.13 kg, including the tower mass of 898,205.13 kg and the nacelle and rotor mass of 169,818 kg (see Table 2).

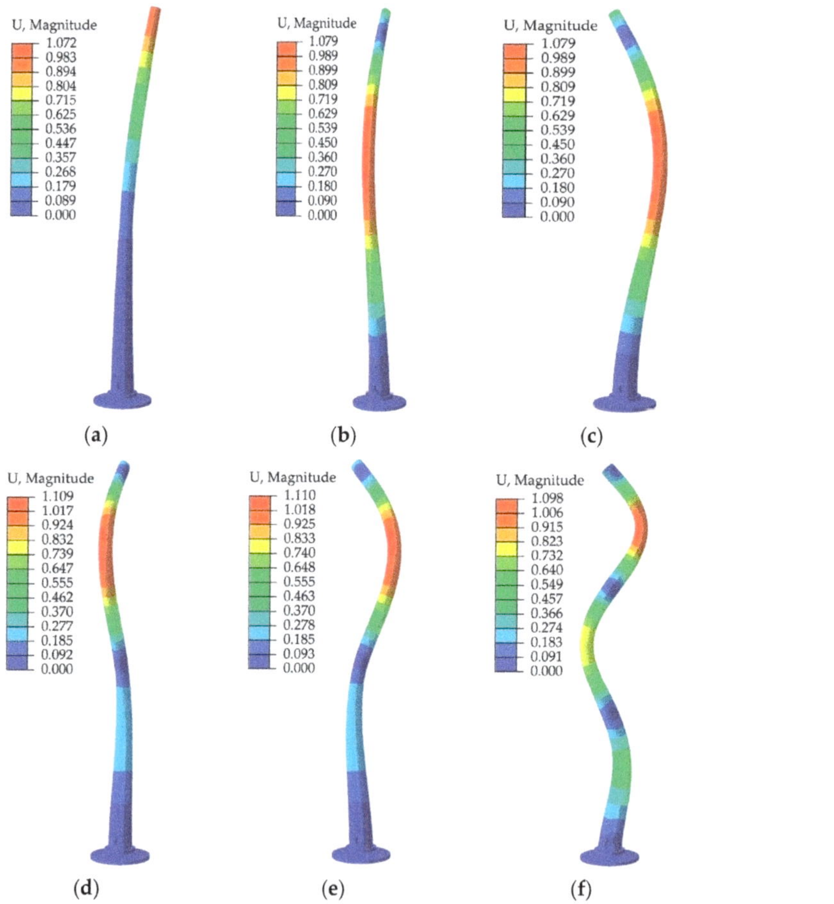

Figure 7. First six order modes of the tower without TMD: (**a**) first order mode; (**b**) second order mode; (**c**) third order mode; (**d**) fourth order mode; (**e**) fifth order mode; (**f**) sixth order mode.

Table 2. Parameters of TMD.

Layout Location	Mass (kg)	Spring Stiffness (N/m)	Damping Ratio	Damping Coefficient c_d(N·s/m)
75 m	5340.12	633,525.12	0.035	4105.13
117 m	5340.12	17,597.92	0.035	684.19

The dynamic characteristics of the tower were evaluated using time history analysis. Gravity, anchor cable prestress, and seismic load were the main actions. The seismic load was generated artificially, and the design peak value of the seismic base acceleration was 0.15 g. The input seismic wave was adjusted according to the following formula to make its peak acceleration value equal to the design value:

$$\alpha'(t) = \frac{\alpha'_{max}}{\alpha_{max}}\alpha(t), \tag{6}$$

where $\alpha'(t)$ represents the time history after adjustment, $\alpha(t)$ represents the time history before adjustment, and α'_{max} is the input seismic wave peak acceleration; α_{max} is the design peak acceleration. Figure 8 shows the adjusted time history of seismic excitation.

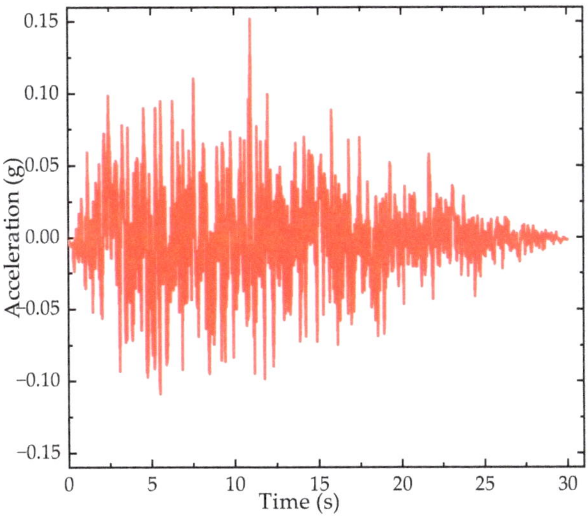

Figure 8. Adjusted time history curve.

A time history analysis of the models with and without TMDs was performed, comparing the displacements, velocities, and accelerations of the two to determine the rate of vibration reduction.

3.2. Results and Discussion

The dynamic responses at the TMD location are compared in Figures 9 and 10, which show the velocities, accelerations, and displacements at the height of 75 m and at the top of the tower. The displacement, velocity, and acceleration were reduced at the top and at the height of 75 m when TMDs were installed, indicating that the TMDs effectively improved the dynamic performance of the tower.

The tower acceleration and velocity were greater at the 75 m height than at the top. The lower steel section of the tower was thicker and stiffer than the upper steel section. Furthermore, the stiffness and frequency at 75 m were greater than those of the entire tower, and the vibration period was shorter. A smaller period led to greater acceleration, greater stiffness, less energy absorbed, and greater velocity, according to the relationships between period and acceleration and between energy and velocity. As a result, the acceleration and velocity in the middle section of the tower were greater than those at the top.

Furthermore, the effect of the TMDs on the accelerations, velocities, and displacements of tower gradually decreased, which is consistent with the results reported in [41]. This demonstrates that controlling the accelerations of the tower and reducing the intensity of vibrations are the key considerations for improving the dynamic performance of the tower when using TMDs.

The vibration attenuation rate that reflects the damping effect of the TMD is defined as follows:

$$\eta = \frac{(D_N - D_T)}{D_N} \times 100\%, \tag{7}$$

where D_N is the peak dynamic response of the structure without a TMD, and D_T is the structure peak dynamic response with a TMD.

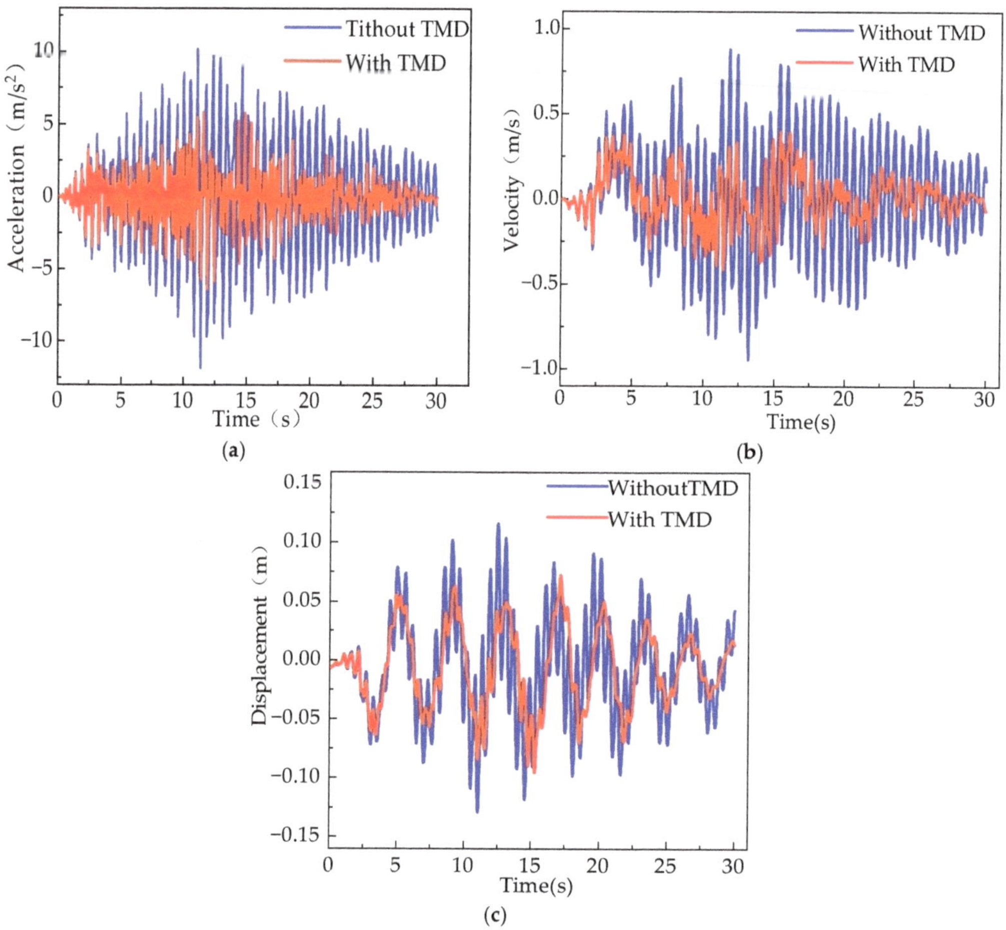

(a) (b) (c)

Figure 9. Comparison of acceleration, velocity and displacement at 75 m with and without TMD towers: (**a**) acceleration comparison; (**b**) velocity comparison; (**c**) displacement comparison.

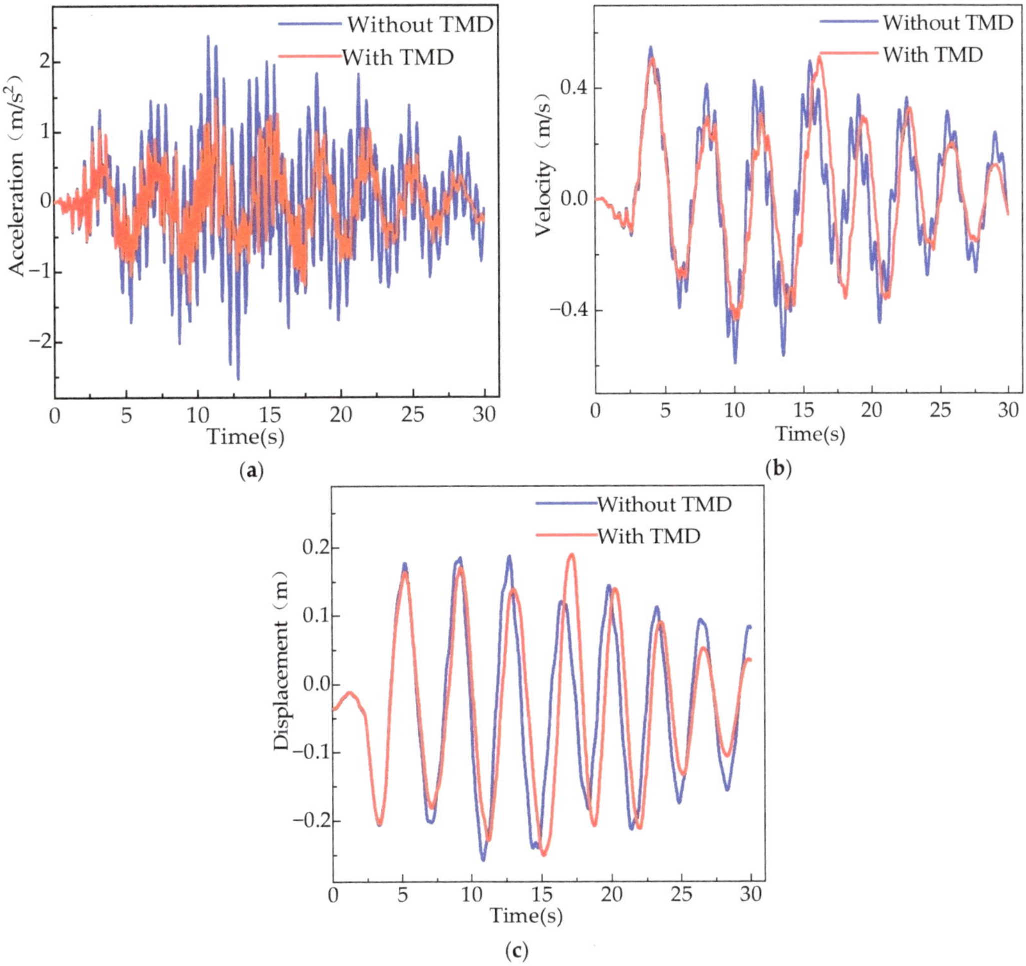

Figure 10. Comparison of acceleration, velocity and displacement at 117 m with and without TMD towers: (**a**) acceleration comparison; (**b**) velocity comparison; (**c**) displacement comparison.

The peak accelerations, velocities, and displacements from the time history analysis were used to calculate the structural vibration attenuation rates shown in Table 3. The results clearly demonstrate that the vibration reduction rate at 75 m was greater than that at the top of the tower after TMDs were installed at 75 m and the top of the tower. This is because the flexibility in the middle of the tower was low, and the vibration energy was high. The energy consumption of the TMD at 75 m was larger than that at the top.

Table 3. Vibration reduction rate at different positions of the tower.

Location	Acceleration Vibration Attenuation Rate	Velocity Vibration Attenuation Rate	Displacement Vibration Attenuation Rate
75 m	70%	64%	60%
Top of tower	51%	35%	9%

4. Optimization of TMD Structural Parameters and Layouts

This section examines the influence of TMD parameters and layout on their vibration reduction effect using the spectral analysis method, optimizes the TMDs on the basis of the seismic performance of the tower, and further investigates whether the selection principle of the TMD optimal parameters is applicable to the steel-concrete tower.

Assuming that the tower was located at a class II site, the design earthquake was the second group, the damping ratio was 0.03, the peak value of the seismic influence coefficient was 0.34, and the characteristic period, Tg, was 0.40 s, the acceleration response spectrum used for the analysis was determined in accordance with the specifications and is shown in Figure 11.

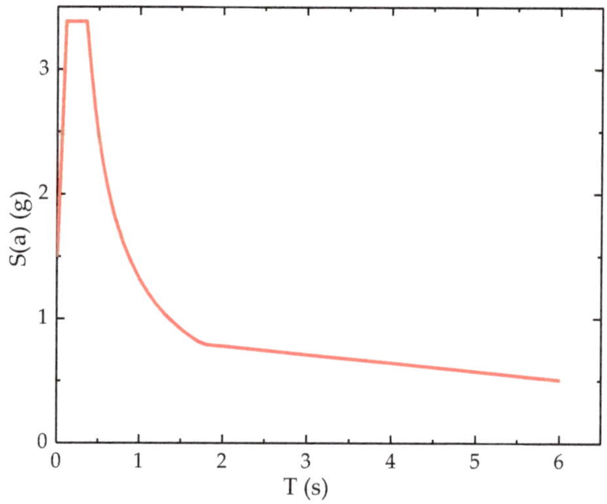

Figure 11. Acceleration response spectrum curve.

4.1. Mass Ratio

Because the TMD mass determines its optimal stiffness and damping, its mass ratio is also the most critical parameter influencing its vibration reduction performance. To investigate the effect of TMD mass on the vibration reduction performance, seven different tower models with TMDs of varying masses were established. The adopted TMD-to-tower structure mass ratios were 0.003, 0.005, 0.007, 0.01, 0.02, 0.03, and 0.04, respectively. The stiffness and damping coefficient of the TMD were the optimal theoretical values corresponding to the mass ratios adopted in the models, and the TMD was assumed to be installed at the top of the tower. Figure 12 shows the vibration reduction rates derived from the spectral analysis.

The acceleration at the top of the tower gradually decreased as the TMD mass ratio increased, but the rate of change also gradually decreased, indicating that, when the TMD mass ratio reached a certain value, the acceleration level gradually stabilized. However, as the TMD mass ratio increased, the velocity and displacement at the tower top first decreased and then increased. The velocity started to rise slowly when the TMD mass ratio reached about 0.02, while the displacement started to rise quickly when the TMD mass ratio reached about 0.01, eventually approaching the largest displacement without a TMD. This demonstrates that, when the TMD mass ratio was within a certain range, increasing the mass could effectively enhance the vibration reduction effect of the TMD; however, when it exceeded a certain value, increasing the TMD mass ratio diminished the inhibitory effect of the TMD on the dynamic responses of the structure, particularly displacements.

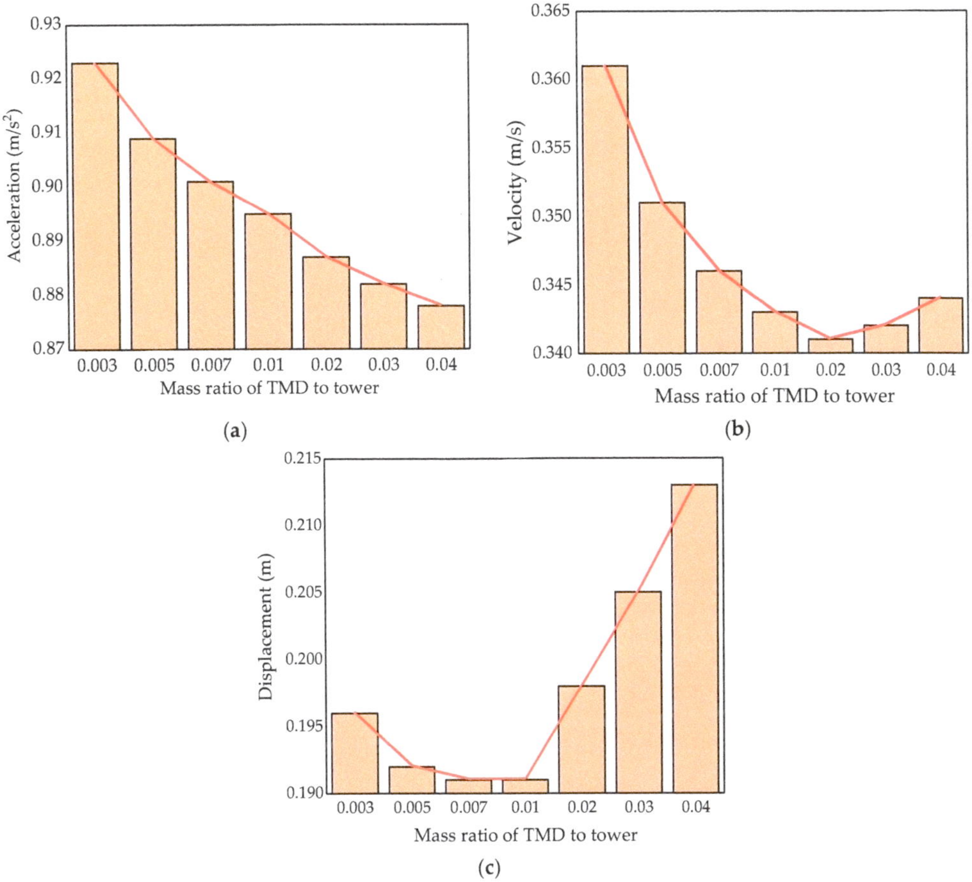

Figure 12. Dynamic response of tower top with different TMD mass ratios: (**a**) acceleration; (**b**) velocity; (**c**) displacement.

The above results were due to the vibration reduction effect of the TMD on the structure on the one hand, and because TMD belonged to the tower structure on the other hand. With the increase in the TMD mass ratio, the acceleration of the structure under constant external load decreased. However, with the increase in the top mass of the tower, the inertia force also increased. When the mass of the TMD reached a certain value, the influence of inertia on the vibration level became greater than the suppression vibrations by the TMD damping and energy consumption. This restricted the TMD effectiveness in reducing the velocities and displacements of the structure only to a certain mass ratio range.

4.2. TMD Layout Schemes

Single or multiple TMDs were installed at the locations of large modal displacements of the first six vibration modes shown in Section 3.2 to further investigate the influence of the TMD location on the tower vibration reduction. Three installation heights were considered: 75 m, 91 m and 117 m, respectively. Table 4 shows the five configurations that were examined. The total TMD mass ratio was 0.01, and the stiffness and damping coefficients were the optimal theoretical values. The impact of various TMD layouts on the dynamic performance of the tower was determined using spectral analysis. Figure 13 shows the accelerations of the various TMD installation positions for different TMD layouts.

Table 4. TMD-specific parameters of different layout schemes.

Layout Scheme	Position of TMD	Mass of TMD (kg)	Stiffness of TMD (N/m)	Damping of TMD (N·s/m)
Scheme 1	117 m	10,680.2313	34,935.77231	1924.379674
Scheme 2	117 m 91 m	5340.11565	17,597.92005 611,867.7529	684.1875557 4034.347311
Scheme 3	117 m 75 m	5340.11565	17,597.92005 611,867.7529	684.1875557 4034.347311
Scheme 4	91 m 75 m	5340.11565	611,867.7529	4034.347311
Scheme 5	117 m 91 m 75 m	3560.0771	11,761.13065 408,926.5413 408,926.5413	373.1213188 2200.129155 2200.129155

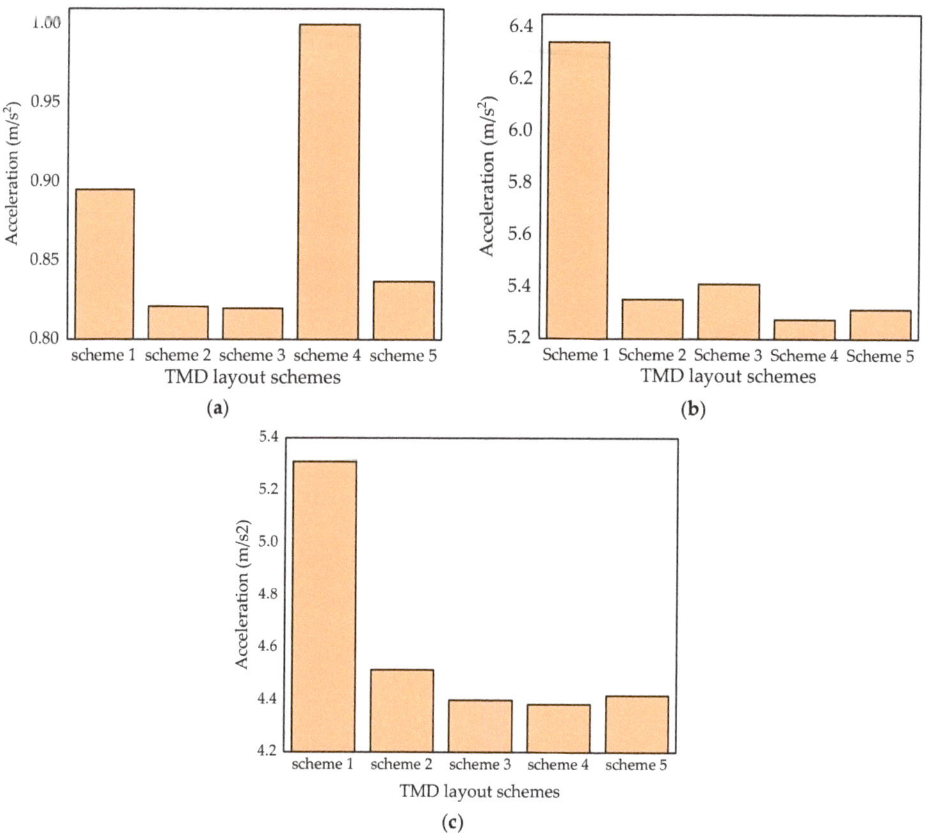

Figure 13. The acceleration of the different installation positions of TMD under various schemes: (**a**) acceleration at 117 m of the tower; (**b**) acceleration at 91 m of the tower; (**c**) acceleration at 75 m of the tower.

At the height of 117 m, the acceleration reduction effects of Layouts 2, 3, and 5 were slightly better than that of Layout 1, while Layout 4 had the weakest acceleration reduction effect. This is because Layouts 1, 2, and 3 all had a TMD at 117 m, whereas, for Layout 1, only one TMD was installed; hence, the vibration reduction effect was slightly worse than the case of another TMD installed at 75 m or 91 m. When TMDs were located at all three locations and the total mass ratio of the TMDs remained constant, the mass of the TMD at the top of the tower decreased and acceleration increased, but the dynamic performance could not be improved further. As a result, the TMDs at the top and in the middle (or upper) part of the tower had the strongest effect on the dynamic response of the tower top.

At the height of 91 m, the control effects of Layouts 2, 3, 4, and 5 on accelerations were clear, whereas the control effect of Layout 1 was extremely poor. Because Layout 1 had only one TMD at 117 m but not at 91m, the dynamic responses could not be effectively suppressed. A TMD was installed at 91 m in Layouts 2, 4, and 5, while two TMDs were located at 75 m and 117 m in Layout 3, i.e., the 91 m location was between the two TMDs and, thus, within their control range. Accordingly, a TMD could be positioned in the middle and upper part, in the middle (or at the top), or in the middle and at the top to effectively control the dynamic responses of the tower middle and upper part.

Layouts 3, 4, and 5 had a good vibration reduction effect at 75 m, while Layouts 1 and 2 had a poor vibration reduction effect, and Layout 1 had the weakest control effect. This is because Layouts 3, 4, and 5 had a TMD at 75 m, which could effectively control the vibrations there, whereas Layouts 1 and 2 had no TMD at 75 m, and Layout 1 had only one TMD at 117 m, which was the furthest away from 75 m, resulting in the worst effect. Therefore, installing a TMD in the middle, as well as in the middle and upper part or at the top, could significantly reduce the dynamic response in the tower middle.

In summary, for the whole steel-concrete tower, setting TMDs in the middle and at the top of the tower (Layout 3) represented the arrangement with a vibration reduction effect.

5. Conclusions

The vibration reduction effect of a TMD on the wind turbine tower was studied using the time history analysis method. Then, the influence of different TMD parameters and layouts on the tower vibrations was studied using the spectrum analysis method to provide suggestions for the optimization of TMDs to control steel-concrete wind turbine towers.

The following conclusions can be drawn from the analyses presented in this study:

1. Comparing the results with and without a TMD, after a TMD was installed, the dynamic responses of the steel-concrete tower were significantly reduced. Moreover, the acceleration reduction effect of the TMD was the most obvious, indicating that controlling the accelerations of the tower was critical for the TMD to improve the dynamic performance of the tower.
2. If a TMD was installed in the middle and at the top of the steel-concrete tower, the vibration reduction effect of the TMD located in the middle was better than that at the top.
3. With the increase in TMD mass ratio, the acceleration at the top of the tower decreased and eventually stabilized, the velocity decreased first and then increased slightly, and the displacement decreased first and then increased significantly. Therefore, the TMD mass ratio is recommended to be 0.01.
4. It was found that the vibration reduction effect of a TMD decreased with the distance from the TMD installation point.
5. The optimal TMD installation locations to reduce the seismic responses of the steel-concrete tower were in the middle and at the top of the tower, respectively. In an actual project, conclusions 3 and 4 can be considered comprehensively.

Author Contributions: Conceptualization, Y.Y. and J.T.; methodology, K.J.; software, C.L.; validation, C.L.; formal analysis, C.L.; investigation, Y.Z.; data curation, K.J.; writing—original draft preparation, C.L.; writing—review and editing, Y.Y.; visualization, Y.Y.; supervision, Y.Y.; project administration, Y.Y. All authors have read and agreed to the published version of the manuscript.

Funding: This research received no external funding.

Conflicts of Interest: The authors declare no conflict of interest.

References

1. Joselin Herbert, G.M.; Iniyan, S.; Sreevalsan, E.; Rajapandian, S. A review of wind energy technologies. *Renew. Sustain. Energy Rev.* **2007**, *11*, 1117–1145. [CrossRef]
2. Yan, K.; Wan, J.; Ren, W.; Wei, Y.; Feng, X.; Wang, H. The status and development of materials of large-scale wind turbine tower. *Wind Energy* **2013**, *3*, 102–105.
3. Agbayani, N.A.; Vega, R.E. The rapid evolution of wind turbine tower structural systems: A historical and technical overview. In Proceedings of the Structures Congress, Chicago, IL, USA, 29–31 March 2012.
4. Cotrell, J.; Stehly, T.; Johnson, J.; Roberts, J.O.; Parker, Z.; Scott, G.; Heimiller, D. Analysis of Transportation and Logistics Challenges Affecting the Deployment of Larger Wind Turbines: Summary of Results. Available online: https://permanent.fdlp.gov/gpo110145/61063.pdf____%60.pdf (accessed on 2 September 2022).
5. Sin, S.; Xu, Q.; Yun, H.; Pan, H. Status of corrosion and protection for offshore wind towers. *Corros. Prot.* **2010**, *31*, 875–877, 885.
6. Ungkurapinan, N. Development of Wind Turbine Towers Using Fiber Reinforced Polymers. Ph.D. Thesis, University of Manitoba, Winnipeg, MB, Canada, 2005.
7. Quilligan, A.; O'Connor, A.; Pakrashi, V. Fragility analysis of steel and concrete wind turbine towers. *Eng. Struct.* **2012**, *36*, 270–282. [CrossRef]
8. Zhang, Y.; Cai, X.; Gao, Q.; Ding, W.X. Research summary of wind turbine tower structure. *Chin. J. Eng. Des.* **2016**, *23*, 108–115, 123.
9. Polyzois, D.J.; Raftoyiannis, I.G.; Ungkurapinan, N. Static and dynamic characteristics of multi-cell jointed GFRP wind turbine towers. *Compos. Struct.* **2009**, *90*, 34–42. [CrossRef]
10. Alshurafa, S.A. Development of Meteorological Towers Using Advanced Composite Materials. Ph.D. Thesis, University of Manitoba, Winnipeg, MB, Canada, 2012.
11. Sami, A.; Hanan, A.; Dimos, P. Finite element method for the static and dynamic analysis of FRP guyed tower. *J. Comput. Des. Eng.* **2019**, *3*, 436–446.
12. Deng, R.; Zhou, X.H.; Deng, X.W.; Ke, K.; Bai, J.L.; Wang, Y.H. Compressive behaviour of tapered concrete-filled double skin steel tubular stub columns. *J. Constr. Steel Res.* **2021**, *184*, 106771. [CrossRef]
13. Rebelo, C.; Moura, A.; Gervásio, H.; Veljkovic, M.; Simões da Silva, L. Comparative life cycle assessment of tubular wind towers and foundations—Part 1: Structural design. *Eng. Struct.* **2014**, *74*, 283–291. [CrossRef]
14. McKenna, R.; Ostman v.d. Leye, P.; Fichtner, W. Key challenges and prospects for large wind turbines. *Renew. Sustain. Energy Rev.* **2016**, *53*, 1212–1221. [CrossRef]
15. Kang, C.; Hartwig, S.; Marx, S. Behavior of externally prestressed segmental towers' dry joint under torsion effects. *Struct. Concr.* **2019**, *20*, 1350–1357. [CrossRef]
16. Kaldellis, J.K.; Zafirakis, D. The wind energy (r)evolution: A short review of a long history. *Renew. Energy* **2011**, *36*, 1887–1901. [CrossRef]
17. Pons, O.; Fuente, A.; Armengou, J. Towards the sustainability in the design of wind towers. *Energy Procedia* **2017**, *115*, 41–49. [CrossRef]
18. Song, H.; Cong, O.; Hao, H.; Xu, Y. Research on mechanical behaviors of horizontal joint connection of prefhbricated concrete towers. *Build. Struct.* **2016**, *46*, 16–20.
19. Song, H.; Cong, O.; Hao, H.; Xu, Y. Research on mechanical behaviors of vertical connection node in prefabricated concrete tower. *Build. Struct.* **2018**, *48*, 679–683.
20. Kim, M.O.; Kim, T.M.; Lee, D.Y.; Kim, M.K. Experimental Investigation of the Steel-Concrete Joint in a Hybrid Tower for a Wind Turbine under Fatigue Loading. *KSCE J. Civ. Eng.* **2019**, *23*, 2971–2982. [CrossRef]
21. Huang, S.; Huang, M.; Lyu, Y.; Xiu, L. Effect of sea ice on seismic collapse-resistance performance of wind turbine tower based on a simplified calculation model. *Eng. Struct.* **2021**, *227*, 111426. [CrossRef]
22. Takeuchi, T.; Nakamura, T.; Matsui, R.; Ogawa, T.; Imamura, A. Seismic performance of tubular truss tower structures focusing on member fracture. *J. Struct. Constr. Eng.* **2011**, *76*, 1971–1980. [CrossRef]
23. Vca, B.; Sns, A. Seismic performance of offshore wind turbine in the vicinity of seamount subduction zone. *Structures* **2021**, *34*, 423–432.
24. Bin, L.I.; Yang, X.; Gao, C. Experimental research and comparative analysis on seismic performance of wind turbine frustum tower drum and latticed tower. *J. Build. Struct.* **2013**, *34*, 161–166.
25. Bian, Y.; Liu, X.; Sun, Y.; Zhong, Y. Optimized Design of a Tuned Mass Damper Inerter (TMDI) Applied to Circular Section Members of Transmission Towers. *Buildings* **2022**, *12*, 1154. [CrossRef]
26. Henriques, J.M.; Barros, R.C. Design of a Lattice Wind Tower and a Comparison of the Structural Response with and without a TMD. In Proceedings of the Fourteenth International Conference on Civil Structural and Environmental Engineering Computing, Cagliari, Italy, 3–6 September 2013.
27. Chen, J.; Georgakis, C.T. Tuned rolling-ball dampers for vibration control in wind turbines. *J. Sound Vib.* **2013**, *332*, 5271–5282. [CrossRef]
28. Wang, L.; Zhao, X.; Zheng, Y.M. A combined tuned damper and an optimal design method for wind-induced vibration control for super tall buildings. *Struct. Des. Tall Spec. Build.* **2016**, *25*, 468–502. [CrossRef]

29. Kalehsar, H.E.; Khodaie, N. Wind-induced vibration control of super-tall buildings using a new combined structural system. *J. Wind. Eng. Ind. Aerodyn.* **2018**, *172*, 256–266. [CrossRef]
30. Fujita, T.; Matsumoto, Y.; Masaki, N.; Suizu, Y. Tuned mass damper using multistage rubber bearing for vibration control of tall buildings (1st report. fundamental study of the vibration characteristics). *Trans. Jpn. Soc. Mech. Eng.* **1990**, *56*, 634–639. [CrossRef]
31. Zhang, Z.; Li, A.; Chen, Y. Research on wind-induced vibration hybrid control of Hefei TV Tower. In Proceedings of the 2011 International Conference on Electric Technology and Civil Engineering (ICETCE), Lushan, China, 22–24 April 2011.
32. Fitzgerald, B.; Sarkar, S.; Staino, A. Improved reliability of wind turbine towers with active tuned mass dampers (ATMDs). *J. Sound Vib.* **2018**, *419*, 103–122. [CrossRef]
33. Kottenstette, R.; Cotrell, J. Hydrogen storage in wind turbine towers. *Int. J. Hydrog. Energy* **2004**, *29*, 1277–1288. [CrossRef]
34. Liu, J.; Liu, H. ABAQUS finite element analysis of reinforced concrete. *Equip. Manufactring Technol.* **2009**, *6*, 69–70.
35. *Code for Design of Concrete Structures*; China Architecture & Building Press: Beijing, China, 2010.
36. Liu, Q.; Fu, W.; Li, M.; Chen, Z. Numerical simulation on the process of multi-point forming for tube. *J. Harbin Inst. Technol.* **2010**, *17*, 858–886.
37. Du, J.; Xu, Y.; Xie, S.; Yang, R. Research on vibration control of wind turbine tower based on TMD. *Acta Energ. Sol. Sin.* **2021**, *42*, 157–162.
38. Li, Y. Application research of TMD in civil engineering field in China. *Shanxi Archit.* **2020**, *46*, 41–43.
39. Luft, R.W. Optimal tuned mass dampers for buildings. *J. Struct. Div.-ASCE* **1979**, *105*, 2766–2772. [CrossRef]
40. Li, B.; Zhang, Y.; Yan, G.; Gu, H. Optimal design method for structures with viscous dampers in base isolation layer. *J. Harbin Inst. Technol.* **2022**, *54*, 101–110.
41. Wei, Y.; Chen, G. Modal damping optimization for general damped structures. *J. Vib. Eng.* **2006**, *19*, 433–437.
42. Hoang, N.; Fujino, Y.; Warnitchai, P. Optimal tuned mass damper for seismic applications and practical design formulas. *Eng. Struct.* **2008**, *30*, 707–715. [CrossRef]
43. Zhang, M.; Xu, F. Tuned mass damper for self-excited vibration control: Optimization involving nonlinear aeroelastic effect. *J. Wind. Eng. Ind. Aerodyn.* **2022**, *220*, 104836. [CrossRef]
44. Huang, Z.; Ni, L.; Qiu, H. Comparative study on four TMD vibration control simulation methods of wind turbine tower. *J. Henan Polytech. Univ.* **2020**, *39*, 85–90.

buildings

MDPI

Article

Experiment Study on the Hysteretic Performance of a Novel Replaceable Beam-to-Column Joint with Energy-Dissipating Steel Hinge

Yongchao Ma [1,2], Ai Qi [1,*], Guiyun Yan [1,*], Lianqiong Zheng [2] and Panrong Xue [3]

1 College of Civil Engineering, Fuzhou University, Fuzhou 350108, China
2 College of Civil Engineering, Fujian University of Technology, Fuzhou 350118, China
3 Fujian Provincial Erjian Construction Group Corp., Ltd., Fuzhou 350001, China
* Correspondence: qikai@fzu.edu.cn (A.Q.); yanguiyun@sina.com (G.Y.)

Abstract: A novel precast replaceable beam-to-column joint with energy-dissipating steel hinges was proposed for the connection of precast structures to improve the seismic performance and post-earthquake resilience. The proposed joint was installed in the predetermined plastic hinge region at beams and the flange segments of the proposed joint were weakened to achieve damage concentration. Cyclic loading tests were conducted on the proposed joint and the steel sleeve confined concrete joint to study the hysteretic performance, including failure mode, load–displacement curves, ductility, and energy-dissipation capacity. Moreover, the hystertic performance of the damage-repaired proposed joint was investigated to verify the post-earthquake resilience. Results demonstrated that the proposed joints could develop favorable failure mode with the necking rupture of the weakened steel plate in steel hinge. The damage of the proposed joint was concentrated in the energy-dissipating hinges while no serious damage was observed in the precast framing components, achieving the objective of damage concentration. Compared with steel sleeve confined concrete joint, the hysteresis curve of proposed joint was more plump while an obvious pinching effect was observed in the steel-confined concrete joint. The bearing capacity and energy-dissipation capacity of the proposed joint were about 1.25 times and 1.55 times of that for the steel sleeve confined concrete joint, respectively. In addition, the hysteretic performance of the repaired specimen was identical to the original one, with the desired failure mode caused by the fracture of the steel hinge. It was noted that the hysteretic performance of the repaired joint was better than the steel sleeve confined concrete joint. The bearing capacity was recovered at up to 96.6% of the original joint while the energy-dissipation capacity was recovered at 96.1%, indicating that the proposed joint achieved the post-earthquake resilience to a great extent.

Keywords: replaceable beam-to-column joint; energy-dissipating steel hinges; post-earthquake resilience; damage concentration; cyclic loading test

Citation: Ma, Y.; Qi, A.; Yan, G.; Zheng, L.; Xue, P. Experiment Study on the Hysteretic Performance of a Novel Replaceable Beam-to-Column Joint with Energy-Dissipating Steel Hinge. *Buildings* **2022**, *12*, 1180. https://doi.org/10.3390/buildings12081180

Academic Editors: Bo Wang, Bo Fu and Xinxin Wei

Received: 7 July 2022
Accepted: 3 August 2022
Published: 7 August 2022

Copyright: © 2022 by the authors. Licensee MDPI, Basel, Switzerland. This article is an open access article distributed under the terms and conditions of the Creative Commons Attribution (CC BY) license (https://creativecommons.org/licenses/by/4.0/).

1. Introduction

Industrialization of the construction industry promotes the development of precast reinforced concrete structures, which have entered the rapid growth stage. Precast reinforced concrete structures can improve construction quality, construction efficiency, and economic benefit, which are conducive to environmental protection [1,2]. However, the mechanical properties of the connection area of the prefabricated members always affect the integrity and seismic performance of the fabricated structure, which is more likely to be damaged and result in the collapse of the whole structure under the action of earthquakes [3,4]. Therefore, the research and development of new beam–column connections with the required mechanical properties is of great significance to improve the seismic performance of prefabricated concrete structures [5–8].

In-depth research works have been conducted on the fabricated beam–column connections or various connection methods for the connection of precast components, such as post-cast precast concrete joints [9,10], bolted connection of prefabricated members [11–13], and prestressed prefabricated joint [14–17]. Bahrami et al. [18] presented two kinds of bending connection precast joints, and carried out numerical research on the mechanical behavior under cyclic loading. The results showed that plastic hinge formed at the precast beam of the frame with two kinds of flexural connection precast joints led to an improved failure mode of bending. Girgin et al. [19] tested the mechanical behavior of five precast hybrid half-scale specimens, and it was observed that the maximum strain developed in the beam bottom flexural reinforcement played an important role in the overall behavior of the connections. Nzabonimpa et al. [20] presented a new assembled mechanical joint with removing the laminate, and established experimental and numerical studies of mechanical properties of joint structures. To improve energy consumption for precast beam–column connection, Tartaglia and Ferrante [21,22] acted on beam–column joints with friction dampers to control plastic deformation and improve node energy consumption.

Moreover, research on improving the seismic performance of precast reinforced concrete structures with prefabricated beam–column connections characterized by plastic damage has attracted more and more attention [23]. The hysteretic performance and damage characteristics of prefabricated, prestressed concrete frame systems with slightly pressed joints were studied [24], and the results demonstrated that slightly pressed joints were conducive to the control of deformation capacity and cracking. Wang et al. [25] arranged a replaceable low-yield-point steel bar in the assembled node to improve energy-dissipation capacity and control damage. Li et al. [26] investigated fabricated RC beam–column plastic controllable steel joint specimens to effectively control concrete damage. Zheng et al. [27] proposed a novel type of steel reciprocating bending energy-dissipation hinge connected by pin shaft to achieve beam-end energy dissipation and plastic hinge outward migration. Teng et al. [28] proposed a novel prefabricated beam-to-column steel joint and compared its hysteretic performance with a monolithic joint. The results showed that the use of buckling energy-dissipation segments was beneficial for concentrating the plastic deformation. Ertas et al. [29] experimentally investigated the performance of four types of precast concrete joints, which showed that, compared with other precast concrete joints, bolted concrete joints had the best ductility, strength, and energy-dissipation capacity. Li et al. [30] developed prefabricated steel joint precast concrete structures to replace cast-in-situ concrete beam-to-column joints. It was found experimentally that this joint exhibited better hysteretic performance and higher energy dissipation and ductility than a monolithic joint. A new hybrid beam–column connection was proposed for precast concrete structures, and the seismic behavior including hysteretic curves, skeleton curves, and dissipation capacity were studied through experimental and numerical study [31]. Moreover, it was also verified that the energy-dissipation capacity of the fabricated beam-to-column joint could be effectively enhanced by employing a friction device or dampers [32–34].

In this paper, a new kind of precast replaceable beam-to-column joint with an energy-dissipating steel hinge was proposed, which can be applied to the connection of precast reinforced concrete structures to improve seismic performance. In addition, through the rational design of the proposed joint, the accumulated plastic damage was concentrated in the energy-dissipating steel hinge and prevented the precast beams and columns from serious damage, improving the post-earthquake resilience performance of the precast reinforced concrete structures. By replacing the damaged upper and lower energy-dissipation steel plates in steel hinge, the proposed joints can be repaired conveniently and efficiently so as to restore the working performance. The hysteretic properties such as failure mode, load–displacement curves, stiffness degradation, and energy-dissipation capacity of the proposed joints were studied and compared with those of prefabricated steel sleeve confined concrete joints. Besides, the post-earthquake resilience performance was discussed by the comparison between the repaired specimen and original specimen.

2. Precast Replaceable Beam-to-Column Joint with Energy-Dissipating Steel Hinge

The configurations of the precast replaceable beam-to-column joint with an energy-dissipating steel hinge are shown in Figure 1. The proposed joint is composed of precast columns, precast beams with embedded steel beam segments, a confined steel sleeve for the core joint, and an energy-dissipating steel hinge. The joint core area was confined with a steel sleeve, with horizontal end plates to enhance the shear resistance. The steel sleeve was welded to a short cantilever beam segment, and then connected to the steel hinge by high-strength bolts. The longitudinal bars of the precast beam were welded to the steel beam with holed-end plates, and then the precast beam was also bolted to the steel hinge through the embedded steel beam segment. The grouting sleeves were embedded at the end of the rebars in the upper precast column, while the upper and lower plate stiffeners inside the steel sleeve were designed with holes to install reinforcements in the bottom column. The energy-dissipating steel hinge consisted of a flange energy-dissipation steel plate and a web-mechanical hinge with a pin shaft. The flange energy-dissipation steel plate in the energy-dissipating steel hinges was designed to sustain the bending moment, while the web-mechanical hinge with the pin shaft connector resists the shear force. The flange energy-dissipation steel plate was composed of an energy-dissipation steel plate and constraint sleeve so as to prevent the out-of-plane buckling behavior. Q235B was used for energy-dissipation steel plates, and the energy-dissipation steel plates were weakened with diamond-shaped holes to achieve yielding prior to other members and achieve the damage concentration.

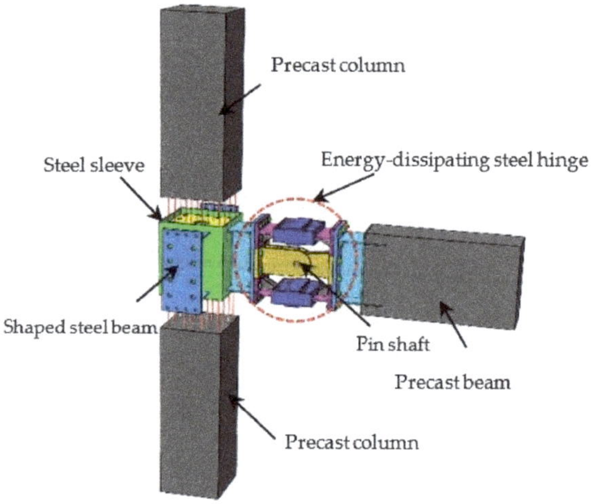

Figure 1. Precast replaceable beam-to-column joint with an energy-dissipating steel hinge.

3. Experimental Investigation

3.1. Design of Tested Specimens

Two tested specimens, including one precast replaceable beam-to-column joint with an energy-dissipating steel hinge (J-R-1) and one steel sleeve confined concrete joint (J-2) were designed. It should be noted that the specimen J-R-1 was repaired after the first test by replacing the damaged member in steel hinge and the repaired specimen (J-R-2) was tested again to verify the post-earthquake resilience. The geometric dimensions, configurations of the precast columns, and joint core area were the same for tested specimens, while the energy-dissipating steel hinges were installed at the predetermined plastic hinge region for J-R-1 and J-R-2. The structural configurations, geometric dimensions of each specimen, and the details of the energy-dissipating steel hinge and join core area are shown in Figures 2–5.

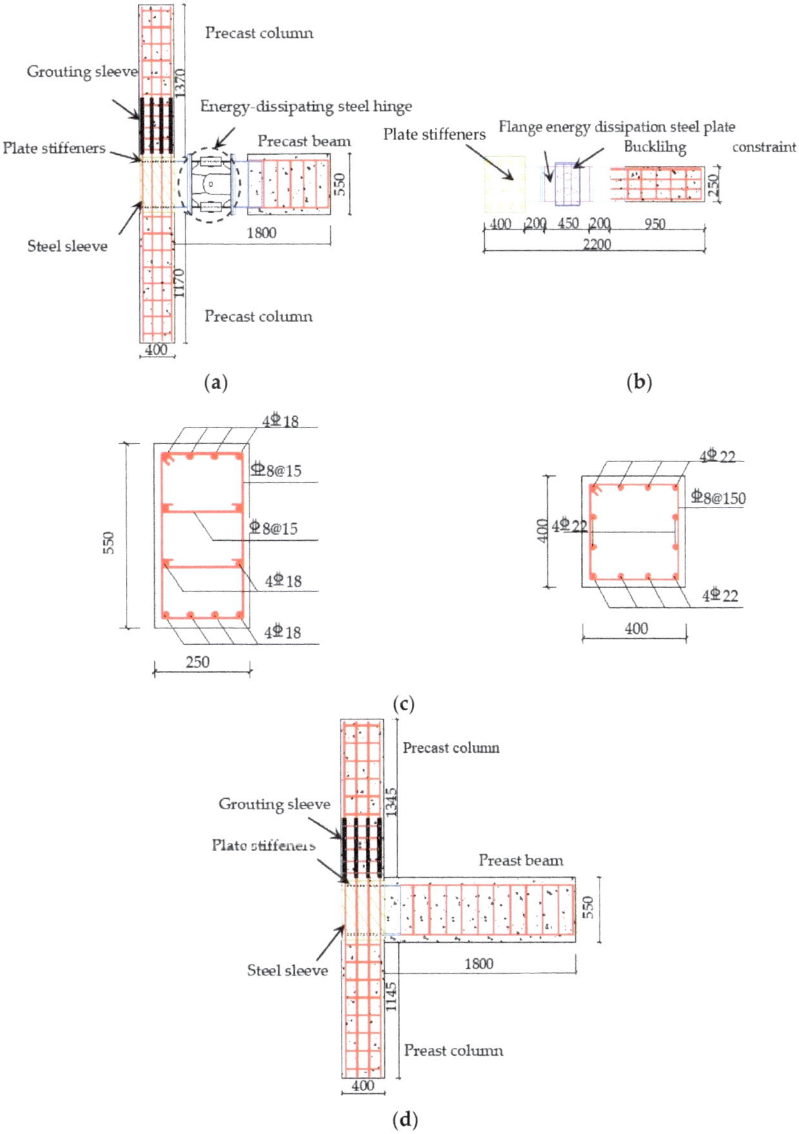

Figure 2. (**a**) Front view of the precast replaceable beam-to-column joint with an energy-dissipating steel hinge; (**b**) vertical view of the precast replaceable beam-to-column joint with an energy-dissipating steel hinge; (**c**) configurations of the beam and column; (**d**) front view of the steel sleeve confined concrete joints (unit: mm).

The concrete strength grade for concrete members was C50, and Q235 was adopted for the energy-dissipation steel plates while Q345 for the other steel members. The thickness of the energy-dissipation steel plates was 10 mm, which was weakened with diamond-shaped openings in the middle area, and the thickness of the steel sleeve for the joint core area was 10 mm.

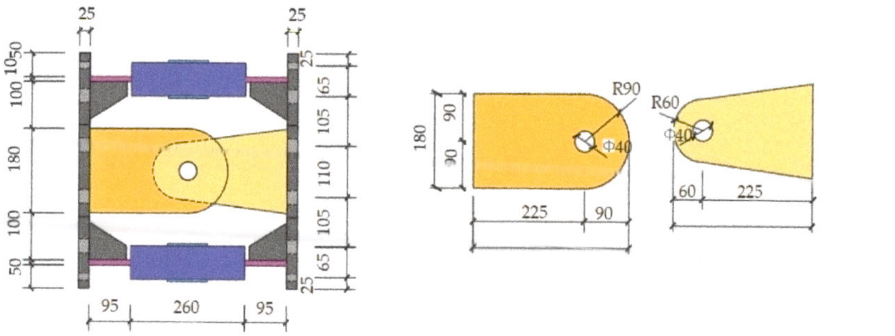

(a) (b)

Figure 3. (**a**) Front view of the energy-dissipating steel hinge; (**b**) front view of the web mechanical hinge (unit: mm).

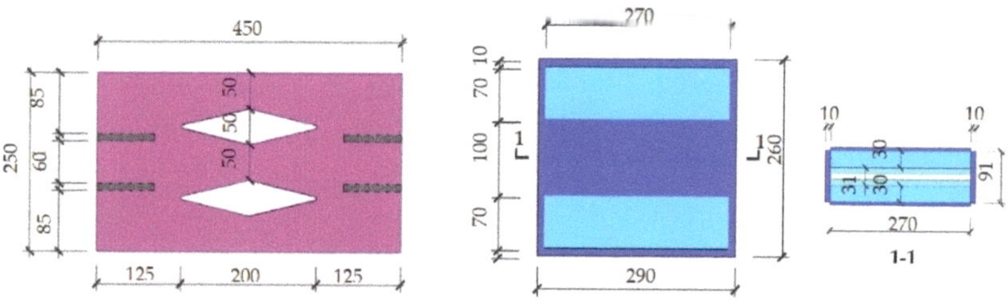

(a) (b)

Figure 4. (**a**) The reduced energy-dissipation steel plate with diamond-shaped holes; (**b**) buckling constraint sleeve construction (unit: mm).

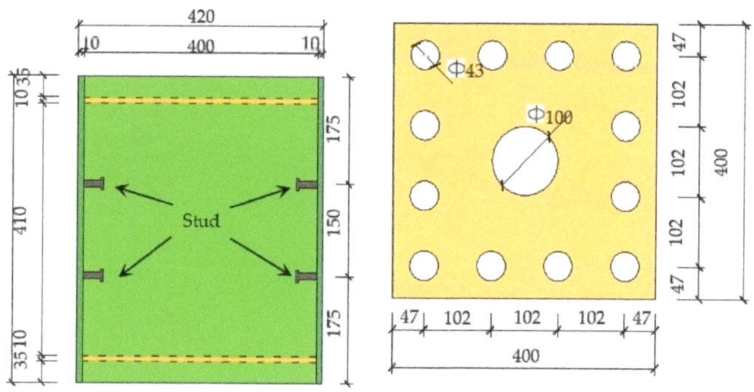

(a) (b)

Figure 5. (**a**) Front view of the steel sleeve in the joint core area; (**b**) vertical view of the steel sleeve in the joint core area (unit: mm).

3.2. Material Properties

C50 commercial concrete was used to pour precast members. The material properties of 150 mm cubic C50 concrete test blocks (same condition curing) were tested according to the Mechanical Properties test Method of Ordinary Concrete (GB50081-2016). Cube compressive strength was 56 MPa, elastic modulus was 3.54 GPa, tensile strength was 7.6 MPa, and the Poisson's ratio was 0.194. According to the tensile test of steel, the mechanical properties of the steel are shown in Table 1.

Table 1. Material properties of the steel.

Steel (bar) Model	Plate Thickness (Diameter) t(d)/mm	Yield Strength f_y/MPa	Yield Strain με	Ultimate Strength f_u/MPa	Elongation Ratio
Q345	10	374.2	2322	489.7	25.8
Q235	10	269.8	1659	373.6	24.5
HRB400	22	414.9	2527	563.7	21.4
HRB400	18	419.9	2446	558.5	22.6
HRB400	8	432.2	2612	572.5	21.9

3.3. Test Device and Loading Scheme

The test device is shown in Figure 6, including sliding support, jack, single-shaft hinged support, the adjustable beam-end support, the horizontal MTS hydraulic servo actuator, and the reaction wall. The MTS hydraulic servo actuator of 500 kN was connected to the reaction wall, and the other end was connected to the top of the upper column to apply horizontal cyclic reciprocating load. An axial load was applied to the top of the column with an axial compression ratio of 0.3. Loading was carried out by load–displacement control, where before yielding, the specimens were loaded with $0.25P_c$, $0.5P_c$, and $0.7P_c$ (P_c was the theoretic ultimate bearing capacity obtained from numerical simulation), and then displacement with the multiple increments of yield displacement (Δ_j) was applied after yielding and repeated three times. The test was stopped when the bearing capacity of the specimens decreased to 85% of the peak load or when the specimens were obviously damaged.

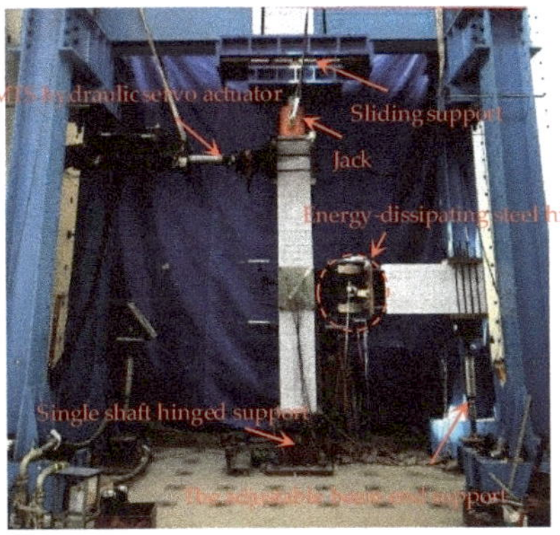

Figure 6. Test set-up.

4. Experiment Phenomenon and Failure Mode

4.1. Specimen J-R-1

When the specimen J-R-1 was loaded to $0.7P_c$, cracks emerged at the top and bottom surface of the precast beam and the cracks developed into shear cracks when loaded up to 20 mm (displacement loading $2\Delta_j$). When the loading reached 50 mm (displacement loading $5\Delta_j$), the lower flange energy-dissipating steel plate slightly inclined due to the rotational behavior of steel hinge and the energy-dissipating steel plate consumed plastic energy; some noises was heard at this stage. The bearing capacity of the specimen increased with the test going on and no obvious damage was observed. Besides, the shear cracks on the side of the beam no longer developed. When loading to 60 mm (displacement loading $6\Delta_j$), the lower flange energy-dissipating steel plate made a big noise at the second loading cycle. At the same time, the bearing capacity dropped significantly and decreased to less than 85% of the peak load, and then the test was terminated. During the test, no cracks appeared in the upper and lower precast columns. Figure 7 shows the failure mode of the precast replaceable beam-to-column joint with energy-dissipating steel hinge.

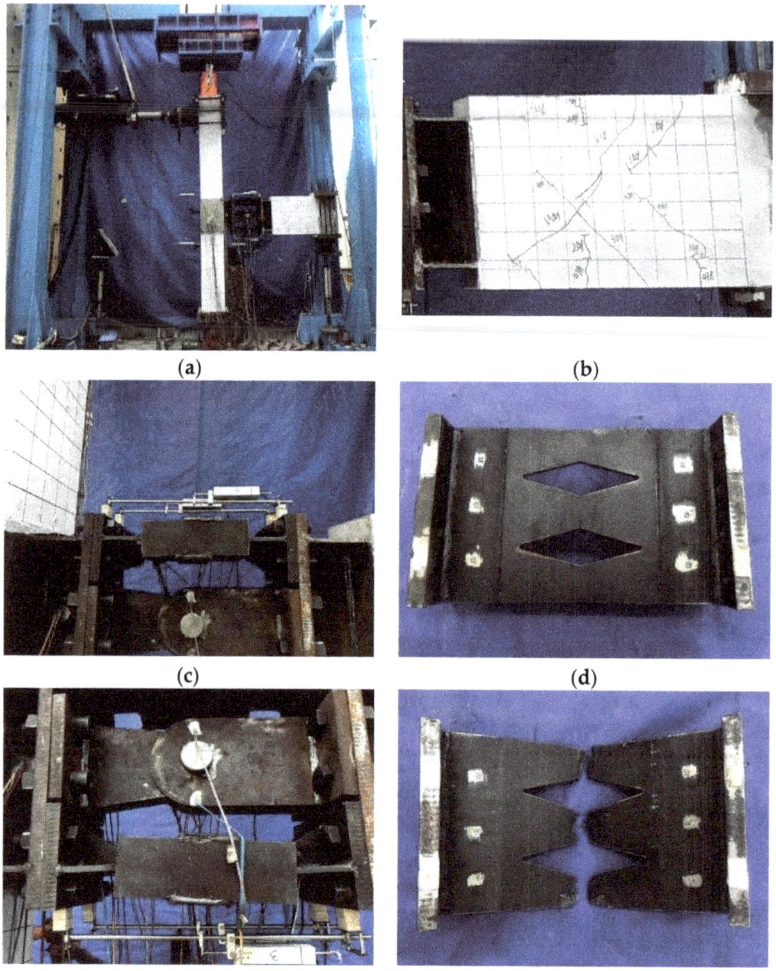

Figure 7. (**a**) The failure mode of the precast replaceable beam-to-column joint; (**b**) crack development in concrete; (**c**) failure mode of the upper connector; (**d**) failure mode of the upper steel plate; (**e**) failure mode of the lower connector; (**f**) failure mode of the lower steel plate.

The failure mode of the reduced energy-dissipating steel plate with diamond-shaped openings in the specimen J-R-1 is shown in Figure 7d,f, respectively. From Figure 7d, it was found that cracking and necking appeared at the smallest connecting area between the openings of the weakened section of the upper energy-dissipating steel plate. Figure 7f showed that the energy-dissipation steel plate ruptured at the reduced section area rather than suffering from out-of-plane buckling failure. The plastic damage of the specimen J-R-1 was concentrated on the reduced energy-dissipation steel plate with diamond-shaped openings in the steel hinge, indicating that the objective of damage concentration was achieved.

4.2. Joint Repair Process

In this experiment, there was no crack development in the precast column and only slight cracks were observed in the precast beam. The overall of J-R-1 was concentrated in the energy-dissipating steel plate in the steel hinge. The damaged flange energy-dissipating steel plates were removed and new steel plates were reinstalled, while the other precast framing components were not replaced, to generate repaired specimen J-R-2. The repair process is illustrated in Figure 8, and the whole process only took about one hour with high efficiency and convenient operation.

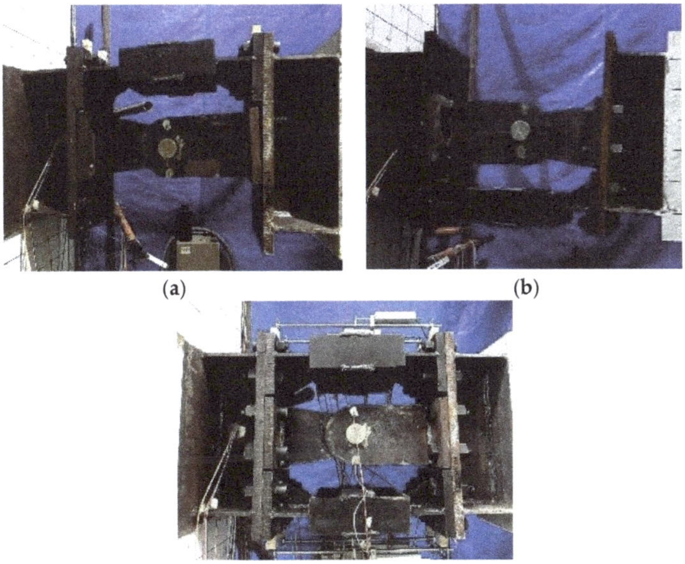

(a) (b) (c)

Figure 8. (**a**) Removing the lower restraint connector; (**b**) removing the upper connector; (**c**) completed joint repair.

4.3. Specimen J-R-2

The low-cycle reciprocating loading test was carried out on specimen J-R-2 with same loading protocol as J-R-1. The experiment phenomenon of specimen J-R-2 was similar to that of specimen J-R-1.

As shown in Figure 9b, it was found there was almost no further development of the previous cracks on the precast beam during the second cyclic test. When the specimen J-R-2 was loaded to 50 mm (the loading displacement of $5\Delta_j$), the upper flange energy-dissipating steel plate inclined slightly and a small noise was heard. The energy-dissipating steel plate started to dissipate plastic energy and the bearing capacity of the specimen kept increasing. When loading to the first cycle of 60 mm (displacement loading $6\Delta_j$), it was found that the hysteresis curve showed a tendency of decreasing. In the second cycle of loading, the upper flange energy-dissipating steel plate made a noticeable noise during deformation, and the hysteretic curve obviously decreased. Finally, at the third cycle, there was a huge sound from the upper flange energy-dissipating steel plate and the bearing capacity of the joint dropped significantly to less than 85% of the peak load. The test was terminated and there were also no cracks in the upper and lower precast columns.

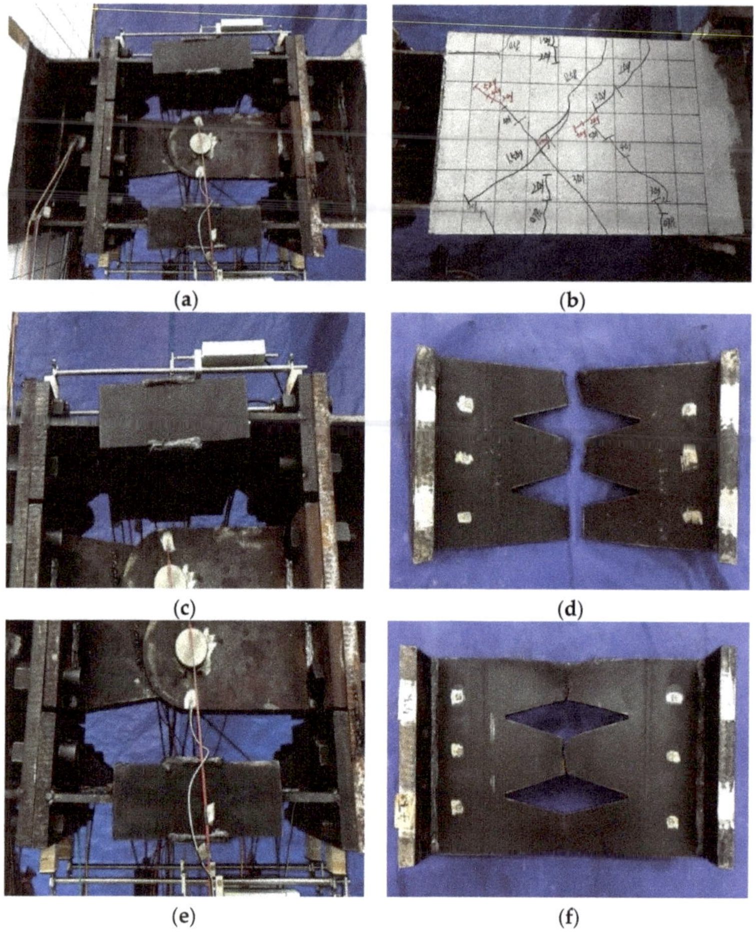

Figure 9. (**a**) Failure mode of the energy-dissipating hinge; (**b**) crack development in concrete; (**c**) failure mode of the upper connector; (**d**) failure mode of the upper steel plate; (**e**) failure mode of lower connector; (**f**) failure mode of the lower steel plate.

The failure mode of the reduced energy-dissipating steel plate with diamond-shaped openings of specimen J-R-2 is shown in Figure 9c,e. It showed that there were also serious cracks at the weakened section of the flange-reduced energy-dissipating steel plate and the necking phenomenon was obvious. No out-of-plane buckling failure occurred in the specimen. In conclusion, the plastic damage accumulated in the reduced energy-dissipating steel plate of the specimen J-R-2, and was similar with specimen J-R-1.

4.4. Specimen J-2

Compared with specimen J-R-1 and J-R-2, there were obvious shear cracks that developed in the side of the beam of specimen J-2 when loaded to 20 mm (displacement loading $2\Delta_j$). With loading increasing, the shear crack ran across the whole side of the beam at 40 mm (displacement loading $4\Delta_j$). The bearing capacity of specimen J-2 decreased when the displacement was approximately 60 mm (displacement loading $6\Delta_j$). The width of the shear crack on the beam side reached 1 cm when loading to 80 mm (displacement loading $8\Delta_j$). The bearing capacity gradually dropped and was reduced to less than 85% of the peak load. At this time, the test was terminated, and Figure 10 demonstrates the failure mode of specimen J-2.

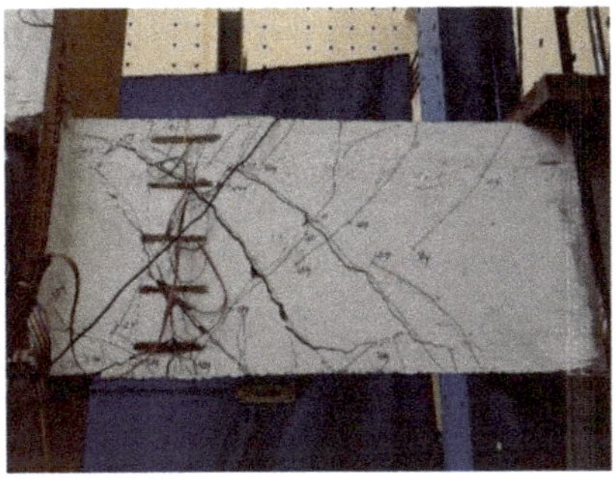

Figure 10. Test phenomenon of specimen J-2.

5. Experimental Results and Discussion

5.1. Hysteretic Behaviors and Strengths

The load–displacement hysteretic curves of each specimen are presented in Figure 11. It can be seen that the hysteretic loops of specimens J-R-1 and J-R-2 were more plump in spindle-shaped, as demonstrated in Figure 11a. Besides, the curves for specimens J-R-1 and J-R-2 were approximately coincident, indicating the mechanical behavior was effectively restored after repair. After the yield of specimens J-R-1 and J-R-2, the plastic damage accumulated in the reduced energy-dissipating steel plate with diamond-shaped openings in the energy-consuming steel hinge, and it exhibited a relatively stable hysteresis behavior. Besides, the bearing capacity still increased with the growth of displacement, indicating that the proposed joint could develop good strength. Moreover, the constraint sleeve effectively protected the reduced energy-dissipating steel plate with diamond-shaped openings from out-of-plane buckling damage, leading to stable bearing capacity without abrupt reduction. Only trivial slight pinching in the hysteretic curve was found under larger deformation, which was caused by the slip of bolt connection.

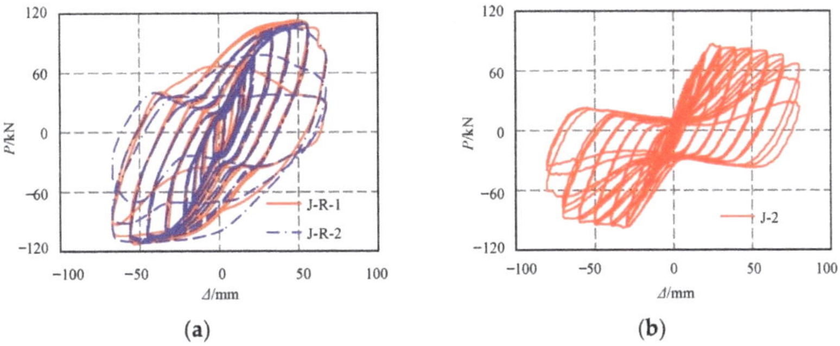

Figure 11. (**a**) Hysteretic curves for specimens J-R-1 and J-R-2; (**b**) hysteretic curve for specimen J-2.

The hysteretic curve of the assembled steel sleeve confined concrete joint specimen J-2 had serious pinching effecting under cyclic loading, as presented in Figure 10b. The bearing capacity of specimen J-2 was smaller than specimens J-R-1 and J-R-2, also resulting in a relatively poor energy-dissipation capacity.

5.2. Skeleton Curves

The skeleton curves of the tested specimens are presented in Figure 12 and Table 2 lists the characteristic value of the hysteresis performance. The parameters in Table 2 are yield displacement Δ_y, the yield load P_y, the peak displacement Δ_m, the peak load P_m, the ultimate displacement Δ_u, the ultimate load P_u, and the ductility u, respectively.

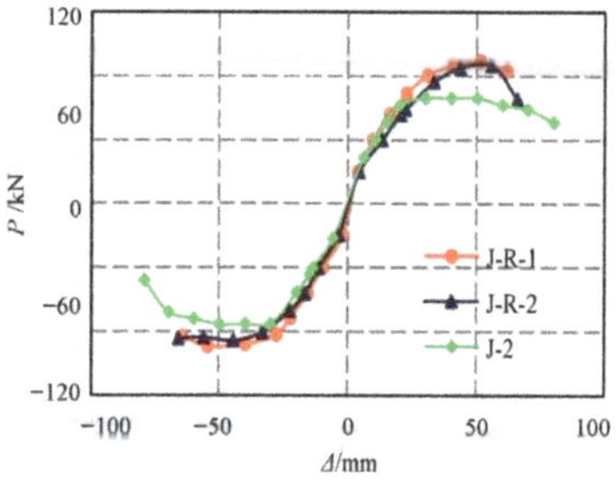

Figure 12. Skeleton curves of different specimens.

Table 2. Characteristic values of the seismic performance of the specimens.

Specimen	Δ_y /(mm)	P_y /(kN)	Δ_m /(mm)	P_m /(kN)	Δ_u /(mm)	P_u /(kN)	u	The Average Ductility
J-R-1	13.73 −13.25	61.83 −62.98	51.28 −54.56	111.6 −111.5	61.84 −64.64	103.9 −103.8	4.50 4.88	4.69
J-R-2	19.17 −19.17	65.77 −77.13	55.6 −44.4	108.4 −107.1	66.16 −66	82 −105.6	3.45 3.44	3.45
J-2	14.26 −14.57	62.64 −56.78	50 −50.06	83.08 −94.20	80.01 −79.64	63.98 −60.03	5.61 5.47	5.54

Note: "−" represents the negative loading direction.

As shown in Figure 12, the stiffness of specimen J-R-2 decreased a bit in the early stage, since specimen J-R-2 only replaced the damaged flange energy-dissipating steel plate while the precast beams with slight cracking and columns remained unchanged. After specimens J-R-1 and J-R-2 yielded, the plastic energy dissipation was mainly taken up by the energy-dissipating steel hinges. The failures of specimens J-R-1 and J-R-2 were all caused by the rupture of the steel plate, leasing to similar development characteristics in the degradation stage in the skeleton curve. The initial stiffness of specimen J-2 was similar to that of specimen J-R-1, indicating that the proposed joint had adequate stiffness. The development of bearing capacity of specimen J-2 was relatively gentle and also degraded more slowly with the damage accumulation caused by the development of cracks. As shown in Table 2, the average peak loads of specimens J-R-1 and J-R-2 were 111.6 kN and 107.8 kN, respectively, indicating the bearing strength was recovered to a great extent. The bearing capacity of specimen J-1 was 88.6 kN, only about 74% of specimen J-R-1, showing that the energy-dissipating steel hinge could significantly improve the bearing capacity. Only the damaged energy-consuming steel plate restraint joints were replaced.

It can also be seen from Table 2 that the average ductility coefficients of specimens J-R-1 and J-R-2 were 4.69 and 3.45, respectively, exhibiting a good deformation capacity. The reduction in ductility of specimen J-R-2 was caused by the slight cracking in precast beam. The damage was accumulated, the initial stiffness decreased, and the yield displacement increased, resulting in a low ductility coefficient. However, the average limit displacements of specimens J-R-1 and J-R-2 were 63.24 mm and 66.08 mm, respectively, indicating that only by replacing the damaged upper and lower energy-dissipating steel plates could the ultimate deformation capacity be recovered.

The average limit displacement of specimen J-2 was 79.82 mm, and the average ductility coefficient of specimen J-2 was 5.54. The average limit displacement and average ductility coefficient of specimen J-2 were larger than those of specimens J-R-1 and J-R-2, since the failure of specimen J-2 was caused by the development of cracks in the precast beam, which was gentler and without serious concrete crushing. However, the failures of specimens J-R-1 and J-R-2 was caused by the necking of weakened steel plates. Once the rupture of the steel plates occurred, the bending moment could not effectively be transferred, leading to abrupt degradation in strength and smaller deformation capacity at test termination.

5.3. Degeneration of Strength

The strength degradation curves of each specimen are presented in Figure 13, while λ_2 and λ_3 are the strength degradation coefficients of the second and third cycles at the same loading level. Before the peak load, the strength degradation coefficient of each specimen was kept at about 1.0, indicating a relatively stable strength development. The strength degradation of specimens J-R-1 and J-R-2 occurred at the loading displacement of $6\Delta_j$, while almost no strength degradation was found before $6\Delta_j$. The reduced energy-dissipating steel plate with diamond-shaped openings was damaged at the reduced section, and the strength degradation curve decreased significantly. However, the strength of specimen J-2 degraded at an earlier stage.

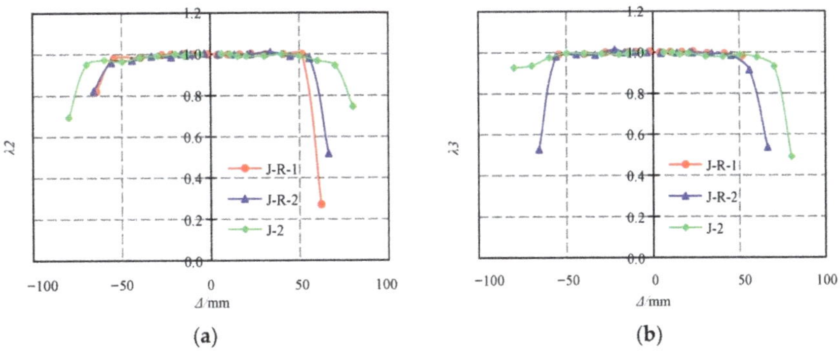

Figure 13. (**a**) Strength reduction curve of λ_2; (**b**) strength reduction curve of λ_3.

5.4. Stiffness Degradation

Figure 14 presents the stiffness degradation curve of the specimen; τ is the normalized stiffness degradation coefficient, defined as $\tau = K_i/K_0$, where K_i is the secant stiffness under various loading levels and K_0 is the initial stiffness. Figure 14 shows that the stiffness degradation rate of specimen J-R-2 was slower than that of specimen J-R-1, because the damage in the precast beam due to crack development was stable after the previous test and the stiffness degradation of specimen J-R-2 was mainly caused by the damage in flange energy-dissipating steel plate. The stiffness degradation rate of specimen J-2 was slower than that of specimens J-R-1 and J-R-2, mainly because specimen J-2 had shear cracks in the beam, and finally the shear failure of the precast beam occurred. Furthermore, the shear stiffness of the precast beam was relatively high, and the stiffness degraded slowly.

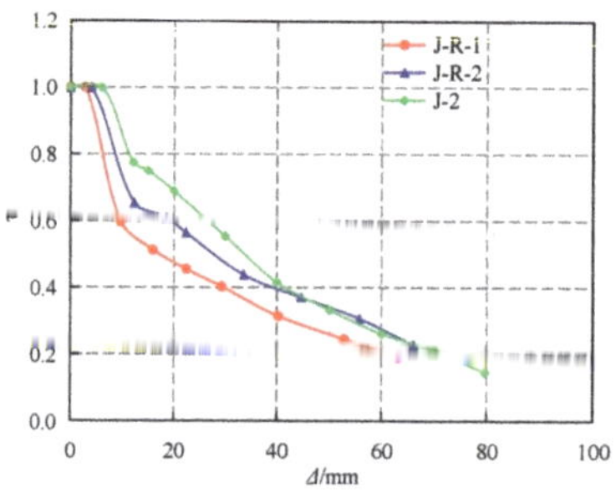

Figure 14. Stiffness reduction curves of specimens.

5.5. Energy-Dissipation Capability

Figure 15 shows the cumulative hysteretic energy dissipation of the tested specimens, and Figure 16 presents the equivalent viscous damping coefficient ζ. From Figure 15, it was analyzed that the total energy dissipation of specimens J-R-1, J-R-2, and J-2 were 46,334 kN·mm, 44,519 kN·mm, and 29,856 kN·mm, respectively. The energy consumption of specimens J-R-1 and J-R-2 were about 1.55 times and 1.49 times of that of specimen J-2, respectively, indicating the adoption of steel hinge could greatly improve the energy-dissipation behavior. The cumulative hysteretic energy-dissipation curves of specimens J-R-1 and J-R-2 were similar but the accumulated hysteretic energy dissipation of specimen J-R-1 was slightly worse than that of specimen J-R-1, since specimen J-R-2 worked with cracks in the precast beam. It was found that the equivalent viscous damping coefficient curves of specimens J-R-1 and J-R-2 were basically the same, as presented in Figure 16. The maximum equivalent viscous damping coefficient of specimens J-R-1, J-R-2, and J-2 were 0.41, 0.44, and 0.23, respectively. In conclusion, it showed that the energy-dissipating steel hinge had good energy dissipation capacity, which could improve the energy-dissipation behavior of the precast replaceable beam-to-column joint with energy-dissipating steel hinge. Moreover, the replacement of damaged steel plates in the steel hinge could effectively restore the energy-dissipating ability.

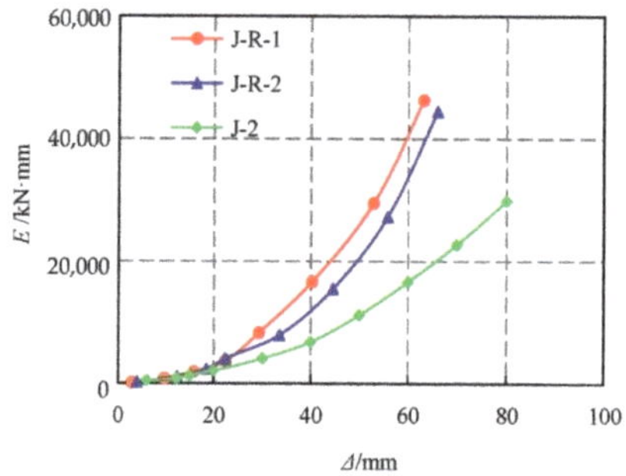

Figure 15. Cumulative hysteretic energy consumption of the specimens.

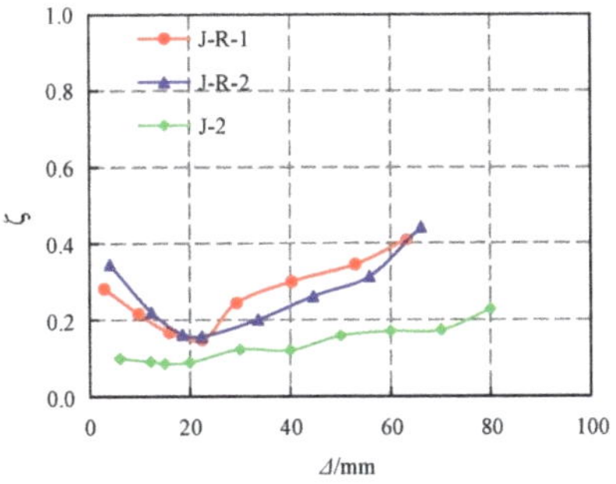

Figure 16. Equivalent viscous damping coefficient.

6. Conclusions

A novel precast replaceable beam-to-column joint with energy-dissipating steel hinges was proposed and the hysteretic performance was investigated through experimental research. Based on the results presented in this paper, some conclusions can be drawn as follows:

(1) Hysteresis performances of the precast beam-to-column joint (e.g., carrying capacity, energy consumption, strength degradation) could be improved by the utilization of an energy-dissipating steel hinge. The failure of the precast replaceable beam-to-column joint with energy-dissipating steel hinges was caused by the necking rupture of weakened flange energy-dissipation steel plates, which exhibited stable mechanical behavior. However, for the steel sleeve confined concrete joint, the failure was caused by the shear cracking in the precast beam.

(2) The precast replaceable beam-to-column joint with energy-dissipating steel hinges exhibited good mechanical behavior under cyclic loading, with plump, spindle shaped hysteresis curves, and an obvious pinching effect was found in the hysteresis curve of the steel-sleeved confined concrete joint. The bearing capacity of specimens J-R-1 and J-R-2 was 26% and 22% higher than that of specimen J-2, while the energy-dissipation capacities were 55% and 49% higher, respectively.

(3) For the precast replaceable beam-to-column joint with the energy-dissipating steel hinge, the damage was concentrated in the weakened steel plates in the energy-dissipating steel joint while cracks in precast beams and columns were not obvious, indicating that the main structure was basically free of damage. The damage of the proposed joint could be repaired by replacing the damaged members, since it was found that the bearing capacity of J-R-2 was recovered at up to 96.6% of J-R-1 while the energy-dissipation capacity was recovered at 96.1%. Besides, the failure process and hysteretic performance of the repaired specimen J-R-2 were similar to those of specimen J-R-1. The precast replaceable beam-to-column joint with energy-dissipating steel hinges had good post-earthquake resilience.

Author Contributions: Conceptualization, Y.M. and A.Q.; methodology, G.Y.; validation, L.Z.; data curation, Y.M. and P.X.; writing—original draft, Y.M.; writing—review and editing, P.X.; visualization, Y.M.; supervision, A.Q. and L.Z.; project administration, G.Y.; funding acquisition, G.Y. All authors have read and agreed to the published version of the manuscript.

Funding: This research was funded by the National Natural Science Foundation of China, grant No. 51878174 and Research and Development Foundation of Fujian University of Technology, GY-Z18184.

Institutional Review Board Statement: Not applicable.

Informed Consent Statement: Not applicable.

Data Availability Statement: Not applicable.

Conflicts of Interest: The authors declare no conflict of interest.

References

1. Khare, R.K.; Maniyar, S.R.; Uma, S.R.; Bidwai, V.B. Seismic performance and design of precast concrete building structures: An overview. *J. Struct. Eng.* **2011**, *38*, 272–284.
2. Breccolotti, M.; Gentile, S.; Tommasini, M.; Materazzi, A.L.; Bonfigli, M.F.; Pasqualini, B.; Bonfigli, M.F.; Pasqualini, B.; Colone, V.; Gianesini, M. Beam-column joints in continuous RC frames: Comparison between cast-in situ and precast solutions. *Eng. Struct.* **2016**, *127*, 129–144. [CrossRef]
3. Esmaeeli, E.; Barros, J.A.O.; Sena-Cruz, J.; Varum, H.; Melo, J. Assessment of the efficiency of prefabricated hybrid composite plates (HCPs) for retroftting of damaged interior RC beam–column joints. *Compos. Struct.* **2015**, *119*, 24–37. [CrossRef]
4. FIB (Fédération Internationale du Béton). *Seismic Design of Precast Concrete Buildings*; Fédération Internationale du Béton: Lausanne, Switzerland, 2003; 262p.
5. Kurama, Y.C.; Sritharan, S.; Fleischman, R.B.; Restrepo, J.I.; Henry, R.S.; Cleland, N.M.; Ghosh, S.K.; Bonelli, P. Seismic-resistant precast concrete structures: State of the art. *J. Struct. Eng.* **2018**, *144*, 03118001. [CrossRef]
6. Choi, H.K.; Choi, Y.C.; Choi, C.S. Development and testing of precast concrete beam-to-column connections. *Eng. Struct.* **2013**, *56*, 1820–1835. [CrossRef]
7. Wu, G.; Feng, D.C. Research progress on fundamental performance of precast concrete frame beam-to-column connections. *J. Build. Struct.* **2018**, *39*, 1–16.
8. Figueira, D.; Ashour, A.; Yıldırım, G.; Aldemir, A.; Şahmaran, M. Demountable connections of reinforced concrete structures: Review and future developments. *Structures* **2021**, *34*, 3028–3039. [CrossRef]
9. Yan, Q.; Chen, T.; Xie, Z. Seismic experimental study on a precast concrete beam-column connection with grout sleeves. *Eng. Struct.* **2018**, *155*, 330–344. [CrossRef]
10. Im, H.J.; Park, H.G.; Eom, T.S. Cyclic Loading Test for Reinforced-Concrete-Emulated Beam-Column Connection of Precast Concrete Moment Frame. *ACI Struct. J.* **2013**, *110*, 115–126. [CrossRef]
11. Latour, M.; Rizzano, G. Seismic behavior of cross-laminated timber panel buildings equipped with traditional and innovative connectors. *Arch. Civ. Mech. Eng.* **2017**, *17*, 382–399. [CrossRef]
12. Lu, X.; Fan, L.; Zhao, B. Pseudo dynamic test on a reduced scale jointed precast concrete frame structure. *J. Build. Struct.* **2008**, *29*, 58–65.
13. Tartaglia, R.; D'Aniello, M.; Landolfo, R. The influence of rib stiffeners on the response of extended end-plate joints. *J. Constr. Steel Res.* **2018**, *148*, 669–690. [CrossRef]
14. Yang, J.; Guo, T.; Chai, S. Experimental and numerical investigation on seismic behaviours of beam-column joints of precast prestressed concrete frame under given corrosion levels. *Structures* **2020**, *27*, 1209–1221. [CrossRef]
15. Pan, P.; Wang, H.S.; Guo, H.S.; Liu, K.; Wang, D.; Qi, H.; Geng, J. Experimental study of seismic performance of unbounded post-tensioned pre-stressed beam-to-column dry connections. *J. Build. Struct.* **2018**, *39*, 46–55.
16. Koshikawa, T. Moment and energy dissipation capacities of post-tensioned precast concrete connections employing a friction device. *Eng. Struct.* **2017**, *138*, 170–180. [CrossRef]
17. Pan, Y.; Chen, X.; Wang, H. Seismic fragility analysis of unbonded post-tensioned fabricated RC frame structures. *J. Harbin Inst. Technol.* **2018**, *50*, 71–77.
18. Bahrami, S.; Madhkhan, M.; Shirmohammadi, F.; Nazemi, N. Behavior of two new moment resisting precast beam to column connections subjected to lateral loading. *Eng. Struct.* **2017**, *132*, 808–821. [CrossRef]
19. Girgin, S.C.; Misir, I.S.; Kahraman, S. Experimental Cyclic Behavior of Precast Hybrid Beam-Column Connections with Welded Components. *Int. J. Concr. Struct. Mater.* **2017**, *11*, 229–245. [CrossRef]
20. Nzabonimpa, J.D.; Hong, W.K.; Kim, J. Experimental and non-linear numerical investigation of the novel detachable mechanical joints with laminated plates for composite precast beam-column joint. *Compos. Struct.* **2018**, *185*, 286–303. [CrossRef]
21. Tartaglia, R.; D'Aniello, M.; Campiche, A.; Latour, M. Symmetric friction dampers in beam-to-column joints for low-damage steel MRFs. *J. Constr. Steel Res.* **2021**, *184*, 106791. [CrossRef]
22. Ferrante Cavallaro, G.; Francavilla, A.B.; Latour, M.; Piluso, V.; Rizzano, G. Cyclic response of low yielding connections using different friction materials. *Soil Dyn. Earthq. Eng.* **2018**, *114*, 404–423. [CrossRef]
23. Elettore, E.; Freddi, F.; Latour, M.; Rizzano, G. Design and analysis of a seismic resilient steel moment resisting frame equipped with damage-free self-centering column bases. *J. Constr. Steel Res.* **2021**, *179*, 106543. [CrossRef]
24. Ryotaro, K.; Hiroyasu, S.; Zhe, Q.; Takashi, S. Precast prestressed concrete frames for seismically retrofitting existing RC frames. *Eng. Struct.* **2019**, *184*, 345–354.
25. Wang, H.; Marino, E.M.; Pan, P.; Liu, H.; Nie, X. Experimental study of a novel precast prestressed reinforced concrete beam-to-column joint. *Eng. Struct.* **2018**, *156*, 68–81. [CrossRef]
26. Li, Z.; Peng, Z.; Qi, Y. Full-scale experimental study on seismic behaviors of plasticity controllable steel joint of prefabricated RC beam column. *J. Struct. Eng.* **2019**, *40*, 43–50.
27. Zheng, L.; Yan, G.; Wei, C. Experimental and numerical investigation of steel energy -dissipating hinge under cyclic loading. *China Civ. Eng. J.* **2020**, *53*, 29–43.

28. Qi, Y.; Teng, J.; Shan, Q.; Ding, J.; Li, Z.; Huang, C.; Xing, H.; Yi, W. Seismic performance of a novel prefabricated beam-to-column steel joint considering buckling behaviour of dampers. *Eng. Struct.* **2021**, *229*, 111591. [CrossRef]
29. Ertas, O.; Ozden, S.; Ozturan, T. Ductile connections in precast concrete moment resisting frames. *PCI J.* **2006**, *51*, 66–76. [CrossRef]
30. Li, Z.; Qi, Y.; Teng, J. Experimental investigation of prefabricated beam-to-column steel joints for precast concrete structures under cyclic loading. *Eng. Struct.* **2020**, *209*, 110217. [CrossRef]
31. Ye, M.; Jiang, J.; Chen, H.M.; Zhou, H.Y.; Song, D.D. Seismic behavior of an innovative hybrid beam-column connection for precast concrete structures. *Eng. Struct.* **2021**, *227*, 111436. [CrossRef]
32. Li, D.; Wu, C.; Zhou, Y.; Luo, W.; Lie, W. A precast beam-column connection using metallic damper as connector: Experiment and application. *J. Constr. Steel Res.* **2021**, *181*, 106628. [CrossRef]
33. Li, Y.; Geng, F.; Ding, Y.; Wang, L. Experimental and numerical study of low-damage self-centering precast concrete frame connections with replaceable dampers. *Eng. Struct.* **2020**, *220*, 111011. [CrossRef]
34. Huang, L.; Clayton, P.M.; Zhou, Z. Seismic design and performance of self-centering precast concrete frames with variable friction dampers. *Eng. Struct.* **2021**, *245*, 112863. [CrossRef]

buildings

MDPI

Article

Effects of Coal Gangue Coarse Aggregate on Seismic Behavior of Columns under Cyclic Loading

Hanqing Liu [1,2,*], Guoliang Bai [1,2,*], Fang Yan [1], Yu Gu [1] and Kefan Zhu [1]

[1] School of Civil Engineering, Xi'an University of Architecture & Technology, Xi'an 710055, China
[2] Key Lab of Structural Engineering and Earthquake Resistance, Ministry of Education (XAUAT), Xi'an 710055, China
* Correspondence: hqliu@xauat.edu.cn (H.L.); guoliangbai@126.com (G.B.)

Abstract: Coal gangue is the rock discharged from coal mining. Using coal gangue as coarse aggregate is one of the solutions for the sustainable development of construction engineering. Five one-half scaled coal gangue concrete (CGC) frame columns with different coal gangue coarse aggregate replacement ratios were designed, and the effect of coal gangue coarse aggregate on seismic behavior of columns under cyclic loading was studied. The test results show that the failure of coal gangue coarse aggregate under cyclic loading is the main reason for the reduction in hysteretic performance of CGC specimens. Compared with natural aggregate concrete (NAC) specimen, the hysteretic behavior, deformation performance, and energy consumption of CGC columns were reduced. However, the seismic performance of CGC specimens with higher replacement ratio was better than that of CGC specimens with a lower replacement ratio. Compared with NAC specimen, the ductility and total energy dissipation of CGC specimen with r = 100% were only reduced by 8.2% and 12.8%. In addition, the test results also found that the higher the replacement ratio, the greater the shear deformation of the specimen. It is recommended to appropriately increase the stirrup ratio of CGC specimens in seismic design.

Keywords: coal gangue coarse aggregate; coal gangue concrete; column; cyclic loading; seismic behavior

Citation: Liu, H.; Bai, G.; Yan, F.; Gu, Y.; Zhu, K. Effects of Coal Gangue Coarse Aggregate on Seismic Behavior of Columns under Cyclic Loading. *Buildings* **2022**, *12*, 1170. https://doi.org/10.3390/buildings12081170

Academic Editor: Antonio Formisano

Received: 23 June 2022
Accepted: 3 August 2022
Published: 5 August 2022

Copyright: © 2022 by the authors. Licensee MDPI, Basel, Switzerland. This article is an open access article distributed under the terms and conditions of the Creative Commons Attribution (CC BY) license (https://creativecommons.org/licenses/by/4.0/).

1. Introduction

Coal gangue is the solid waste discharged in the process of coal mining, and its emission accounts for about 10–20% of the coal production in that year [1]. As of 2012, the cumulative stock of coal gangue in China are more than 5 billion tons [2], which is the largest solid waste discharged [3,4]. At present, coal gangue is mainly treated by stacking, which not only occupies a lot of land, but also pollutes the surrounding soil and water in the long-term stacking process [5–7]. Therefore, how to effectively use coal gangue is an urgent matter to be solved [8].

With the development of construction engineering in recent years, the demand for natural aggregates has increased day by day [9], which is in sharp contrast to the massive stockpiling of coal gangue. The mining of natural aggregates will not only consume a lot of energy, but at the same time, high-quality natural aggregates in some areas have been exhausted [10], and the sustainable development of the construction engineering is facing severe challenges.

Coal gangue is essentially a kind of rock, which is similar to natural aggregate in terms of physical and chemical composition [11,12]. Using coal gangue as coarse aggregate partly or completely replacing natural aggregate to produce coal gangue concrete (CGC) will effectively reduce the mining amount of natural gravel [13–15]. Additionally, the application of CGC can also relieve the supply pressure of natural aggregate [16–18], which is one of the effective solutions to realize the sustainable development of the construction engineering.

To promote the utilization of coal gangue in construction engineering, many scholars have conducted research on CGC. Zhang et al. [19] and Li et al. [20] studied spontaneous combustion coal gangue concrete and found that coal gangue has adverse effects on the working and mechanical properties of CGC. Wang et al. [21] and Liu et al. [22] analyzed the elastic modulus of CGC through experiments and found that the higher the replacement ratio of CGC, the lower the elastic modulus. Based on the experimental data, the calculation models of elastic modulus of CGC were proposed. Zhou et al. [23] analyzed the influence of the types of gangues on the mechanical behavior of CGC and established the stress–strain relationship model of CGC. Guan et al. [24] and Qiu et al. [25] analyzed the mechanical properties of CGC in a freeze–thaw environment and found that the antifreeze performance of CGC is worse than that of natural aggregate concrete (NAC). Bai et al. [26] and Wang et al. [27] conducted experimental studies on the shear and flexural properties of CGC beams and found that coal gangue has little effect on the flexural and shear capacity of CGC beams, but the deformation of CGC beams was significantly greater than that of NAC beams.

Obviously, the existing literature mostly stays on the basic properties of CGC materials, while the research on the mechanical properties of CGC components, especially the seismic performance of CGC components, is relatively less. For CGC structures, seismic performance is also an inevitable problem. Therefore, five specimens are designed in this paper, and the seismic performance of CGC columns is analyzed. The effects of coal gangue on the failure process, hysteretic behavior, deformation performance, and energy consumption of CGC columns are studied. In addition, the deformation properties of CGC columns are also discussed. The research in this paper can provide a reference for the seismic design of CGC structures.

2. Materials and Methods

2.1. Materials

Coarse aggregate was divided into coal gangue coarse aggregate (CGCA, Figure 1) and natural coarse aggregate (NCA), with gradation of 5–31.5 mm. The mechanical properties of aggregates are shown in Table 1. The fine aggregate is medium sand with a bulk density of kg·m^{-3} and an apparent density of 2780 kg·m^{-3}. The cement is P·O 42.5.

Figure 1. Coal gangue coarse aggregate (CGCA).

Table 1. Properties of aggregates.

Type	CGCA [1]	NCA [2]
Crush index (%)	21.8	10.7
Bulk density (kg·m^{-3})	1160	1590
Apparent density (kg·m^{-3})	2321	2656
Water absorption (%)	5.4	0.8
Water content (%)	0.6	0.5

[1] CGCA: coal gangue coarse aggregate; [2] NCA: natural coarse aggregate.

To compensate for the high absorption of mixing water by CGCA, additional water was added based on the difference between absorption rate and water content of CGCA [28]. Finally, the calculated mix proportions of each group of concrete are shown in Table 2. By adjusting the amount of water reducer, the slump of each group of concrete was controlled at about 80 mm. Three cubes with a side length of 150 mm were reserved in each group during concrete pouring to test the cube compressive strength (f_{cu}).

Table 2. Mix proportion of concretes.

ID	r [1] (%)	Unit Weight (kg·m^{-3})							f_{cu} [3] (MPa)
		Cement	FA [2]	NCA	CGCA	Mixing Water	Additional Water	Water Reducer	
KZA-1	0	318.2	873.8	1068.0	0.0	140.0	0.0	2.54	50.5
KZA-2	30	353.7	855.6	732.0	313.7	145.0	14.2	3.18	42.8
KZA-3	50	394.7	834.9	510.2	510.2	150.0	23.0	3.16	36.7
KZA-4	70	442.9	811.0	297.4	693.8	155.0	31.3	4.43	35.9
KZA-5	100	500.0	783.0	0.0	957.0	160.0	43.1	5.00	32.4

[1] r: replacement ratio of coal gangue; [2] FA: fine aggregate; [3] f_{cu}: cube compressive strength.

The longitudinal reinforcement in the column uses a grade 400 hot-rolled ribbed steel bar (HRB 400), and the stirrup uses a grade 300 hot rolled plain steel bar (HPB 300). The mechanical properties of the two kinds of steel bars are presented in Table 3.

Table 3. Material properties of steel bar.

Type	Diameter (mm)	f_y [1] (MPa)	f_u [2] (MPa)	E_s [3] (MPa)	ε_y [4] (10^{-6})
HRB 400	20	439	635	2.00×10^5	2195
HPB 300	8	367	431	2.10×10^5	1748

[1] f_y: yield strength; [2] f_u: ultimate strength; [3] E_s: elasticity modulus; [4] ε_y: yield strain.

2.2. Specimen Details

To analyze the effect of CGCA on the seismic behavior of columns, five one-half scaled columns were designed based on the different replacement ratios (r = 0%, 30%, 50%, 70%, and 100%). The detailed parameters of each specimen are shown in Table 4.

Table 4. Specimen parameters.

Specimen	r [1] (%)	n [2]	λ [3]	ρ_c [4] (%)	ρ_{sv} [5] (%)	H_0 [6] (mm)
KZA-1	0					
KZA-2	30					
KZA-3	50	0.30	3.17	1.40	0.34	950
KZA-4	70					
KZA-5	100					

[1] r: replacement ratio of coal gangue; [2] n: ratio of axial compression stress to strength; [3] λ: shear span ratio; [4] ρ_c: column reinforcement ratio; [5] ρ_{sv}: volume reinforcement ratio of the stirrups; [6] H_0: effective height from the loading point to the bottom of the column.

The geometric dimensions and reinforcement of each specimen are the same, as shown in Figure 2. In addition, to avoid local damage to the top of the column during the loading process, the stirrups within 300 mm of the top of the column were densified.

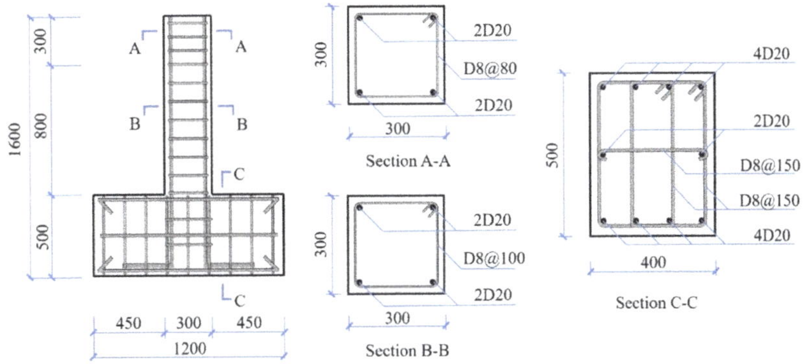

Figure 2. Specimen details (unit: mm). Note: section A-A, section B-B and section C-C in the figure are the reinforcement details of the corresponding positions of the specimen respectively.

2.3. Test Setup and Measurements

The specimen is fixed on the ground beam by bolts; the loading device is shown in Figure 3. The axial load is applied by the 200 t vertical jack, and its value remains unchanged during the whole loading process, while the horizontal cyclic load is applied by the 100 t MTS electro-hydraulic servo actuator and the load is recorded automatically. The vertical jack is connected with the reaction girder through the directional support. Due to the small friction coefficient of the roller, it can ensure that the jack moves horizontally with the specimen without relative displacement [29]. When loading, the axial load is first applied, and then the horizontal cyclic load is applied [30].

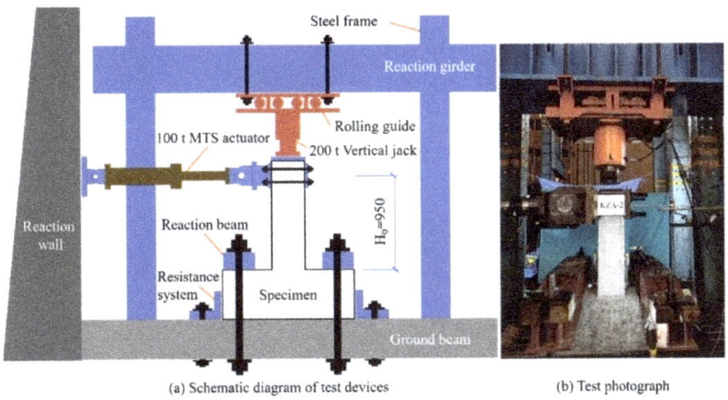

Figure 3. Test apparatus.

The load-deformation double control method is used for horizontal cyclic load [31], and the loading steps are shown in Figure 4. Before the specimen has yielded, the force-controlled mode is adopted, the increment is 10 kN, and each load is pushed and pulled once. After the specimens yields, the displacement-controlled method is used for loading, and the yield displacement (Δ_y) is used as the increment, and each displacement is repeated for three cycles [32]. When the load drops to 85% of the peak load, it indicates that the specimen has been damaged [33].

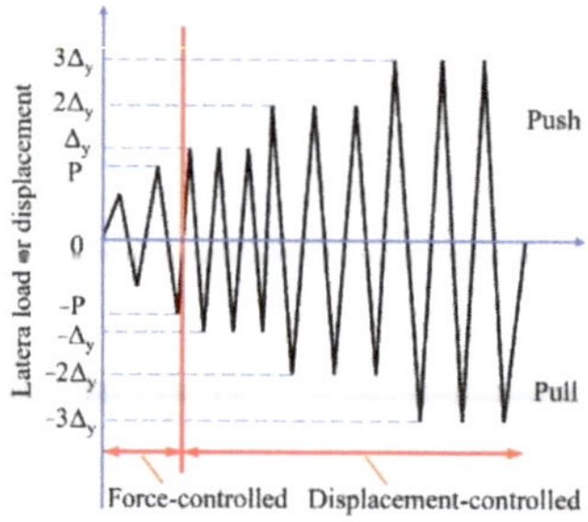

Figure 4. Cyclic lateral loading procedure.

The layout of the linear variable differential transducers (LVDTs) on the surface of the specimen is shown in Figure 5a. The displacement at the loading point of the column top is measured by LVDT-1, and LVDT-2 is used to record the horizontal displacement of the base to correct the column top displacement. LVDT-3/4 are two orthogonally arranged transducers used to measure the shear deformation in the plastic hinge area of the specimen [34]. The strains of the longitudinal bars and stirrups were measured by strain gauges (SG) glued to the surface of reinforcement, and the layout is shown in Figure 5b–d.

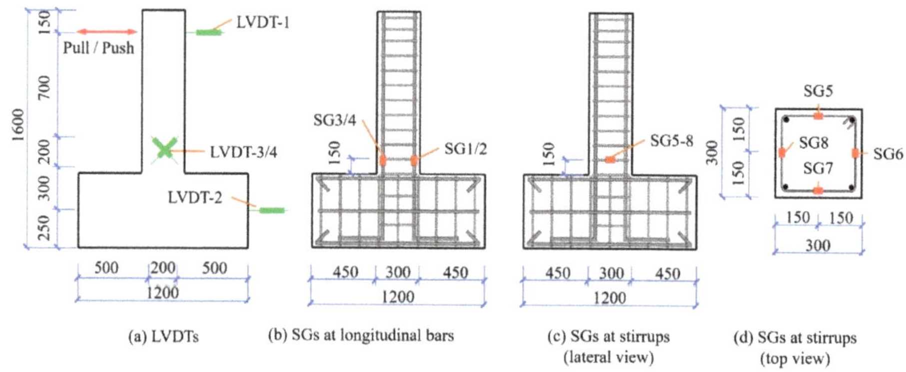

Figure 5. Layout of measuring points (unit: mm).

3. Results

3.1. General Observations and Failure Modes

The failure process of CGC specimens with different replacement ratios is similar to that of the NAC specimen, and they all go through four stages: cracking, yield, ultimate, and failure. The typical specimen KZA-5 was used as an example to illustrate the failure process.

When the load P = 80 kN, a horizontal crack with a length of about 50 mm was first observed at a distance of 150 mm from the base of the column. After continuing to load for a period of time, the test loading was changed from force-controlled to displacement-controlled. When the displacement Δ = 6.0 mm, at a distance of 300 mm from the base of the column, the original horizontal crack expanded to form an oblique crack about 80 mm, the direction of which was roughly at a 45-degree angle to the horizontal. When the displacement Δ = 12.0 mm, a vertical crack with a length of about 100 mm appeared at a distance of 400 mm from the base of the column. When the displacement Δ = 14.8 mm, the bearing capacity of the specimen reached the peak load. Continuing to load, the concrete cover at the column foot peeled off. When the displacement Δ = 32.1 mm, the bearing capacity decreased to 85% of the maximum load, indicating that the specimen was damaged, and the loading stopped. The failure modes of specimens are shown in Figure 6.

(**a**) KZA-1 (**b**) KZA-2 (**c**) KZA-3 (**d**) KZA-4 (**e**) KZA-5

Figure 6. Failure modes of specimens.

It can be seen from Figure 6 that many oblique cracks appear directly on the surface of the NAC column after cyclic load, while the oblique cracks on the surface of CGC specimens are mainly formed by the expansion of horizontal cracks. The main reason for this phenomenon is that compared with NAC, the elastic modulus of CGC is small, so the bending deformation of the CGC specimen is greater than that of the NAC specimen at the initial stage of loading, resulting in more horizontal bending cracks on the surface of the CGC specimen, and then further expand into oblique cracks during the cyclic action of load.

According to whether the reinforcement yields after the column is damaged, there are three failure modes of the column under cyclic loading, including bending, shear, and bending–shear [35]. The reinforcement strains of typical specimens KZA-1 and KZA-5 are plotted in Figure 7. According to the measured yield strain of reinforcement in Table 3, the longitudinal reinforcement and stirrup in the NAC and CGC columns yielded before the specimen was damaged, and then the concrete was crushed. Therefore, the failure mode of CGC specimens with different replacement ratios is consistent with that of the NAC specimen, which belongs to bending–shear failure.

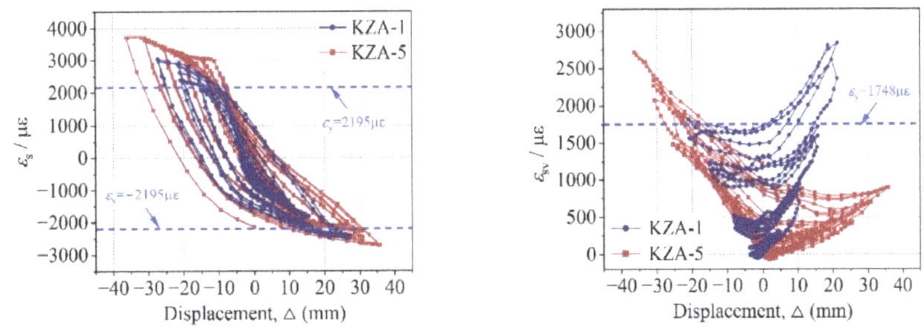

(**a**) Strain of longitudinal reinforcement (**b**) Strain of stirrup

Figure 7. Strain of reinforcement.

3.2. Hysteretic Curves

The hysteretic curves of specimens are shown in Figure 8. Affected by the coal gangue coarse aggregate, compared with the NAC specimen, the bearing capacity of the CGC specimens decreased by 1.7%, 4.6%, 24.7%, and 26.9% with the increase in the coal gangue replacement ratio. Moreover, compared with the NAC specimen, the biggest difference in the hysteresis curve of the CGC specimen is that the unloading curve is flatter, resulting in a significant reduction in the area enclosed by the hysteresis curve, the residual deformation after complete unloading is small, and the recovery deformation lag phenomenon is not significant.

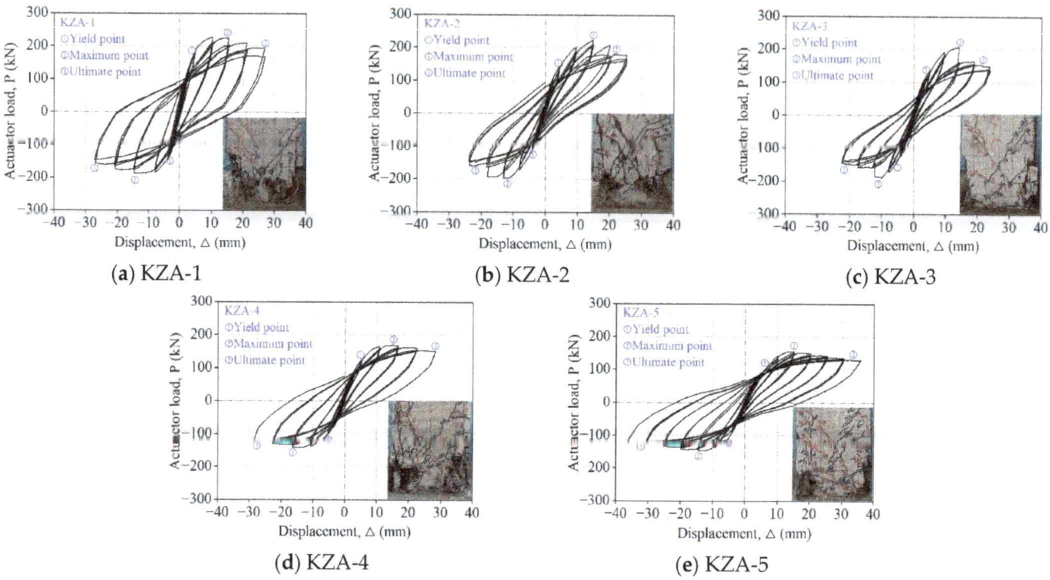

Figure 8. Hysteretic curves of specimens.

Observation of the crack interface for the specimen during loading showed that the different crack interface characteristics of the CGC specimen and NAC specimen are the main reasons for the difference in hysteresis performance. Figure 9 is the typical crack interface of the CGC specimen. Due to the low strength of CGCA, it cannot prevent the cracks from propagating, resulting in the crack directly passing through the CGCA. Therefore, the crack interface of the CGC specimen is relatively linear, the mechanical bite force between cracks is poor, and the crack can be closed quickly after unloading, resulting in poor energy consumption and small residual deformation of the specimen.

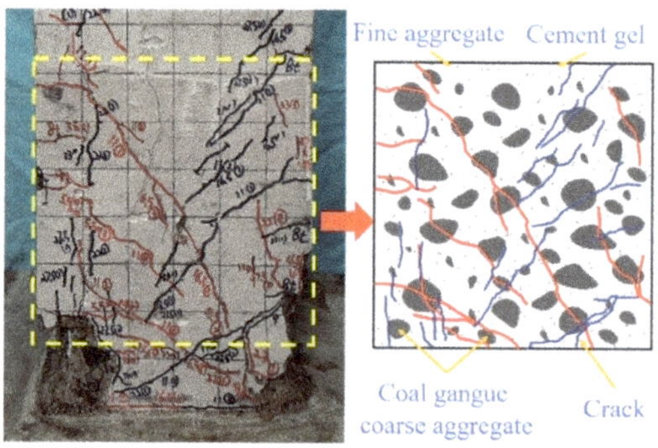

Figure 9. Crack interface of CGC specimen.

However, it is worth noting that due to the large deformation of CGCA, the higher the replacement ratio of the CGC specimen, the more the cyclic times of the hysteretic curve, and the larger the total area under the hysteretic cycle. Therefore, it can be found from Figure 8 that the energy dissipation of the CGC specimen with a higher replacement ratio is better.

3.3. Skeleton Curves

The skeleton curve of each specimen is obtained by connecting the trajectories of the maximum load reached by each cyclic loading [36], as shown in Figure 10.

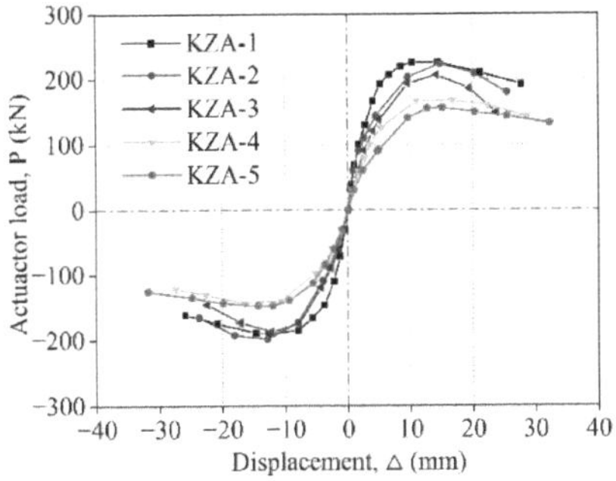

Figure 10. Skeleton curves of specimens.

Obviously, in the ascending section of the skeleton curve, the stiffness of CGC specimens is less than that of NAC specimens due to the influence of CGCA. In the descending section of the skeleton curve, the curves of specimens KZA-1, KZA-4, and KZA-5 are relatively soft, while the curves of specimens KZA-2 and KZA-3 are relatively steep. Obviously, the ductility of CGC specimens with a high replacement ratio is close to that of the NAC specimen, and is significantly higher than that of CGC specimens with low replacement ratio.

3.4. Displacement Ductility

The seismic performance of a structure depends not only on the bearing capacity of the component, but also on its deformation performance to a large extent. The ductility coefficient (μ) is an important index to evaluate the deformation performance of specimens [37], and the calculation formula is shown in Equation (1).

$$\mu = \Delta_u / \Delta_y \tag{1}$$

where Δ_u is the displacement when the load is reduced to 85% of the peak load; Δ_y is the yield displacement of the specimen, which is calculated by the energy equivalent method [38].

According to the ductility coefficient of specimens in Table 5, a number of findings can be reported. (a) Compared with KZA-1, the yield displacements of KZA-2, KZA-3, KZA-4, and KZA-5 are increased by 5.2%, 14.9%, 19.4%, and 30.0%, while the ductility coefficients are reduced by 28.6%, 24.4%, 12.6%, and 8.2%, indicating that the deformation characteristics of CGC specimens have a large deformation ability before yield but a poor deformation ability after yield. (b) Compared with KZA-2, the ultimate displacements of KZA-3, KZA-4, and KZA-5 are increased by 15.8%, 38.8%, and 59.1%, and the ductility coefficients are increased by 5.9%, 22.3%, and 28.5%, indicating that the deformation capacity of CGC specimens is positively correlated with the coal gangue replacement ratio. Combined with the hysteretic curve of specimens (see Figure 8), it can be found that due to the large deformation of CGCA, the hysteretic performance of CGC specimens with high replacement ratio is improved, so the deformation performance is increased. (c) The ductility coefficient of CGC specimens is between 3.05–3.92, which is lower than that of the NAC specimen (μ = 4.27), but it meets the requirement that the ductility coefficient of reinforced concrete seismic structure is not less than 3.0, indicating that the CGC specimen has good ductility performance under cyclic load.

Table 5. The ductility coefficients of specimens.

Specimen	P_y (kN)	Δ_y (mm)	Δ_y/Δ_u	P_m (kN)	Δ_m (mm)	Δ_m/Δ_u	P_u (kN)	Δ_u (mm)	μ
KZA-1	186.16	6.30	23.4%	216.83	12.52	46.6%	184.31	26.89	4.27
KZA-2	174.75	6.63	32.8%	210.50	11.86	58.7%	178.80	20.19	3.05
KZA-3	159.71	7.24	31.0%	197.38	13.92	59.5%	167.78	23.39	3.23
KZA-4	133.03	7.52	26.8%	155.76	15.69	56.0%	132.40	28.03	3.73
KZA-5	129.97	8.19	25.5%	151.19	14.79	46.0%	128.51	32.13	3.92

Note: the values in the table are the average values of the push and pull directions; P_y, P_m, and P_u respectively represent the yield load, maximum load, and ultimate load of the specimen; Δ_y, Δ_m, and Δ_u are the displacements corresponding to P_y, P_m, and P_u, respectively; μ is the ductility coefficient.

3.5. Energy Dissipation Capacity

The area under the hysteretic curve reflects the energy dissipated by the specimen under the cyclic loading [39]. The relationship between the energy dissipation of each specimen and the displacement is shown in Figure 11.

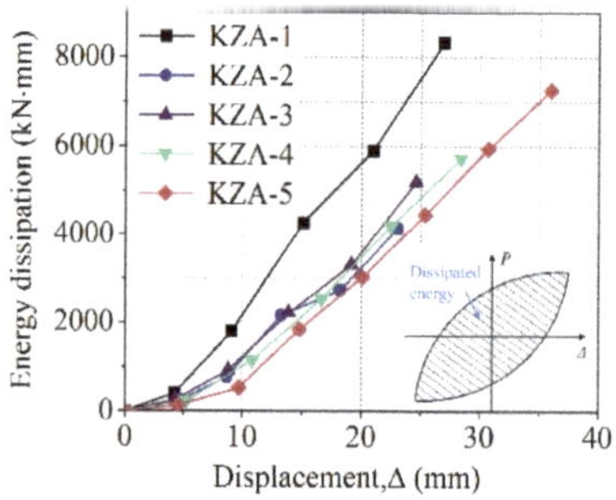

Figure 11. Energy dissipation of specimens.

Due to the damage of CGCA at the crack interface of CGC specimen (see Figure 9), the energy consumption of CGC specimen is less than that of NAC specimen whether it is the energy consumption of single cyclic loading or the total energy consumption. Compared with KZA-1, the energy dissipation of the CGC specimens with different replacement ratios decreased by 50.3%, 37.7%, 31.2%, and 12.8%.

However, it is worth noting that the deformation capacity of CGC specimen is improved due to the large deformation of coal gangue coarse aggregate (see Table 5). Therefore, compared with KZA-2, the total energy dissipation of KZA-3, KZA-4, and KZA-5 increased by 25.2%, 38.2%, and 75.4%, respectively.

3.6. Shear Deformation

Through the two transducers (LVDT-3,4) arranged in Figure 5, the deformation at the plastic hinge area of the specimen under cyclic load can be measured. The shear deformation (γ) calculation diagram is shown in Figure 12, and the calculation formulas are shown in Equations (2) and (3).

$$\gamma = \gamma_1 + \gamma_2 = (\frac{\sin\theta}{h_c} + \frac{\cos\theta}{b_c})\overline{X} = \frac{\sqrt{{h_c}^2 + {b_c}^2}}{h_c b_c}\overline{X} \tag{2}$$

$$\overline{X} = \frac{|\delta_A + \delta'_A| + |\delta_B + \delta'_B|}{2} \tag{3}$$

where h_c and b_c represent the measured height and width of the plastic hinge region, as shown in Figure 12; $\overline{X}$ is the average deformation along the diagonal direction; $\delta_A + \delta'_A$ and $\delta_B + \delta'_B$ are measured by LVDT-3 and LVDT-4.

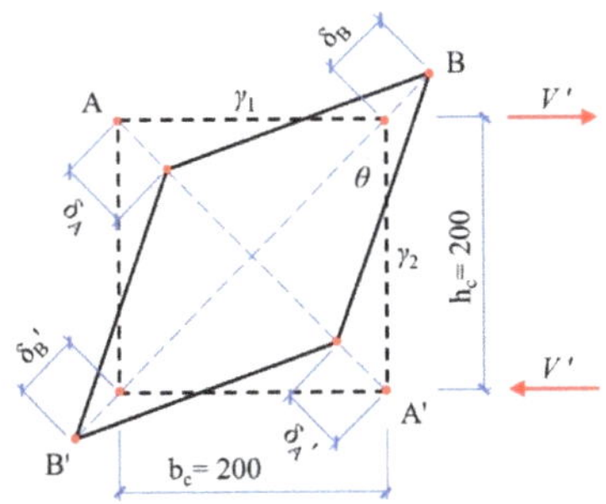

Figure 12. Shear deformation calculation diagram (unit: mm).

The shear deformation at the plastic hinge area of each specimen is shown in Figure 13, and its value is the average value in the push and pull directions. In terms of the overall trend, the development of the shear deformation of the specimen can be divided into two stages: the load rising stage and the load falling stage.

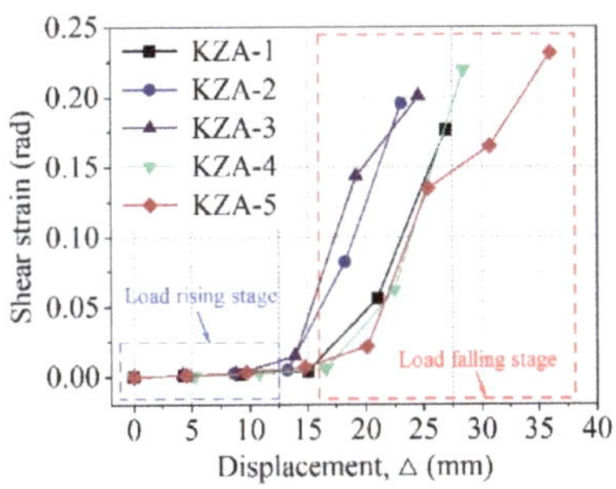

Figure 13. Shear deformation at the plastic hinge region.

In the load rising stage, the overall deformation of the specimen is small, and the oblique cracks on the surface of the specimen have not significantly expanded, so the shear deformation of each specimen is very small. The shear deformation curves of the CGC specimens and NAC specimens roughly coincide, and the coal gangue coarse aggregate has little effect on the shear deformation of the specimen.

In the load falling stage, due to the continuous expansion of oblique cracks on the surface of the specimen, the shear deformation of each specimen increases rapidly with the increase in displacement until the specimen is damaged. Compared with NAC specimens, CGC specimens have greater shear deformation because there is a large amount of damage of coal gangue coarse aggregate at the crack interface, resulting in the reduction in mechanical bite force at the crack interface. When the specimens are damaged, the shear deformation of KZA-2, KZA-3, KZA-4, and KZA-5 increased by 10.9%, 13.7%, 24.2%, and 31.2%, respectively, compared with KZA-1.

The destruction of coal gangue coarse aggregate leads to the increase in the shear deformation of the CGC specimen, and the increase in the shear deformation of CGC specimen will further reduce the energy dissipation capacity of the crack interface. Therefore, it is suggested that increasing the stirrup ratio in CGC specimens can not only effectively restrict the deformation of CGC and enhance the friction at the crack interface, but also improve the hysteretic performance and energy dissipation capacity of CGC specimens, which is an effective way to improve the seismic performance of CGC specimens. Moreover, the existing research [40] shows that the use of the steel section will effectively improve the seismic performance of concrete structures. Therefore, in addition to appropriately increasing the reinforcement ratio of coal gangue concrete, it can also be considered to arrange the steel section in the coal gangue concrete to improve the seismic performance of the coal gangue concrete structure.

It should be noted that this paper mainly studies the mechanical behavior of coal gangue concrete frame joints under repeated load, and whether the addition of coal gangue coarse aggregate will affect the corrosion resistance of reinforcement [41]; the durability of concrete remains to be further studied.

4. Conclusions

Using solid waste coal gangue as coarse aggregate can significantly reduce the mining of natural aggregates and help realize the sustainable development of construction engineering. Through the cyclic loading test of five concrete frame columns with different coal gangue replacement ratios, the following conclusions can be drawn.

The coal gangue coarse aggregate cannot block the propagation of cracks, resulting in the decrease in friction at the crack interface. Therefore, CGC specimens have two typical characteristics under cyclic loads: most of the deformation can be recovered quickly after each unloading, and the residual deformation of the specimen is small; the deformation of the specimen is large before yield, while the deformation capacity is poor after yield.

Both CGC and NAC specimens show a bending–shear type of failure, but affected by the coal gangue coarse aggregate, the ductility coefficients and total energy consumption of CGC specimens are reduced by 8.2–28.6% and 12.8–50.3%, respectively compared with KZA-1.

Affected by the large deformation of coal gangue coarse aggregate, the CGC specimens with higher replacement ratio have better seismic performance. Compared with KZA-2, the ductility coefficients of KZA-3, KZA-4, and KZA-5 increased by 15.8%, 38.8%, 59.1%, and the total energy consumption increased by 25.2%, 38.2%, and 75.4%, respectively.

Compared with KZA-1, the shear deformation at the plastic hinge area of CGC specimens increased by 10.9–31.2%. Larger shear deformation directly increases the width of the diagonal crack, which is also one of the reasons for the decrease in energy dissipation capacity of the CGC specimens. In the design, it is suggested to appropriately increase the stirrup ratio of the CGC specimen to improve its seismic performance.

In general, although the seismic performance of CGC specimens is reduced by the influence of coal gangue coarse aggregate compared with NAC specimens, the displacement ductility coefficient of CGC specimens with different replacement ratios is between 3.05–3.92, indicating that CGC specimens can be applied to structures in areas with seismic fortification requirements.

Author Contributions: Conceptualization, H.L. and G.B.; methodology, H.L. and G.B.; software, H.L. and Y.G.; validation, G.B.; formal analysis, G.B.; investigation, F.Y. and Y.G.; resources, G.B.; data curation, H.L., F.Y. and K.Z.; writing—original draft preparation, H.L.; writing—review and editing, G.B.; visualization, G.B.; supervision, G.B.; project administration, G.B.; funding acquisition, G.B. All authors have read and agreed to the published version of the manuscript.

Funding: This research was funded by the National Natural Science Foundation of China, grant number 52078410; Shaanxi Provincial Science and Technology Plan Achievement Promotion Project, grant number 2020CGHJ-017; and Key Laboratory Project of Shaanxi Provincial Department of Education, grant number 20JS071.

Institutional Review Board Statement: Not applicable.

Informed Consent Statement: Not applicable.

Data Availability Statement: Not applicable.

Conflicts of Interest: The authors declare no conflict of interest.

References

1. Querol, X.; Izquierdo, M.; Monfort, E.; Álvarez, E.; Font, O.; Moreno, T.; Alastuey, A.; Zhuang, X.; Lu, W.; Wang, Y. Environmental characterization of burnt coal gangue banks at Yangquan, Shanxi Province, China. *Int. J. Coal Geol.* **2008**, *75*, 93–104. [CrossRef]
2. Wang, C.; Ni, W.; Zhang, S.; Wang, S.; Gai, G.; Wang, W. Preparation and properties of autoclaved aerated concrete using coal gangue and iron ore tailings. *Constr. Build. Mater.* **2016**, *104*, 109–115. [CrossRef]
3. Li, D.; Song, X.; Gong, C.; Pan, Z. Research on cementitious behavior and mechanism of pozzolanic cement with coal gangue. *Cem. Concr. Res.* **2006**, *36*, 1752–1759. [CrossRef]
4. Zhang, Y.; Ling, T. Reactivity activation of waste coal gangue and its impact on the properties of cement-based materials–A review. *Constr. Build. Mater.* **2020**, *234*, 117424. [CrossRef]
5. Okagbue, C.; Ochulor, O. The Potential of Cement-Stabilized Coal-Reject as a Construction Material. *Bull. Eng. Geol. Environ.* **2007**, *66*, 143–151. [CrossRef]
6. Jabło´nska, B.; Kityk, A.; Busch, M.; Huber, P. The structural and surface properties of natural and modified coal gangue. *J. Environ. Manag.* **2017**, *190*, 80–90. [CrossRef]
7. Moghadam, M.; Ajalloeian, R.; Hajiannia, A. Preparation and application of alkali-activated materials based on waste glass and coal gangue: A review. *Constr. Build. Mater.* **2019**, *221*, 84–98. [CrossRef]
8. Dong, Z.; Xia, J.; Fan, C.; Gao, J. Activity of calcined coal gangue fine aggregate and its effect on the mechanical behavior of cement mortar. *Constr. Build. Mater.* **2015**, *100*, 63–69. [CrossRef]
9. Zaben, A.; Maslehuddin, M.; Omar, A.; Al-Dulaijan, S. Influence of mix composition on the properties of recycled aggregate concrete. *Struct. Concr.* **2021**, *22*, 2939–2951. [CrossRef]
10. Piccinali, A.; Diotti, A.; Plizzari, G.; Sorlini, S. Impact of Recycled Aggregate on the Mechanical and Environmental Properties of Concrete: A Review. *Materials* **2022**, *15*, 1818. [CrossRef]
11. Li, C.; Wan, J.; Sun, H. Investigation on the activation of coal gangue by a new compound method. *J. Hazard. Mater.* **2010**, *179*, 515–520. [CrossRef] [PubMed]
12. Makul, N.; Fediuk, R.; Amran, M.; Zeyad, A.M.; de Azevedo, A.R.G.; Klyuev, S.; Vatin, N.; Karelina, M. Capacity to Develop Recycled Aggregate Concrete in South East Asia. *Buildings* **2021**, *11*, 234. [CrossRef]
13. Xu, D.; Yang, Y.; Chen, Z. Experimental Study and Damage Model on the Seismic Behavior of Reinforced Concrete L-Shaped Columns under Combined Torsion. *Appl. Sci.* **2020**, *10*, 7008. [CrossRef]
14. Bravo, M.; Duarte, A.P.C.; de Brito, J.; Evangelista, L.; Pedro, D. On the Development of a Technical Specification for the Use of Fine Recycled Aggregates from Construction and Demolition Waste in Concrete Production. *Materials* **2020**, *13*, 4228. [CrossRef] [PubMed]
15. Silva, F.A.N.; Delgado, J.M.P.Q.; Azevedo, A.C.; Lima, A.G.B.; Vieira, C.S. Preliminary Analysis of the Use of Construction Waste to Replace Conventional Aggregates in Concrete. *Buildings* **2021**, *11*, 81. [CrossRef]
16. Díaz, M.; Almendro-Candel, M.B.; Blanco, D.; Jordan, M.M. Aggregate Recycling in Construction: Analysis of the Gaps between the Chilean and Spanish Realities. *Buildings* **2019**, *9*, 154. [CrossRef]
17. Marsh, E. Civil infrastructure systems materials research support at the National Science Foundation. *Cem. Concr. Comp.* **2003**, *25*, 575–583. [CrossRef]
18. Gao, S.; Zhang, S.; Guo, L. Application of Coal Gangue as a Coarse Aggregate in Green Concrete Production: A Review. *Materials* **2021**, *14*, 6803. [CrossRef]
19. Zhang, Y.; Wang, Q.; Zhou, M.; Fang, Y.; Zhang, Z. Mechanical properties of concrete with coarse spontaneous combustion gangue aggregate (SCGA): Experimental investigation and prediction methodology. *Constr. Build. Mater.* **2020**, *255*, 119337. [CrossRef]
20. Li, S.; Zhou, M.; Zhang, L. Properties of spontaneous combustion coal gangue coarse aggregate and its influence on concrete. *J. Build. Mater.* **2020**, *23*, 334–340, 380.
21. Wang, Q.; Li, Z.; Zhang, Y.; Zhang, H.; Zhou, M.; Fang, Y. Influence of coarse coal gangue aggregates on elastic modulus and drying shrinkage behaviour of concrete. *J. Build. Eng.* **2020**, *32*, 101748. [CrossRef]
22. Liu, H.; Xu, Q.; Wang, Q.; Zhang, Y. Prediction of the elastic modulus of concrete with spontaneous-combustion and rock coal gangue aggregates. *Structures* **2020**, *28*, 774–785. [CrossRef]
23. Zhou, M.; Dou, Y.; Zhang, Y.; Zhang, Y.; Zhang, B. Effects of the variety and content of coal gangue coarse aggregate on the mechanical properties of concrete. *Constr. Build. Mater.* **2019**, *220*, 386–395. [CrossRef]
24. Guan, X.; Qiu, J.; Song, H.; Qin, Q.; Zhang, C. Stress–strain behaviour and acoustic emission characteristic of gangue concrete under axial compression in frost environment. *Constr. Build. Mater.* **2019**, *220*, 476–488. [CrossRef]
25. Qiu, J.; Zhou, Y.; Vatin, N.; Guan, X.; Sultanov, S.; Khemarak, K. Damage constitutive model of coal gangue concrete under freeze-thaw cycles. *Constr. Build. Mater.* **2020**, *264*, 120720. [CrossRef]
26. Bai, G.; Zhu, C.; Wang, J. Experimental study on shear behavior of coal gangue concrete beams. *J. Build. Struct.* **2020**, *41*, 49–55. [CrossRef]
27. Wang, Q.; Li, Z.; Zhou, M. Effects of spontaneous-combustion coal gangue aggregate (SCGA) replacement ratio on flexural behavior of SCGA concrete beams. *J. Build. Struct.* **2020**, *41*, 64–74. [CrossRef]
28. Huda, S.; Alam, M. Mechanical behavior of three generations of 100% repeated recycled coarse aggregate concrete. *Constr. Build. Mater.* **2014**, *65*, 574–582. [CrossRef]

29. Choi, M.; Lee, C. Seismic Behavior of Existing Reinforced Concrete Columns with Non-Seismic Details under Low Axial Loads. *Materials* **2022**, *15*, 1239. [CrossRef]
30. Hashemi, B.; Bonab, A. Experimental investigation of the behavior of laced columns under constant axial load and cyclic lateral load. *Eng. Struct.* **2013**, *57*, 536–543. [CrossRef]
31. Karen, E.; Caballero-Morrison, J.; Bonet, J.; Pedro, S. An experimental study of steel fiber-reinforced high-strength concrete slender columns under cyclic loading. *Eng. Struct.* **2013**, *57*, 565–577. [CrossRef]
32. Hyun, J.; Bang, J.; Lee, B.; Kim, Y. Effects of the Replacement Length of Concrete with ECC on the Cyclic Behavior of Reinforced Concrete Columns. *Materials* **2021**, *14*, 3542. [CrossRef] [PubMed]
33. Cassese, P.; Menna, C.; Occhiuzzi, A.; Asprone, D. Experimental Behavior of Existing RC Columns Strengthened with HPFRC Jacket under Concentric and Eccentric Compressive Load. *Buildings* **2021**, *11*, 521. [CrossRef]
34. El-Mandouh, M.; Omar, M.; Elnaggar, M.; Abd El-Maula, A. Cyclic Behavior of High-Strength Lightweight Concrete Exterior Beam-Column Connections Reinforced with GFRP. *Buildings* **2022**, *12*, 179. [CrossRef]
35. Wang, J.; Yi, X.; Liu, Q.; Fang, X. Seismic Performance of Steel-Reinforced Concrete Columns with Q690 High-Strength Steel. *Materials* **2022**, *15*, 2979. [CrossRef] [PubMed]
36. Shi, K.; Zhang, M.; Zhang, T.; Li, P.; Zhu, J.; Li, L. Seismic Performance of Steel Fiber Reinforced High–Strength Concrete Beam–Column Joints. *Materials* **2021**, *14*, 3235. [CrossRef]
37. Rodrigues, H.; Arêde, A.; Varum, H.; Costa, A.G. Experimental evaluation of rectangular reinforced concrete column behaviour under biaxial cyclic loading. *Earthq. Eng. Struct. Dyn.* **2013**, *42*, 239–259. [CrossRef]
38. Jiuru, T.; Chaobin, H.; Kaijian, Y.; Yongcheng, Y. Seismic behavior and shear strength of framed joint using steel–fiber reinforced concrete. *J. Struct. Eng.* **1992**, *118*, 341–358. [CrossRef]
39. Chopra, A. *Dynamics of Structures: Theory and Applications to Earthquake Engineering*; Prentice Hall: Hoboken, NJ, USA, 2017.
40. Montava, I.; Irles, R.; Pomares, J.C.; Gonzalez, A. Experimental Study of Steel Reinforced Concrete (SRC) Joints. *Appl. Sci.* **2019**, *9*, 1528. [CrossRef]
41. Bautista, A.; Pomares, J.C.; González, M.N.; Velasco, F. Influence of the microstructure of TMT reinforcing bars on their corrosion behavior in concrete with chlorides. *Constr. Build. Mater.* **2019**, *229*, 116899. [CrossRef]

buildings

MDPI

Article

Performance Evaluation of Reinforced Concrete Columns under Simultaneously Combined Fire and Cyclic Loads

Qingjun Chen [1,2] and Yu Jiang [1,2,*]

1 State Key Laboratory of Disaster Reduction in Civil Engineering, Tongji University, Shanghai 200092, China; chenqj@tongji.edu.cn
2 College of Civil Engineering, Tongji University, Shanghai 200092, China
* Correspondence: jiangyu0512@tongji.edu.cn; Tel.: +86-021-6598-6185

Abstract: Reinforced concrete (RC) structures could suffer from the combined action of fires, earthquakes, and other loads during their life cycle; more importantly, coupled disasters lead to further deterioration and damage to structural performance. This paper investigated the multiple performances and distinguished the safe working conditions of the RC column subjected to simultaneously combined fire and cyclic loads. The numerical model considered the degradation of the mechanical properties of steel and concrete and the bond-slip performance between steel and concrete at high temperatures. The results show that the performance of RC columns with different section sizes, longitudinal reinforcement ratios, cover thicknesses, axial load ratios, and cyclic loads differs greatly under simultaneously combined fire-cyclic loads. In specific, when the cyclic load application time is less than 2 h, the cyclic load has little effect on the response of the RC column. According to the different characteristics of RC columns when subjected to combined fire-cyclic loads, the firing process of RC columns is divided into four stages. To avoid the excessive performance degradation of RC columns, the minimum designed fire resistance time of RC columns is recommended to be 2.5 times the fire resistance time of the RC column under static loads.

Keywords: fire and dynamic loads; seismic engineering; reinforced concrete column; numerical simulation; bond-slip behavior

Citation: Chen, Q.; Jiang, Y. Performance Evaluation of Reinforced Concrete Columns under Simultaneously Combined Fire and Cyclic Loads. *Buildings* **2022**, *12*, 1062. https://doi.org/10.3390/buildings12071062

Academic Editors: Bo Wang, Bo Fu and Xinxin Wei

Received: 18 June 2022
Accepted: 18 July 2022
Published: 21 July 2022

Copyright: © 2022 by the authors. Licensee MDPI, Basel, Switzerland. This article is an open access article distributed under the terms and conditions of the Creative Commons Attribution (CC BY) license (https://creativecommons.org/licenses/by/4.0/).

1. Introduction

Reinforced concrete (RC) structures may suffer fire, earthquake [1–4], or dynamic loads during the life cycle. In case of fire, the mechanical properties of steel bars and concrete, which are the constituent materials of the RC structure, will degrade [5–7], the concrete may burst at high temperatures [8–10], and serious redistribution of internal forces will occur in the structures, resulting in a significant decline in the bearing capacity of RC structures [11–13]. It is common for earthquakes/dynamic loads to occur simultaneously or sequentially with fires, such as the Loma Prieta earthquake in 1989 [14], the Northridge earthquake in 1994 [15], and the Kobe earthquake in 1995 [16]. Aftershocks may occur during the continuous fire process. Under the condition of coupling of these disasters, the deformation, bearing capacity, and damage mechanism of RC structures are more complicated, which will cause more damage to RC structures than one disaster alone (fire, earthquake, and other dynamic loads). RC components are the important parts of the structure, and the combined action of fire and earthquake/dynamic loads will lead to aggravated performance degradation of components, which will lead to the partial and overall collapse of the structure. Therefore, studying the performance of RC components under the combined action of fire and dynamic loads has great engineering significance and can provide a reference for the design of engineering structures under the combined action of fire and earthquake/dynamic loads.

At present, research on the performance of RC components under the combined action of fire and earthquake/dynamic loads mainly focuses on post-earthquake fires (PEFs) or post-fire earthquakes (PFEs). For example, Shah et al. [17] proposed that non-ductile plates and shells of RC frames are prone to brittle failure in PEF, while RC frames with ductile design have good lateral resistance before and after a fire. Wen et al. [18] indicated that the concrete spalling width has a greater influence on the deformation of the seismically damaged RC column when it is exposed to fire, while the spalling length has less of an effect. Wang et al. [19] presented that the lateral resistance, effective stiffness, and ductility of RC columns decreased under PEF. Ni et al. [20] pointed out the fire would reduce the bearing capacity and stiffness of RC walls. Wang et al. [21] showed that the load level of beams and columns has little effect on the residual bearing capacity of the joint, but the load level has a significant effect on the ductility of the joint, and the ductility coefficient can more truly reflect the response of the joint after the fire than the limit displacement. Liu et al. [22] proposed that the failure mode of RC frame joints changed from plastic hinge failure at ambient temperature to shear failure in the joint area after the fire. Lu et al. [23] found that except for the core area failure of the plane nodes after exposure to fire, the rest of the nodes were all beam end bending failures. Jin et al. [24] simulated the seismic performance of 3D mesoscale RC columns after exposure to fire and proposed that axial compression ratios in the range of 0.2 to 0.6 are beneficial for improving the lateral force resistance and initial stiffness of RC short columns, but have a negative impact on ductility. Demir et al. [25] reported that the time after exposure to fire had a limited influence on the residual lateral resistance and ductility of the column. Han et al. [26] proposed that factors such as the fire exposure time, cross-sectional size, slenderness ratio, and concrete strength have significant effects on the residual strength of steel-tube RC columns after exposure to fire.

In short, most of the existing studies are on the performance of RC structures or components under exposure to fire and earthquake/dynamic load successively, but there are few reports on the simultaneous fire and earthquake/dynamic load, which could be the dominant design condition for the RC columns. The RC column is an important vertical load member of the structure, and its static bearing capacity and deformation capacity are significantly reduced during fire exposure [27–32]. The empirical model of Shah et al. [31] showed that tie spacing, confinement of columns, and axial load ratio have a marked influence on the fire resistance of the RC column. When the RC column is subjected to axial dynamic loads (such as vertical seismic motion or vertical mechanical vibration [33–35]) during fire exposure, the performance of the RC column will further deteriorate. Therefore, it is necessary to study the performance of RC columns under axial cyclic loads and exposure to fire which can provide a reference for the design of RC structures under the combined action of fire and dynamics.

In this paper, the RC column subjected to the simultaneously combined fire and cyclic loads is taken as the research object. Firstly, the nonlinear bond-slip constitutive of the connector element is used for simulating the nonlinear bond-slip behavior between the steel bar and the concrete, and the effectiveness of the numerical simulation method was verified by comparing the deformation and damage. Secondly, the effects of amplitudes and the number of cycles of cyclic loads on the axial deformation, stiffness, and fire resistance time of RC columns were analyzed. Finally, the specific cyclic characteristics (amplitude and number of cycles) were selected as the basic cyclic loads, and the effect of cyclic loads on the response of RC columns with different column characteristics (section size, cover thickness, longitudinal reinforcement ratio, tie spacing, and axial load ratio) were analyzed.

2. Description of Models and Verification of Numerical Method

2.1. Description of Column

Raut and Kodur [36] conducted a fire resistance test on several RC columns, and this paper takes the test specimen NSC1 as the basic research object. NSC1 is 3350 mm long with a cross-section of 203 mm × 203 mm and 50 mm cover thickness. NSC1 has four Φ20-mm longitudinal reinforcements and a Φ10-mm stirrup with 200 mm spacing. Figure 1 shows the elevation and cross-section details of the NSC1. The top of NSC1 is under a concentric load with an axial compression ratio of 0.4. The boundary conditions are shown in Figure 2a, the bottom 100 mm range of the column is fixed displacement in all directions, and the top 100 mm range of the column is fixed displacement except axial displacement. The middle 1.7 m of the column height is exposed to a four-sided fire. The heating curve is shown in Figure 2a and the ambient temperature is 25 °C.

Due to physicochemical properties changes and microstructural degradation of reinforcing steel and concrete during fire exposure [37,38], concrete and reinforcing steel have different mechanical properties and thermophysical properties at elevated temperatures. Therefore, this paper adopts the thermophysical parameters (Section 2.2) and mechanical parameters (Section 2.3) of reinforcing steel and concrete at elevated temperatures. The temperature field calculation needs to consider the actual thermal boundary conditions to set the corresponding heat transfer calculation parameters (Section 2.4).

In this paper, ABAQUS is used for the thermal-mechanical coupling calculation of RC column at high temperatures, in which the concrete was modeled by eight-node thermally coupled brick with trilinear displacement and temperature element (C3D8T) with an approximately uniform size of 25 mm × 25 mm × 50 mm, and the reinforcing steel is simulated by two-node 3-D thermally coupled truss elements (T3D2T) with a size of 25 mm. The RC column mesh and reinforcement details are shown in Figure 2b,c, respectively.

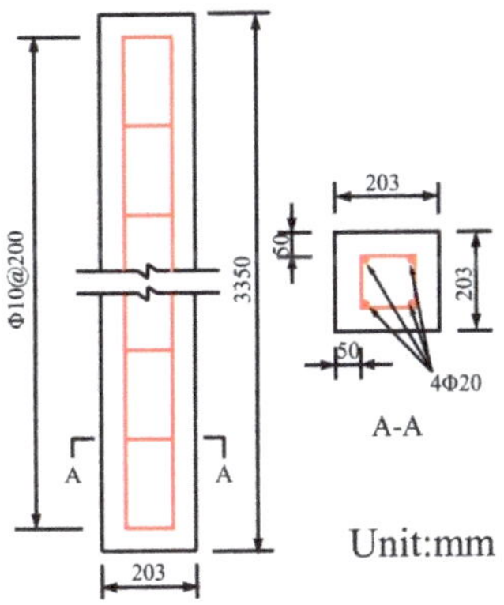

Figure 1. The evolution and cross-section and reinforcing details of NSC1.

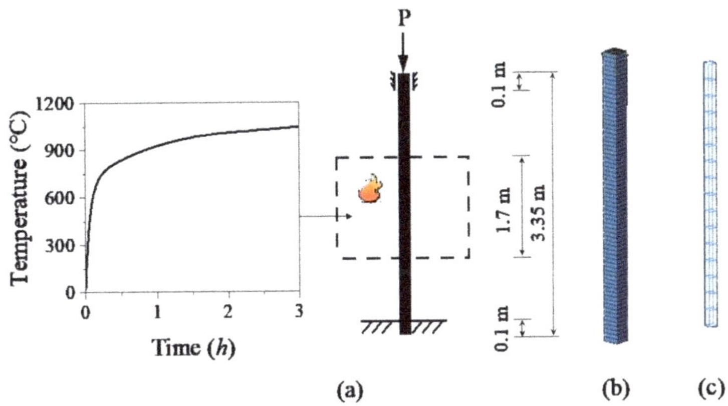

Figure 2. Numerical model details: (**a**) boundary conditions and temperature rise curve; (**b**) concrete mesh; and (**c**) reinforcing steel details.

2.2. Thermophysical Parameters

Because the density of the reinforcing steel can be considered to be independent of the temperatures [39], the value $\rho_s = 7850\ \mathrm{kg/m^3}$ is taken. The specific heat capacity $C_{p,s}$ and thermal conductivity λ_s of reinforcing steel at high temperatures refer to EN 1994-1-2:2005 [39], and the thermal expansion coefficient α_s follows the suggested formula values by Lie [40]. Thermophysical parameters of steel are shown in Figure 3a.

The density ρ_c of concrete at high temperatures is taken according to EN 1992-1-2:2004 [41], and the specific heat capacity $C_{p,c}$, thermal conductivity λ_c, and thermal expansion coefficient α_c are taken according to the recommended values of Lie [40]. All the thermophysical parameters of concrete are shown in Figure 3b.

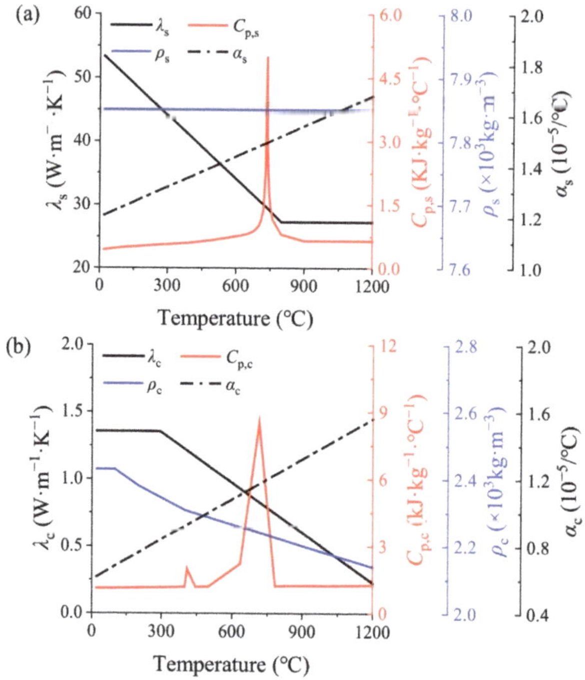

Figure 3. Thermophysical parameters of (**a**) reinforcement steel and (**b**) concrete.

2.3. Constitutive Model and Mechanical Parameters

The elastic, linear hardening model is used for the reinforcing steel. Figure 4a shows the stress-strain curve of reinforcement at elevated temperatures, where $f_{sy,T}$, $f_{sy,T}$, $\varepsilon_{sy,T}$, and $\varepsilon_{su,T}$ are the yield strength, ultimate strength, yield strain, and ultimate strain at elevated temperatures, and the ratio of ultimate strength to yield strength is assumed to be constant at elevated temperatures. Young's modulus $E_{s,T}$ and yield strength $f_{sy,T}$ decline gradually with temperature increases, and the corresponding attenuation parameters are selected according to the recommendations of EN1994-1-2:2005 [39], as shown in Figure 5a, where $E_{s,0}$ is Young's modulus at ambient temperature. The yield strength, ultimate strength, ultimate strain, and the Poisson's ratio of both longitudinal reinforcement and stirrups at ambient temperature are 450 MPa, 705 MPa, 0.17, and 0.3, respectively.

The concrete adopts the plastic-damage model [42,43] which is widely used for describing the dynamic behavior of concrete, and the constitutive curve based on GB50010-2010 [44] is depicted in Figure 4b. The compressive strength $f_{c,0}$, tensile strength $f_{t,0}$, Young's modulus $E_{c,T}$, and Poisson's ratio of the concrete at ambient temperature are 51 MPa, 3.5 MPa,

32.5 GPa, and 0.2, respectively. The degradations of the mechanical properties are shown in Figure 5b, where the degradation of the compressive strength $f_{c,T}$ and the peak strain $\varepsilon_{cr,T}$ are selected according to EN1992-1-2:2004 [41], whereas the deterioration of Young's modulus $E_{c,T}$ is taken from AISC-360-10 [45]. The ratio of the peak tensile strain to the compressive peak strain and the ratio of the tensile strength to the compressive strength at elevated temperatures are constant.

The bond-slip behavior [46] between the reinforcing steel and the concrete is simulated by adding the connector element between the reinforcing steel node and the concrete node and giving the connector element nonlinear bond-slip constitutive properties. To simplify the calculation, the connector is only subjected to axial force, i.e., the slip between the reinforcement and the concrete is assumed to be along the axial direction of the reinforcement.

The bond-slip constitutive which references GB50010-2010 [44] is shown in Figure 4c, where *cr*, *u*, *un*, and *r* represent the cracking point, the peak point, the unloading point, and the residual point, respectively. Figure 5c shows the peak bond stress and the peak slip at elevated temperatures according to the test of Özkal [47], and $\tau_{u,0}$, $\tau_{u,T}$, $S_{u,0}$, and $S_{u,T}$ are the bond strength at ambient temperature, the bond strength at elevated temperatures, the peak slip at ambient temperature and the peak slip at elevated temperatures, respectively. Table 1 shows the values of bond-slip parameters at ambient temperature, where d is the diameter of the reinforcing steel, and $f_{t,T}$ is the tensile strength of the concrete.

Table 1. Bond stress-slip constitutive of concrete and reinforcing steel at ambient temperature.

Feature Point	Crack Point (*cr*)	Peak Point (*u*)	Residual Point (*r*)
Bond stress (N/mm^2)	2.5 $f_{t,T}$	3 $f_{t,T}$	$f_{t,T}$
Slip (mm)	0.025	0.04 d	0.55 d

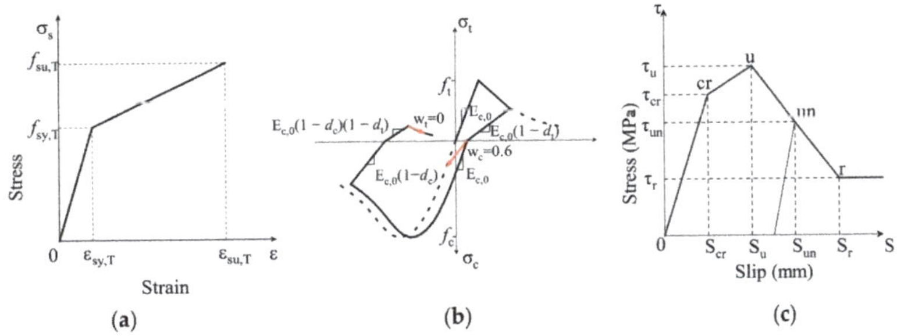

Figure 4. Constitutive models of (**a**) reinforcing steel (**b**) concrete and (**c**) bond strength-slip.

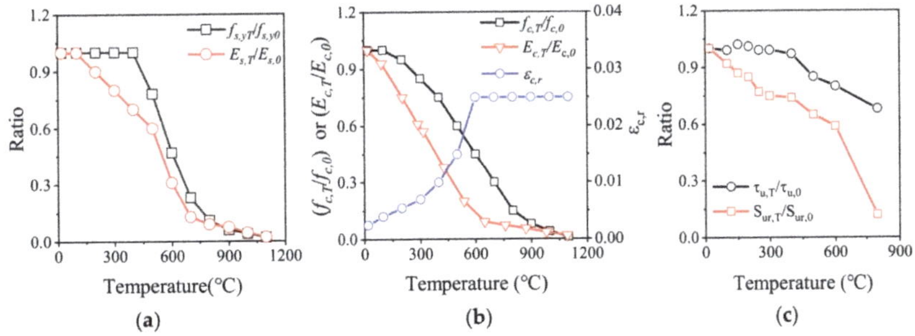

Figure 5. Degradation of mechanical performance of (**a**) reinforcing steel (**b**) concrete and (**c**) bond-slip at high temperature.

2.4. Equation and Calculation Parameter of Heat Transfer

The heat transfer inside the reinforced concrete during fire exposure can be calculated according to the Fourier heat transfer theory, and the heat exchange between the surface of the concrete and the outside occurs through heat convection and heat radiation [48]. Convective heat transfer heat flux density q_{ch}, radiant heat transfer heat flux density q_{rh}, and composite heat transfer heat flux density q_h are defined as Equations (1)–(3). The internal heat conduction equation of the RC column is calculated according to Equation (4).

$$q_{ch} = h_{ch}\left(T_{wh} - T_{fh}\right) \quad (1)$$

$$q_{rh} = \varepsilon_h C_b\left[{T_{wh}}^4 - {T_{am}}^4\right] \quad (2)$$

$$q_h = q_{ch} + q_{rh} \quad (3)$$

$$\rho c\frac{\partial T}{\partial t} = \frac{\partial}{\partial x}\left(\lambda\frac{\partial T}{\partial x}\right) + \frac{\partial}{\partial y}\left(\lambda\frac{\partial T}{\partial y}\right) + \frac{\partial}{\partial z}\left(\lambda\frac{\partial T}{\partial z}\right) \quad (4)$$

where T_{wh} is the surface temperature of the wall, T_{fh} is the temperature of the thermal fluid (this parameter is the input temperature curve in ABAQUS), and T_{am} is the ambient temperature. According to EN1992-1-2:2004 [41], the convective heat transfer coefficient h_{ch} is 25 $W/(m^2{\cdot}K)$ for the fire-exposed surface and 9 $W/(m^2{\cdot}K)$ for the unexposed surface. The emissivity ε_h is 0.7 and 0 for fire exposed surface and unexposed surface, respectively. The Stefan-Boltzmann constant C_b is 5.67×10^{-8} $W/\left(m^2{\cdot}K^4\right)$. In Equation (4), ρ, c, λ, T, and t are the density, specific heat compacity, thermal conductivity, temperature, and time, respectively.

2.5. Verification of Models

The mesh size selection is comprehensively considered through parametric analysis. Table 2 shows the peak deformation, fire resistance time, calculation time ratio, and connector generation of RC columns with different mesh sizes. The approximate mesh size of 25 mm × 25 mm × 50 mm shows a medium error of the peak deformation, lowest error of the fire resistance time, and lower calculation time ratio than other mesh sizes. Additionally, the approximate mesh size of 25 mm × 25 mm × 50 mm can easily generate the connector, so it is suitable for numerical simulation.

Table 2. Parametric analysis of mesh size.

Approximate Mesh Size (mm)	Peak Deformation		Fire Resistance Time		Calculation Time Ratio	Connector Generation
	(mm)	Err%	(h)	Err%		
Test	5.14	-	3.05	-	-	-
50 × 50 ×50	5.83	13.42	1.87	−38.65	0.34	easy
40 × 40 × 50	5.77	12.26	2.49	−18.40	0.90	hard
34 × 34 × 50	5.73	11.48	2.74	−10.17	1.60	hard
25 × 25 × 50	6.07	18.09	3.33	9.18	1.00	easy
20 × 20 × 50	6.20	20.62	3.46	13.61	11.73	hard

Axial deformation and damage of RC columns under fire exposure were used to compare the test results (Test) with the numerical simulation results (Perfect bond means that the reinforcement and concrete showed a perfect bond with no slip between them, whereas Bond-slip means that nonlinear bond-slip behavior between the reinforcement and concrete were considered) to verify the validity of the numerical model.

Figure 6 compares the axial deformation-time curve of the numerical simulation with the test. It can be found that the peak value and the descending value of the bond-slip are closer to the test value than the value of the perfect bond, but the value of the perfect bond is closer to the test value than the value of the bond-slip at the early stage of fire exposure. To quantitatively evaluate the simulation quality of the test, the error analysis is used between the simulated value y_s and the experimental value y_e, and Mean Absolute Percentage Error ($MAPE$) and coefficient of determination R^2 are selected as evaluation indexes [49], where:

$$MAPE = \frac{100\%}{n}\sum_{i=1}^{n}\left|\frac{y_e - y_s}{y_e}\right| \quad (5)$$

$$R^2 = 1 - \frac{\sum_{i=1}^{n}(y_s - \overline{y_e})^2}{\sum_{i=1}^{n}(y_e - \overline{y_e})^2} \quad (6)$$

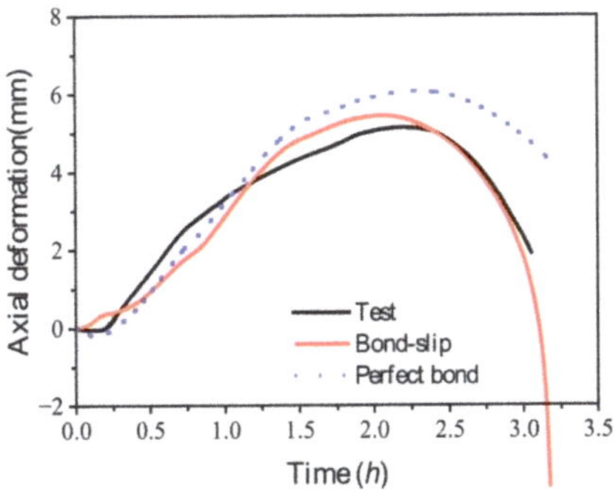

Figure 6. The comparison of the axial deformation-time curve of numerical simulation and test.

Through the error analysis comparing the numerical simulation results and the test values (due to the deformation of 0~0.25, h is an excessive relative error of the test value near 0 mm, so this part of the error is not considered, and the error analysis calculation starts after 0.25 h), it is found that the $MAPE$ of the perfect bond is 34.94% while it is 14.42% of the bond-slip, and the coefficients of determination R^2 of the perfect bond and bond-slip are 0.923 and 0.987, respectively. The above results show that the bond-slip can better reflect the axial deformation of the RC column during fire exposure.

To compare the damage of the test and numerical model, the quality of the numerical simulation is determined by the volume loss ratio and the morphology. In the literature [36], the fire duration was 3.05 h, and the volume loss ratio was 15%. In the numerical model, this paper defines the percentage of the element volume with damage greater than 0.9 as the volume loss ratio. Figure 7 shows the numerically simulated damage diagram with the element of damage factor greater than 0.9 removed and damage after the test. The volume loss ratio of the perfect bond and bond-slip are 6.46 and 14.77%, respectively. Comparing the morphology of the perfect bond and bond-slip with the test, the damage morphology of the bond-slip can better reflect the damage of the test.

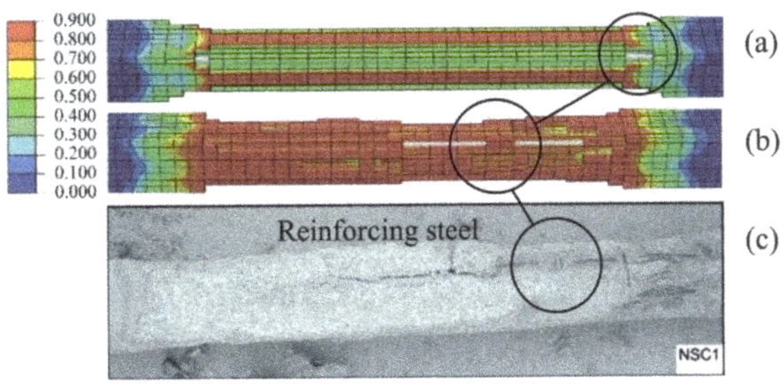

Figure 7. FEM damage and test damage of RC column. (**a**) Perfect bond; (**b**) Bond-slip; (**c**) Test.

To sum up, the bond-slip model can better reflect the axial deformation during fire exposure and damage morphology of RC columns after exposure to fire. On this basis, the axial deformation, stiffness, and fire resistance time of RC columns subjected to cyclic loads during fire exposure can be studied.

3. Performance Evaluation of RC Column Subject to Various Cyclic Loads and Fire

In this section, the response of RC columns under different cyclic loads and exposure to fire are investigated. The deformation, stiffness, and fire resistance time of RC columns are quantitatively analyzed by axial deformation, cyclic load effect coefficient, stiffness degradation ratio, and fire resistance time ratio. Firstly, the response of RC columns subjected to fire and cyclic loads of different amplitudes are studied.

3.1. Working Condition

The vertical cyclic load is taken as 30 s according to the common duration of the earthquake and is applied to different time points (*TP*) during fire. Figure 8 shows the vertical load $\widetilde{P}$ applied during the cyclic load duration, and $\widetilde{P}$ is determined by Equation (7).

$$\widetilde{P} = P \cdot \widetilde{A} \tag{7}$$

where P is the axial compression load, and the cyclic load coefficient $\widetilde{A}$ is shown in Figure 9. The amplitude (A in Figure 9) of the cyclic load is set according to the maximum acceleration under the earthquake action, which is 0.05, 0.10, 0.15, 0.20, 0.30, and 0.40, respectively. The ratio of vertical ground motion to horizontal ground motion is 0.65 [50], and the cyclic load of the reference column is 2 s in a cycle, that is, 15 cycles in 30 s. The working conditions are shown in Table 3. Taking F040T05 as an example, F040 indicates that the cyclic amplitude is 0.4, and T05 indicates that the insertion time point of cyclic load is 0.5 h (abbreviated to h).

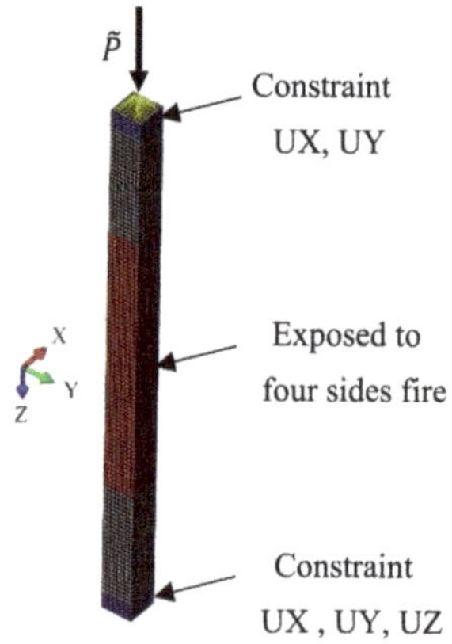

Figure 8. Cyclic load application and boundary conditions.

Table 3. Description of working condition.

A	TP (h)						
	0.0	0.5	1.0	1.5	2.0	2.5	3.0
0.05	F005T00	F005T05	F005T10	F005T15	F005T20	F005T25	F005T30
0.10	F010T00	F010T05	F010T10	F010T15	F010T20	F010T25	F010T30
0.15	F015T00	F015T05	F015T10	F015T15	F015T20	F015T25	F015T30
0.20	F020T00	F020T05	F020T10	F020T15	F020T20	F020T25	F020T30
0.30	F030T00	F030T05	F030T10	F030T15	F030T20	F030T25	F030T30
0.40	F040T00	F040T05	F040T10	F040T15	F040T20	F040T25	F040T30

3.2. Axial Deformation

The axial deformation-time curves of RC columns under different amplitudes are summarized in Figure 10.

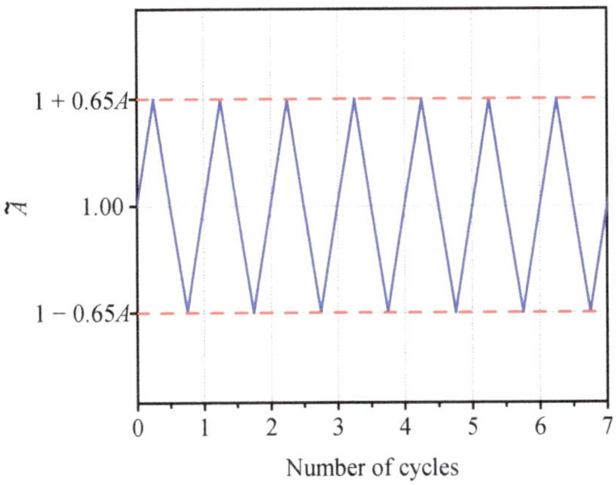

Figure 9. Cyclic load coefficient curve.

In general, cyclic loads with different TP and different amplitudes have various degrees of effect on the axial deformation-time curves of RC columns. The axial deformation-time curves of RC columns are affected within a certain period after the cyclic load is applied, while the subsequent axial deformation-time curve almost overlaps; some RC columns are damaged during the process of cyclic load or after the cyclic load is applied for a while (with the continuous decline of axial deformation as the failure symbol). When $TP = 3.0$ h, except for the condition of $A = 0.05$, the other conditions are damaged during the loading process. To quantitatively analyze the effect of cyclic load on the axial deformation of the RC column during fire exposure, the axial deformation-time curve of the RC column under static load during fire exposure is taken as the benchmark, and the axial deformation-time curve of the RC column under cyclic load during fire exposure is analyzed by error analysis parameter ($MAPE$ is chosen as a representative parameter). The $MAPE$ of each working condition is shown in Figure 11. There is no subsequent axial deformation in the working condition that is damaged during the loading process, so it can be seen from the definition of $MAPE$ that there is no $MAPE$.

As shown in Figure 11, when $TP \leq 1.5$ h and the amplitude is between 0.05~0.40, the effect on the subsequent axial deformation-time curve is less than 10%; when $TP \geq 2.0$ h and the amplitude is greater than 0.3, the cyclic load has a large impact on the subsequent axial deformation-time curve, and the impact increases as the amplitude increases.

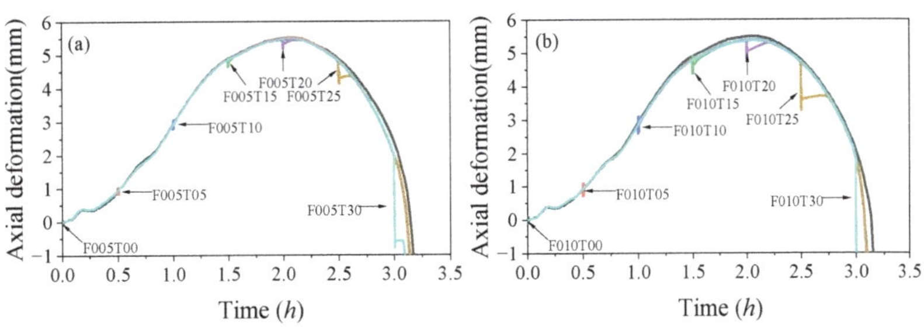

Figure 10. *Cont.*

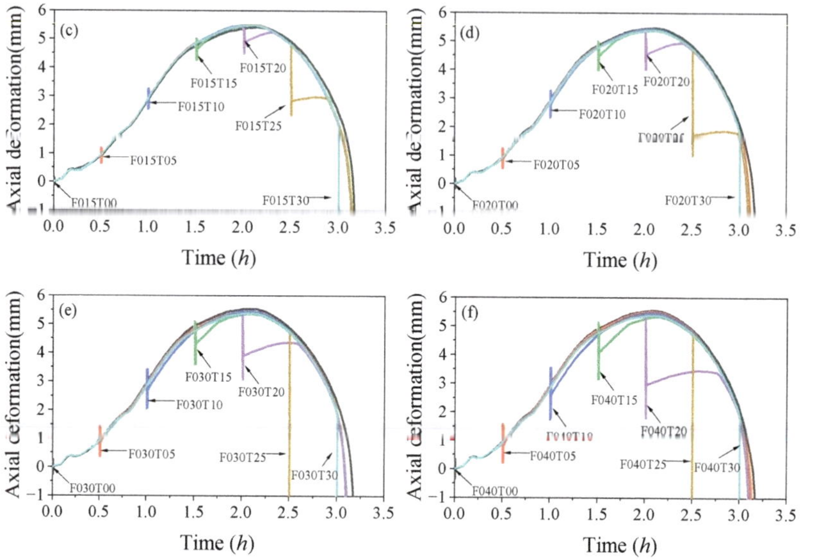

Figure 10. Axial deformation-time curve of different working conditions. (**a**) $A = 0.05$; (**b**) $A = 0.10$; (**c**) $A = 0.15$; (**d**) $A = 0.20$; (**e**) $A = 0.30$; (**f**) $A = 0.40$.

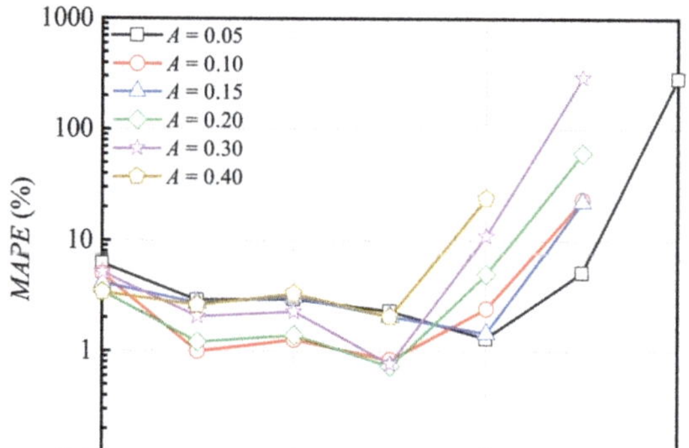

Figure 11. *MAPE* of working conditions.

To quantitatively analyze the effect of the cyclic load on the axial deformation of RC column under cyclic load during fire exposure, the cyclic load effect coefficient η is chosen. η is defined by Equation (1) and the calculation parameters of η are shown in Figure 12, with the deformation Δ_d at t_0 at the end of the application of cyclic load, the deformation Δ_0 at t_0 and the peak deformation Δ_{max} under static load.

$$\eta = \frac{\Delta_0 - \Delta_d}{\Delta_{max}} \times 100\% \tag{8}$$

According to the definition of η, there is no such parameter in the case when the RC column is damaged during the cyclic load. When $\eta > 10\%$, it is considered to have a large impact, and $\eta < 10\%$ means a small impact. The cyclic load effect coefficient η under different TP and different amplitudes is listed in Figure 13. Overall, under the same TP, the larger the amplitude, the greater the effect on the axial deformation of the column. When $TP \leq 1.5$ h, the effect of different amplitude cyclic loads on the axial deformation of the RC column is small, and when $TP > 1.5$ h, the η of the amplitudes of 0.20~0.40 exceeds 10%. The cyclic load with the amplitude of 0.05 exerted a great effect on the axial deformation of the RC column at 3 h, while it had little effect on the axial deformation of the RC column when $TP \leq 2.5$ h.

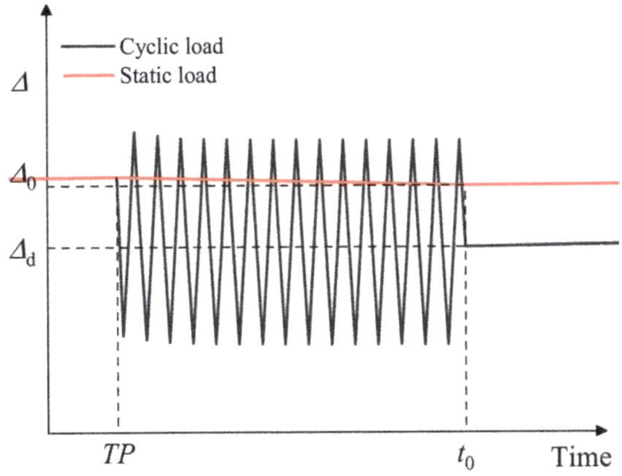

Figure 12. Calculation parameters diagram of cyclic load effect coefficient (η).

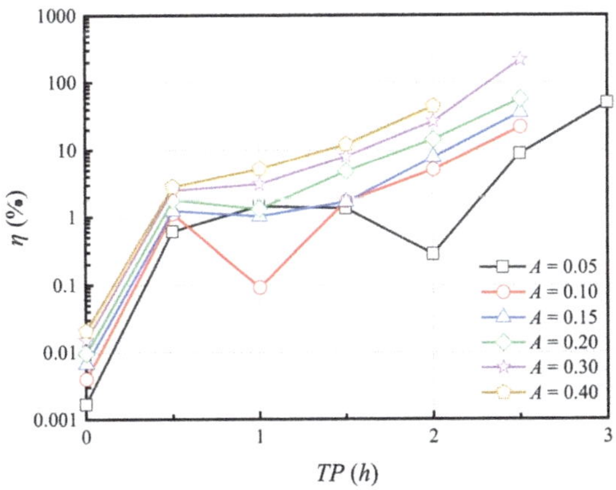

Figure 13. Cyclic load effect coefficient (η) with different working conditions.

3.3. *Stiffness*

The secant stiffness K_i at the maximum displacement of the nth cycle under the vertical cyclic load is calculated according to Equation (9), where ΔP_{i+} and ΔP_{i-} are the difference between the cyclic load and the static load at the corresponding moments of t_{i+} and t_{i-}, respectively; i represents the number i cycle; other parameters are shown in Figure 14. The stiffness degradation ratios under different working conditions are summarized in Figure 15.

$$K_i = \frac{|\Delta P_{i+}| + |\Delta P_{i-}|}{|\Delta_{i+}| + |\Delta_{i-}|} = \frac{2P}{|\Delta_{i+}| + |\Delta_{i-}|} \tag{9}$$

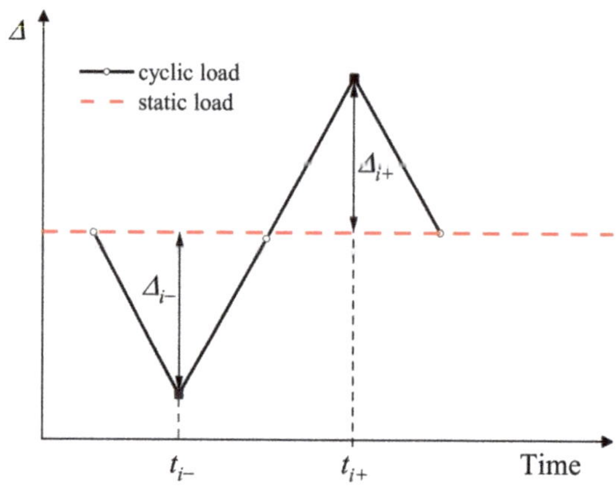

Figure 14. Stiffness degradation parameter diagram.

As shown in Figure 15, when $TP \leq 1.5$ h, the stiffness degradation ratio K_i/K_1 of each cycle in each working condition is between 0.99 and 1.05, indicating that the stiffness degradation caused by cyclic loading has little effect, and the existence of cyclic load leads to slight stiffness strengthening (less than 5%) within a certain cycle number range; when $TP = 2.0$ h, the stiffness of the working conditions with an amplitude greater than or equal to 0.2 dropped sharply in the first and second cycles, and the stiffness decreased slowly after the third cycle. However, with the increase of the cycle, the stiffness decreases slowly for the working conditions with the amplitude less than 0.2, and the cyclic load has little effect on these working conditions. When $TP \geq 2.5$ h, the cyclic load has a significant effect on the stiffness, and the larger the amplitude, the greater the impact. In some cases, failure occurs due to excessive stiffness attenuation, and the larger the amplitude, the earlier the failure.

3.4. Fire Resistance Time and Damage

The time from the start of the fire to the failure of the RC column under different working conditions is defined as the fire resistance time. The fire resistance time of the RC column under static load is $T_{f,0}$, and the fire resistance time under the cyclic load is $T_{f,d}$. The fire resistance time ratios ($T_{f,d}/T_{f,0}$) of different working conditions are displayed in Figure 16. It can be seen that the cyclic load applied within 2 h has little effect on the fire resistance time of the RC column, and the reduction in the fire resistance time is within 5%. When $TP = 2.5$ h, the fire resistance time of the amplitudes of 0.30 and 0.40 is reduced by more than 5%, and the fire resistance time of the amplitude of 0.40 is reduced by more than 20% due to the failure during the loading process; when $TP = 3.0$ h, the fire resistance time of different amplitude conditions is only reduced within 7%, because the cyclic load applied time is close to the failure time of the static load fire condition.

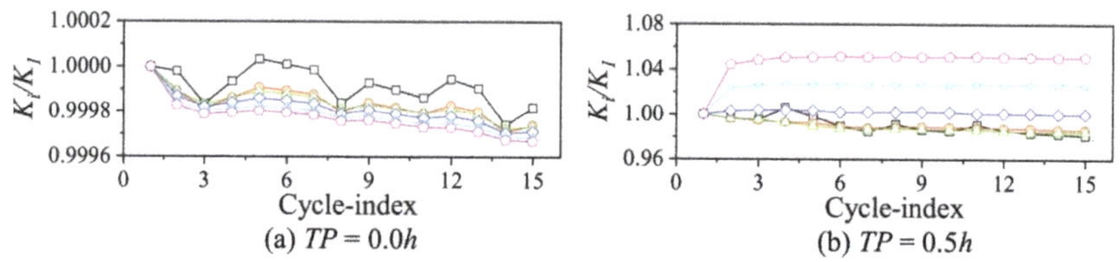

Figure 15. *Cont.*

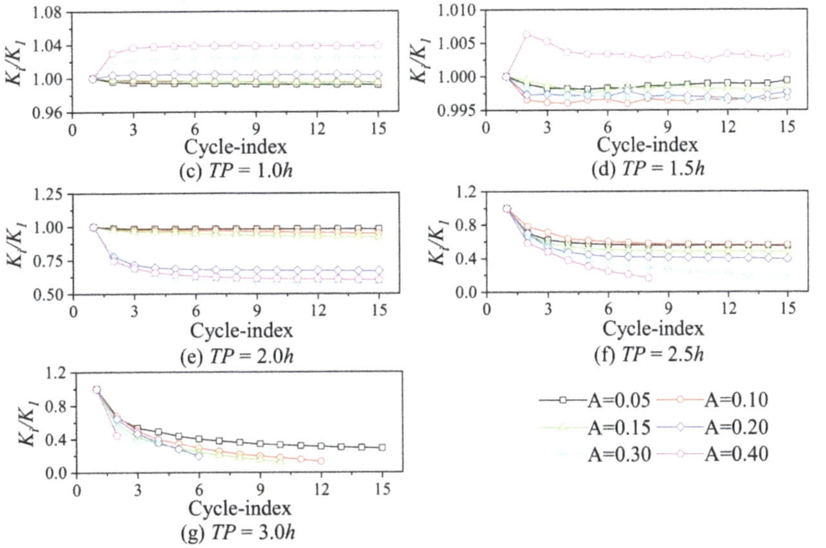

Figure 15. Variation curve of stiffness degradation ratio (K_i/K_1) with cycle index under different working conditions.

To compare and analyze the damage under different working conditions, the damage time consistent with Section 0 is selected. When the fire resistance time is greater than 3.05 h, the damage at 3.05 h is selected, otherwise the damage at failure time is chosen. Figure 17 shows the comparative analysis of damage in working conditions with an amplitude of 0.40, which has a greater impact on the fire resistance time. The elements with damage greater than 0.9 increase with the increases in TP, resulting in a significant decrease in the number of elements that can bear the load. When the TP is greater than 2.5 h, the buckling failure time of the RC column is less than 3.05 h.

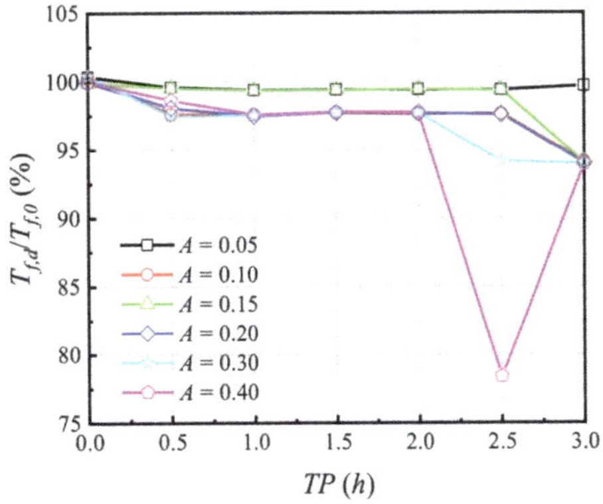

Figure 16. Effect of amplitude and TP of cyclic load on fire resistance time.

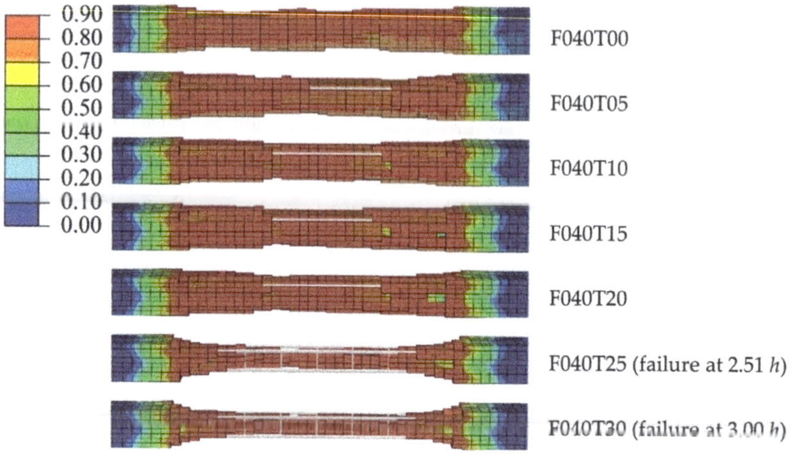

Figure 17. Damage of RC column with different TP and amplitude of 0.40.

3.5. Response Analysis of RC Column under Cyclic Load with Different Cycle Times during Fire Exposure

Through the analysis in Sections 3.2–3.4, when the cycle amplitude is 0.4, the deformation, stiffness, and fire resistance time of the RC column under different TPs are significantly different. Therefore, when the cyclic amplitude is 0.4, the cyclic load duration is 30 s, and the number of cycles (*Noc*) is taken as 15, 30, 60, and 150 times, respectively, the effects of cyclic loads on the deformation, stiffness, and fire resistance time of RC columns under different TPs are investigated. The representative parameters are the cyclic load effect coefficient η, $MAPE$, stiffness degradation ratio (represented by the ratio K_{-1}/K_1 of the last cyclic stiffness K_{-1} to the first cyclic stiffness K_1, the same below), and the fire resistance time ratio $T_{f,d}/T_{f,0}$). The effect of the number of cycles on the deformation, stiffness, and fire resistance time ratio of the RC column during fire exposure is shown in Figure 18. Different cycle times have similar effects on the axial deformation and subsequent axial deformation, stiffness, and fire resistance time ratio of the RC column during the cyclic load duration and fire exposure. The later the cyclic load is added, the greater the effect on the axial deformation. When $TP \leq 1.5$ h, different cycle times have little effect on the stiffness of the RC column during fire exposure. When $TP > 1.5$ h, the more cycles, the greater the effect of cyclic load on the stiffness of the RC column. When the cyclic load is added for the same time, the fire resistance time of different cycle times has little difference.

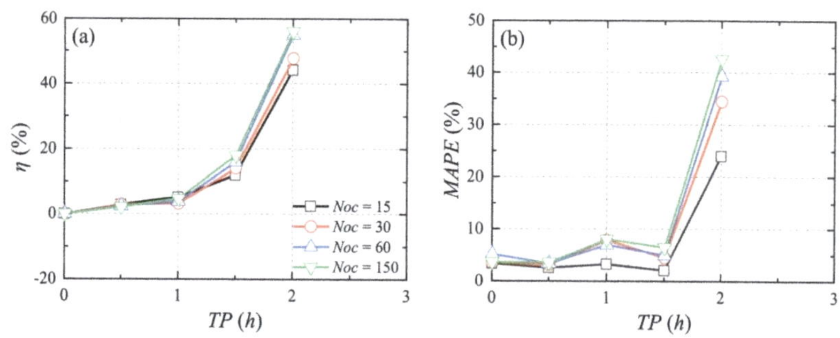

Figure 18. *Cont.*

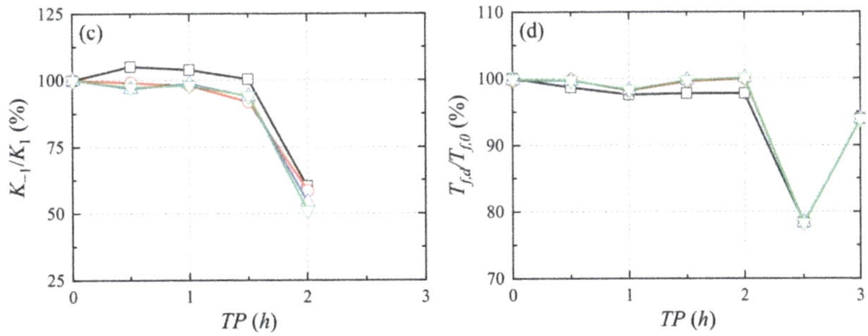

Figure 18. Impact of cyclic loads with different number of cycles (Noc) on the response of RC column during fire exposure: (**a**) cyclic load effect coefficient η; (**b**) $MAPE$; (**c**) stiffness degradation ratio; (**d**) fire resistance time ratio.

4. Multi-Performance Comparison of Various RC Columns Subject to Coupling Effect of Cyclic Loads and Fire

Given the obtained results for one typical RC column, the multi-performance evaluation is further performed for a series of RC columns with various application scenarios. The section size, protective cover thickness, longitudinal reinforcing ratio, tie spacing, and axial load ratio of the RC columns are considered to simulate the inherent characteristics of columns. The cyclic amplitude and circles of the cyclic load are assumed as 0.40 and 15 times for reference. The η, $MAPE$, stiffness degradation ratio K_{-1}/K_1, and fire resistance time ratio $T_{f,d}/T_{f,0}$ were used as representative parameters to investigate the influence of cyclic load on the deformation and damage of RC column with different characteristics. Table 4 summarizes the working conditions of reference columns and columns with different characteristics. For all the columns, the material mechanical properties, thermal mechanical properties, and boundary conditions are the same. Table 4 summarizes the fire resistance time of the reference column and different column characteristics. When improving the section size, cover thickness, and longitudinal reinforcement ratio or reducing the axial load ratio and tie spacing, the fire resistance time of RC columns increases. The fire resistance time of RC columns with different characteristics is consistent with previous research [31,51]. When confinement increases by 50%, the fire resistance time of the RC column increases by 6 and 3.5 to 12% respectively in Table 4 and previous research [31].

Table 4. Column characteristic and fire resistance time.

Description	Section Size $a \times b$ (mm)	Cover Thickness c (mm)	Longitudinal Reinforcement Ratio ρ_l	Tie Spacing sp (mm)	Axial Load Ratio p	Fire Resistance Time (h)
Reference column	203.0 × 203.0	50	3.05%	200	0.4	3.20
$a \times b$	304.5 × 304.5	50	3.05%	200	0.4	4.63
	406.0 × 406.0	50	3.05%	200	0.4	9.97
c	203.0 × 203.0	30	3.05%	200	0.4	1.89
	203.0 × 203.0	40	3.05%	200	0.4	2.86
ρ_l	203.0 × 203.0	50	0.76%	200	0.4	0.86
	203.0 × 203.0	50	1.72%	200	0.4	2.08
	203.0 × 203.0	50	2.47%	200	0.4	2.81
	203.0 × 203.0	50	4.76%	200	0.4	4.57
sp	203.0 × 203.0	40	3.05%	100	0.4	3.52
	203.0 × 203.0	40	3.05%	160	0.4	3.33
	203.0 × 203.0	40	3.05%	320	0.4	3.22

Table 4. *Cont.*

Description	Section Size $a \times b$ (mm)	Cover Thickness c (mm)	Longitudinal Reinforcement Ratio ρ_l	Tie Spacing sp (mm)	Axial Load Ratio p	Fire Resistance Time (h)
p	203.0 × 203.0	50	3.05%	200	0.2	7.90
	203.0 × 203.0	50	3.05%	200	0.3	4.87
	203.0 × 203.0	50	3.05%	200	0.5	2.29
	203.0 × 203.0	50	3.05%	200	0.6	1.77

4.1. Section Size

The effects of cyclic loading on the deformation, stiffness, and fire resistance time of RC columns with different cross-sectional sizes during fire exposure are shown in Figure 19. The larger the cross-sectional size is, the less the cyclic loads affect the RC columns. When $TP < 2.0$ h, the cyclic load has little effect on the axial deformation of RC columns with different cross-sectional sizes during fire exposure, and the cyclic load has a strengthening effect on the stiffness of RC columns with different cross-sectional sizes; when $TP \geq 2.0$ h, the larger the section size, the smaller the effect of the cyclic load on the subsequent axial deformation, and the effect of the cyclic load on the stiffness of the RC column decreases with the increase in the section size. For an RC column with a section size of 406 mm × 406 mm, when the cyclic load is applied within 3 h, the cyclic load has little effect on the stiffness of the RC column. Except for some working conditions that occur failure during cyclic load, the effect of the cyclic load on the fire resistance time of RC columns with a large section size is smaller under the same TP.

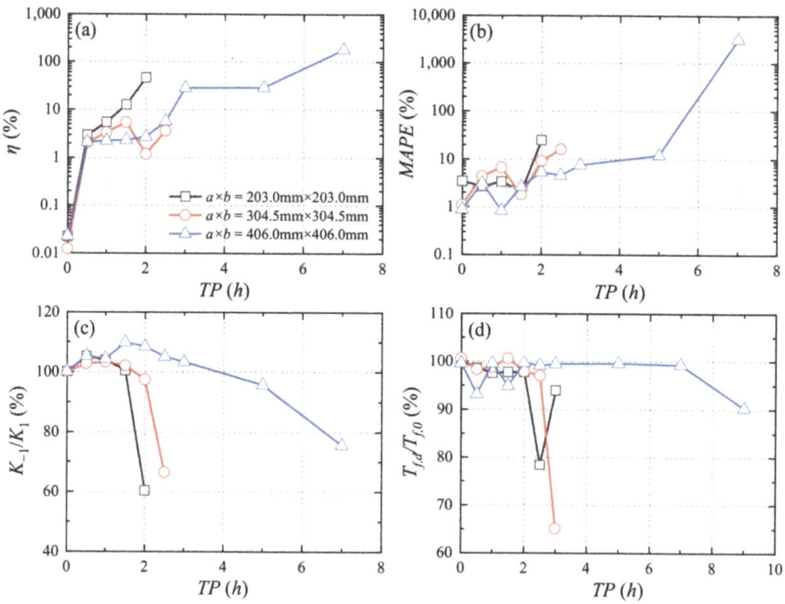

Figure 19. Impact of cyclic load on the response of different section size ($a \times b$) of RC column during fire exposure: (**a**) cyclic load effect coefficient η; (**b**) $MAPE$; (**c**) stiffness degradation ratio; (**d**) fire resistance time ratio.

4.2. Cover Thickness

The effects of cyclic loading on the deformation, stiffness, and fire resistance time of RC columns with different cover thicknesses during fire exposure are shown in Figure 20. The larger the cover thickness, the smaller the effect of the cyclic load on the axial deformation of the RC column, and the greater effect on the fire resistance time. When $TP \leq 1$ h, the cyclic load can improve the stiffness of RC columns with the cover thickness of 40 mm and 50 mm during fire exposure. When $TP > 1$ h, the thicker the cover thickness, the smaller the effect of cyclic load on the stiffness of the RC column during fire exposure, but when the cyclic load is applied at the same TP, the stiffness ratios of different cover thickness shows little difference.

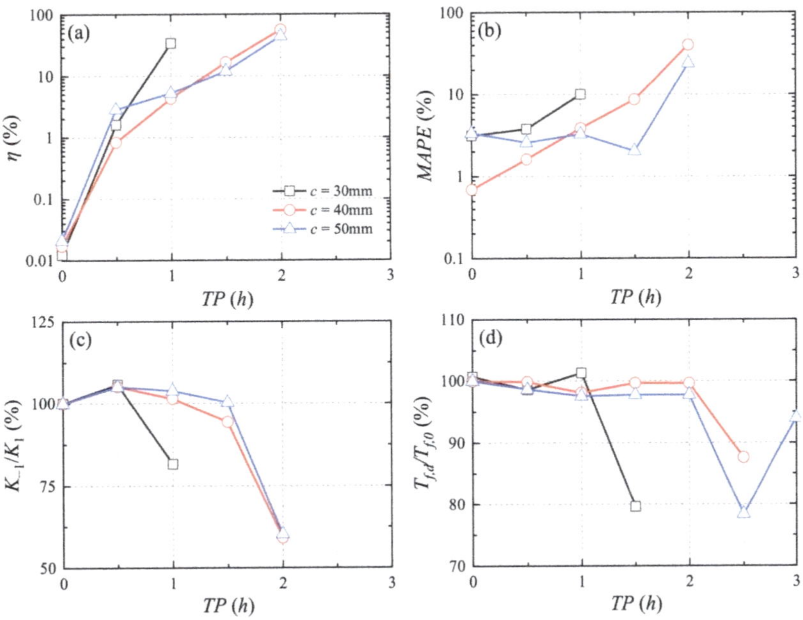

Figure 20. Impact of cyclic load on the response of different cover thickness (c) of RC column during fire exposure: (**a**) cyclic load effect coefficient η; (**b**) $MAPE$; (**c**) stiffness degradation ratio; (**d**) fire resistance time ratio.

4.3. Longitudinal Reinforcement Ratio

The effects of the cyclic loads on the axial deformation, stiffness, and fire resistance time of RC columns with different reinforcement ratios during fire exposure are shown in Figure 21. Because larger reinforcement ratios can supply a larger capacity to RC columns during the same fire conditions, there is a smaller effect of the cyclic load on the axial deformation, subsequent axial deformation, stiffness ratio, and fire resistance time of the RC column with larger reinforcement ratios under the cyclic load and fire exposure at the same TP.

4.4. Tie Spacing

The effects of the cyclic loads on the deformation, stiffness, and fire resistance time of RC columns with different tie spacings during fire exposure are shown in Figure 22. Under the same TP, the effects of cyclic loads on the axial deformation, subsequent axial deformation, and stiffness of RC columns with different tie spacings are not significantly different. Except for the RC column with a 100 mm tie spacing, the effect of the cyclic load on the fire resistance time ratio of RC columns with different tie spacing shows little difference, and the fire resistance time is reduced by more than 20% when the cyclic load is applied for 2.5 h. When the cyclic load is applied for 3 h, the fire resistance time of the RC column with 100 mm tie spacing decreases significantly.

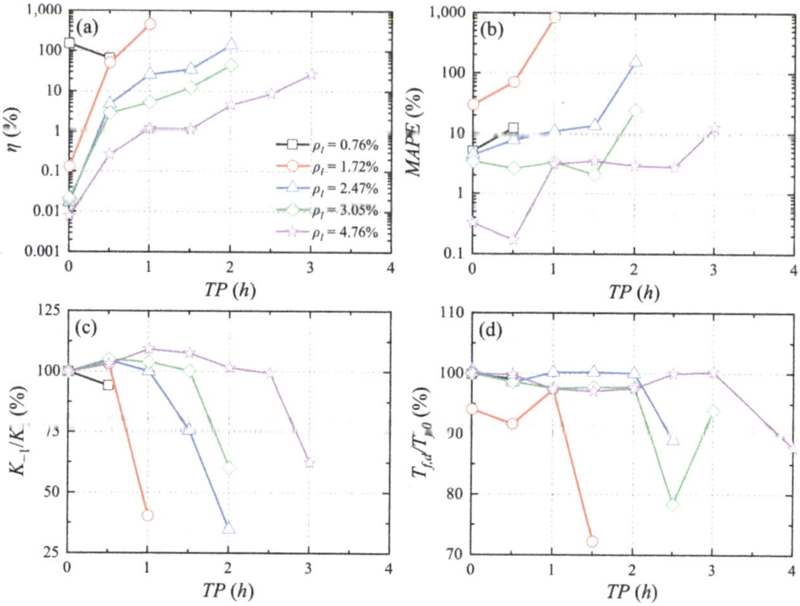

Figure 21. Impact of cyclic load on the response of different longitudinal reinforcement ratio (ρ_l) of RC column during fire exposure: (**a**) cyclic load effect coefficient η; (**b**) $MAPE$; (**c**) stiffness degradation ratio; (**d**) fire resistance time ratio.

4.5. Axial Load Ratio

The effects of cyclic loads on the deformation, stiffness, and fire resistance time of RC columns with different axial compression ratios during fire exposure are shown in Figure 23. The larger the axial compression ratio, the smaller the effect of cyclic load on the axial deformation, subsequent axial deformation, stiffness, and fire resistance time of the RC column with the same TP during fire exposure. When the axial compression ratio is 0.2 and 0.3, the cyclic load has little effect on the fire resistance time of the RC column, and the fire resistance time is more than 4 h, which is significantly improved compared to the reference column.

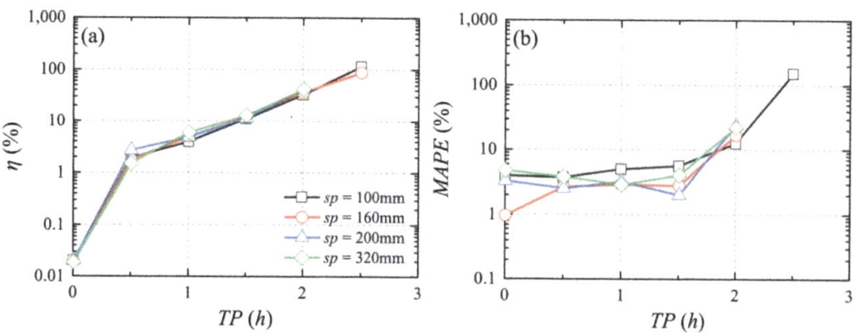

Figure 22. *Cont.*

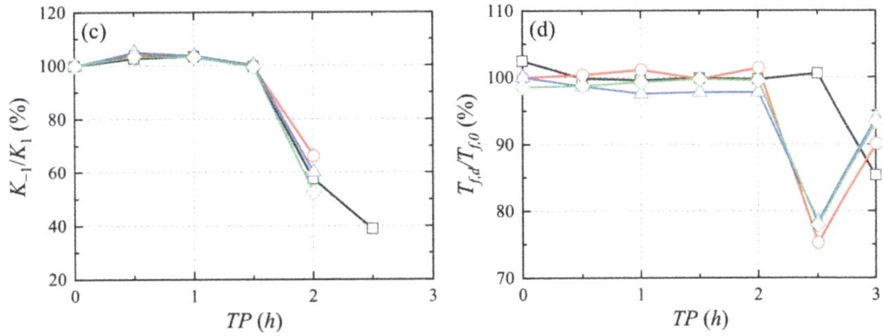

Figure 22. Impact of cyclic load on the response of RC column with different tie spacing (sp) during fire exposure: (**a**) cyclic load effect coefficient η; (**b**) $MAPE$; (**c**) stiffness degradation ratio; (**d**) fire resistance time ratio.

4.6. Stage Division of RC Columns under Fire and Cyclic Loads

The normalized time ratio $TP/T_{f,0}$ is obtained by comparing the cyclic load application time TP with the fire resistance time $T_{f,0}$ of the RC column of the same column characteristic, and the time-normalized data of all working conditions in Sections 4.1–4.5 are summarized to Figure 24. As seen in Figure 24, the effect of the cyclic load on the response of RC columns during fire exposure can be divided into four stages.

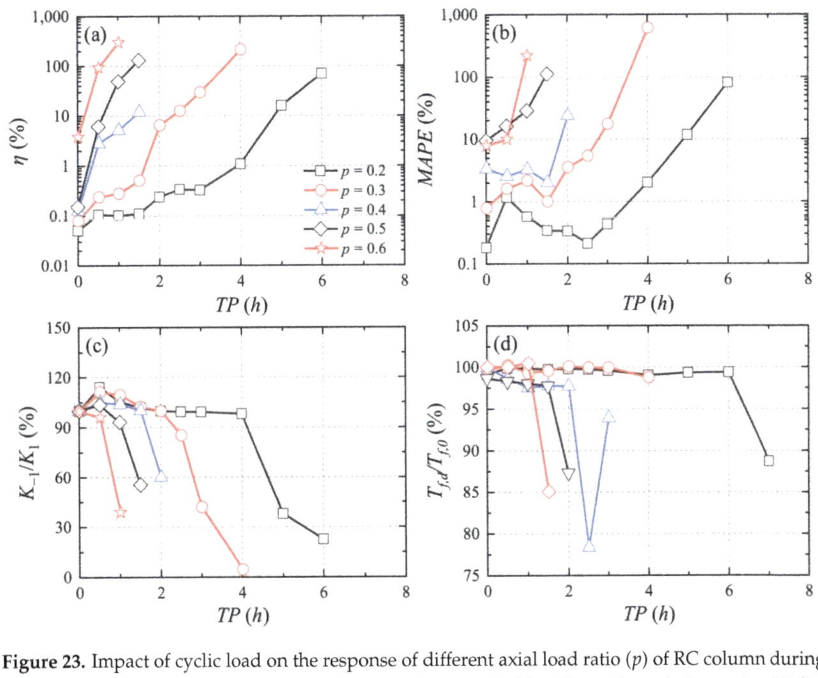

Figure 23. Impact of cyclic load on the response of different axial load ratio (p) of RC column during fire exposure: (**a**) cyclic load effect coefficient η; (**b**) $MAPE$; (**c**) stiffness degradation ratio; (**d**) fire resistance time ratio.

The first stage is the low-impact stage, and $TP/T_{f,0} \in [0, 0.4)$. In this stage, the effects of cyclic loads on the axial deformation, subsequent deformation, stiffness ratio, and fire resistance time of RC columns are predominantly within 10%. The second stage is the medium impact stage, and $TP/T_{f,0} \in [0.4, 0.6)$. In this stage, the effect of the cyclic load on the axial deformation and subsequent deformation of RC columns are chiefly in the range of 1 to 100%, while the effect on the stiffness is 10~60% and the effect on the fire resistance time is slight. The third stage is the high impact stage, and $TP/T_{f,0} \in [0.6, 0.8)$. In this stage, the effect of the axial deformation and subsequent deformation of RC columns are 10–1000%, the effect on the stiffness is more than 20–80%, and the effect on the fire resistance is 0–40%. The fourth stage is the failure stage, and $TP/T_{f,0} \in [0.8, 1]$. In this stage, the cyclic load has a great effect (more than 100%) on the axial deformation, subsequent deformation, and stiffness of the RC column; the failure of the RC column occurs when the cyclic load is applied or after being applied for a few minutes; the fire resistance time of the RC column is about 0–15%.

When designing an RC column under cyclic load and fire exposure, we need to avoid the excessive effect of the cyclic load on the deformation and stiffness of the RC column during fire exposure. Based on the analysis of the above stages, the cyclic load should be applied during the low impact stage within the target fire resistance time ($T_{f,0}$). The fire resistance time is set to $T'_{f,0}$. Taking the upper bound ($TP = 0.4T'_{f,0}$) of the low impact stage, and taking the maximum TP as $T_{f,0}$, then the minimum fire resistance time of the RC column under cyclic load is $T'_{f,0} = 2.5T_{f,0}$; that is, the minimum fire resistance time of the RC column under cyclic load should be designed to be 2.5 times the fire resistance time under static load.

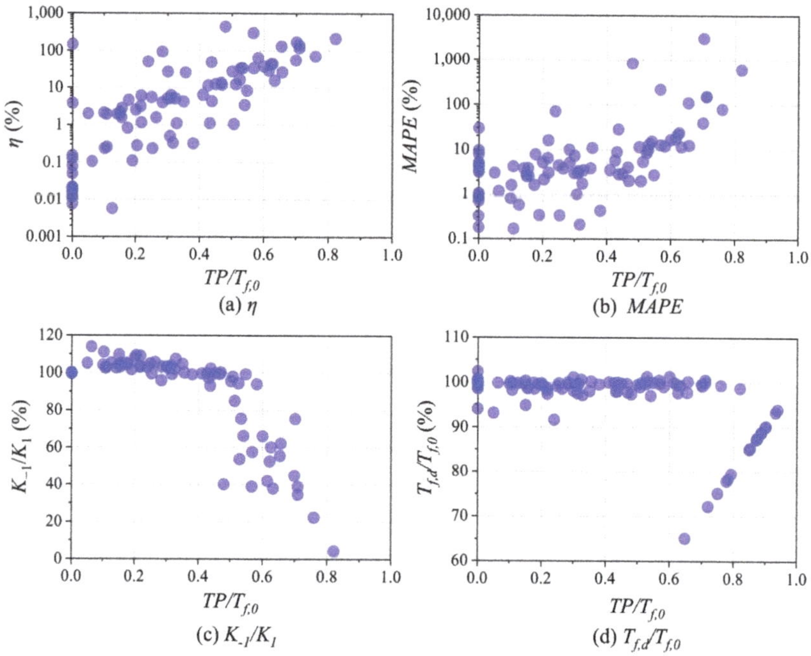

Figure 24. The effect of cyclic load on different columns during fire exposure: (**a**) cyclic load effect coefficient η; (**b**) $MAPE$; (**c**) stiffness degradation ratio; (**d**) fire resistance time ratio.

5. Conclusions

In this study, after the FE model effectively simulated the axial deformation during fire exposure and damage morphology of RC columns after exposure to fire, multiple dynamic performances of RC columns were evaluated in the simultaneously coupling effect of fire and cyclic loads, where the deformation and damage regularity of RC columns were systematically quantified by considering different cyclic load characteristics and structural features. The main conclusions can be summarized as follows:

1. The effect of cyclic loads with different amplitudes at different time points on the deformation-time curve, stiffness degradation ratio, and fire resistance time of RC columns increase with increase in cyclic load addition time points and amplitudes. For the benchmark column, when the cyclic load application time is less than 2 h and the cyclic amplitude is less than 0.30, the cyclic load has little effect on the response of the RC column. The cyclic load with a large amplitude applied later results in buckling failure.
2. Different number of cycles have similar influences on the axial deformation, stiffness degradation ratio, the fire resistance time of RC columns during fire exposure, and axial deformation; stiffness degradation ratio of RC columns is greatly affected by the applied time point of cyclic loading.
3. The reinforcement ratio and the cover thickness have a significant effect on the deformation, stiffness degradation ratio, and fire resistance time of the reinforced concrete column. Increasing the cover thickness, the reinforcement ratio, and the section size can effectively reduce the effect of the cyclic load on the RC column during fire exposure.
4. The effect of cyclic loads on RC columns with different column characteristics during fire exposure can be divided into four stages: low impact stage, medium impact stage, high impact stage, and failure stage. To avoid the excessive deformation and stiffness of RC columns under the cyclic load and fire exposure, it is recommended that the minimum fire resistance time of RC columns under cyclic load be designed to be 2.5 times the fire resistance time under static load. The pertinent experiments will be conducted in the future.

Author Contributions: Conceptualization, Y.J. and Q.C.; methodology, Y.J.; software, Y.J.; validation, Y.J.; formal analysis, Y.J.; investigation, Y.J.; writing—original draft preparation, Y.J.; writing—review and editing, Y.J. and Q.C.; supervision, Q.C. All authors have read and agreed to the published version of the manuscript.

Funding: This research was funded by Basic Research Project of State Key Laboratory of Ministry of Science and Technology (grant number SLDRCE19A-02).

Institutional Review Board Statement: Not applicable.

Informed Consent Statement: Not applicable.

Data Availability Statement: Not applicable.

Acknowledgments: Portions of this research were conducted with the computing resources provided by Tongji University.

Conflicts of Interest: The authors declare no conflict of interest.

Nomenclature

ρ	density
c_p	specific heat compacity
λ	thermal conductivity
α	thermal expansion coefficient
f	strength
ε	strain
E	Young's modulus
f_t	tensile strength of concrete
τ	bond strength
S	slip
d	diameter of the reinforcing steel
q_{ch}	convective heat transfer heat flux density
q_{rh}	radiant heat transfer heat flux density
q_h	composite heat transfer heat flux density
T_{wh}	surface temperature of the wall
T_{fh}	temperature of the thermal fluid
T_{am}	ambient temperature
h_{ch}	convective heat transfer coefficient
ε_h	emissivity
C_b	Stefan-Boltzmann constant
T	temperature
y_s	simulated value
y_e	experimental value
R^2	coefficient of determination
$\widetilde{P}$	cyclic load

P	axial compression load
$\widetilde{A}$	cyclic load coefficient
η	cyclic load effect coefficient
Δ	deformation
K	stiffness
T_f	fire resistance time
Noc	the number of cycles
$a \times b$	section size
c	cover thickness
ρ_l	longitudinal reinforcement ratio
sp	tie spacing
p	axial load ratio
	Subscript
d	under cyclic load
0	at ambient temperature or under static load
y	yield
u	ultimate or peak
T	at elevated temperatures
s	reinforcing steel
c	concrete or compressive
	Abbreviation
RC	reinforced concrete
TP	time point
A	amplitude
$MAPE$	Mean Absolute Percentage Error
h	hours/hour

References

1. Fu, B.; Jiang, H.; Wu, T. Experimental study of seismic response reduction effects of particle damper using substructure shake table testing method. *Struct. Control Health Monit.* **2019**, *26*, e2295. [CrossRef]
2. Jiang, H.; Fu, B.; Lu, X.; Chen, L. Seismic Damage Assessment of RC Members by a Modified Park-Ang Model. *Adv. Struct. Eng.* **2015**, *18*, 353–364. [CrossRef]
3. Cherif, S.; Chourak, M.; Abed, M.; Douiri, A. Potential Seismic Damage Assessment of Residential Buildings in Imzouren City (Northern Morocco). *Buildings* **2018**, *8*, 179. [CrossRef]
4. Fujii, K. Pushover-Based Seismic Capacity Evaluation of Uto City Hall Damaged by the 2016 Kumamoto Earthquake. *Buildings* **2019**, *9*, 140. [CrossRef]
5. Xiao, J.; Xie, Q.; Xie, W. Study on high-performance concrete at high temperatures in China (2004–2016)—An updated overview. *Fire Saf. J.* **2018**, *95*, 11–24. [CrossRef]
6. Hassan, A.; Arif, M.; Shariq, M. Mechanical Behaviour and Microstructural Investigation of Geopolymer Concrete After Exposure to Elevated Temperatures. *Arab. J. Sci. Eng.* **2020**, *45*, 3843–3861. [CrossRef]
7. Liu, Y.Z.; Jin, B.; Huo, J.S.; Li, Z. Effect of microstructure evolution on mechanical behaviour of concrete after high temperatures. *Mag. Concr. Res.* **2018**, *70*, 770–784. [CrossRef]
8. Hertz, K.D. Concrete strength for fire safety design. *Mag. Concr. Res.* **2005**, *57*, 445–453. [CrossRef]
9. Naser, M.Z. Heuristic machine cognition to predict fire-induced spalling and fire resistance of concrete structures. *Automat. Constr.* **2019**, *106*, 102916. [CrossRef]
10. Liu, J.; Tan, K.H.; Yao, Y. A new perspective on nature of fire-induced spalling in concrete. *Constr. Build. Mater.* **2018**, *184*, 581–590. [CrossRef]
11. Zhang, B.; Zhu, H.; Chen, J.; Yang, O. Evaluation of bond performance of corroded steel bars in concrete after high temperature exposure. *Eng. Struct.* **2019**, *198*, 109479. [CrossRef]
12. Caetano, H.; Ferreira, G.; Rodrigues, J.P.C.; Pimienta, P. Effect of the high temperatures on the microstructure and compressive strength of high strength fibre concretes. *Constr. Build. Mater.* **2019**, *199*, 717–736. [CrossRef]
13. Liu, Z.; Wang, Y.; Li, G.; Jiang, J.; Fu, C. Mechanical behavior of cross-shaped steel reinforced concrete columns after exposure to high temperatures. *Fire Saf. J.* **2019**, *108*, 102857. [CrossRef]
14. Levenson, L.M. Residential water heater damage and fires following the Loma Prieta and Big Bear Lake earthquakes. *Earthq. Spectra* **1992**, *8*, 595–603. [CrossRef]

15. Trifunac, M.D.; Todorovska, M.I. The Northridge, California, earthquake of 1994: Fire ignition by strong shaking. *Soil Dyn. Earthq. Eng.* **1998**, *17*, 165–175. [CrossRef]
16. Scawthorn, C. Fire following the Northridge and Kobe earthquakes. In Proceedings of the Thirteenth Meeting of the UJNR Panel on Fire Research and Safety, Gaithersburg, MD, USA, 13–20 March 1996.
17. Shah, A.H.; Sharma, U.K.; Bhargava, P. Outcomes of a major research on full scale testing of RC frames in post earthquake fire. *Constr. Build. Mater.* **2017**, *155*, 1224–1241. [CrossRef]
18. Wen, B.; Wu, B.; Niu, D. Post-earthquake fire performance of reinforced concrete columns. *Struct. Infrastruct. E.* **2016**, *12*, 1106–1126. [CrossRef]
19. Wang, J.; Zhang, X.; Kunnath, S.; He, J.; Xiao, Y. Post-Earthquake Fire Resistance and Residual Seismic Capacity of Reinforced Concrete Columns. *ACI Struct. J.* **2021**, *118*, 123–135. [CrossRef]
20. Ni, S.; Birely, A.C. Post-fire seismic behavior of reinforced concrete structural walls. *Eng. Struct.* **2018**, *168*, 163–178. [CrossRef]
21. Wang, Y.; Kodur, V.; Fu, C.; Liu, C.; Zhou, H.; Naser, M. Seismic Performance of Reinforced Concrete Frame Joints after Exposure to Fire. *ACI Struct. J.* **2021**, *118*, 3–14. [CrossRef]
22. Liu, X.; Gernay, T.; Li, L.; Lu, Z. Seismic performance of post-fire reinforced concrete beam-column joints strengthened with steel haunch system. *Eng. Struct.* **2021**, *234*, 111978. [CrossRef]
23. Lu, Z.; Chen, Y.; Li, L.; Liu, X.; Wei, K. Experimental Study on Seismic Behavior of Plane and Spatial Concrete Beam-column Joints After Exposure to Fire. *J. Tongji Univ. Nat. Sci.* **2020**, *48*, 340–348. [CrossRef]
24. Jin, L.; Li, X.; Zhang, R.; Du, X. Meso-scale modelling the post-fire seismic behavior of RC short columns. *Eng. Fail. Anal.* **2021**, *120*, 105117. [CrossRef]
25. Demir, U.; Goksu, C.; Binbir, E.; Ilki, A. Impact of time after fire on post-fire seismic behavior of RC columns. *Structures* **2020**, *26*, 537–548. [CrossRef]
26. Han, L.; Zhou, K.; Tan, Q.; Song, T. Performance of steel reinforced concrete columns after exposure to fire: Numerical analysis and application. *Eng. Struct.* **2020**, *211*, 110421. [CrossRef]
27. Chen, Y.; Chang, Y.; Yao, G.C.; Sheu, M. Experimental research on post-fire behaviour of reinforced concrete columns. *Fire Saf. J.* **2009**, *44*, 741–748. [CrossRef]
28. Tan, K.H.; Yao, Y. Fire Resistance of Reinforced Concrete Columns Subjected to 1-, 2-, and 3-Face Heating. *J. Struct. Eng.* **2004**, *130*, 1820–1828. [CrossRef]
29. Bengar, H.A.; Shahmansouri, A.A. Post-fire behavior of unconfined and steel tube confined rubberized concrete under axial compression. *Structures* **2021**, *32*, 731–745. [CrossRef]
30. Jaszczak, B.; Kuczma, M.; Szymkuć, W. Comparison of the load-bearing capacity of reinforced concrete columns under fire conditions using the method A, zone method and isotherm 500 method. *Fire Saf. J.* **2021**, *124*, 103396. [CrossRef]
31. Shah, A.H.; Sharma, U.K. Fire resistance and spalling performance of confined concrete columns. *Constr. Build. Mater.* **2017**, *156*, 161–174. [CrossRef]
32. Pul, S.; Atasoy, A.; Senturk, M.; Hajirasouliha, I. Structural performance of reinforced concrete columns subjected to high-temperature and axial loading under different heating-cooling scenarios. *J. Build. Eng.* **2021**, *42*, 102477. [CrossRef]
33. Saljoughian, A.; Mostofinejad, D. Grooving methods in square RC columns strengthened with longitudinal CFRP under cyclic axial compression. *Eng. Struct.* **2018**, *174*, 724–735. [CrossRef]
34. Shao, Y.; Zhu, Z.; Mirmiran, A. Cyclic modeling of FRP-confined concrete with improved ductility. *Cem. Concr. Comp.* **2006**, *28*, 959–968. [CrossRef]
35. Lam, L.; Teng, J.G. Stress–strain model for FRP-confined concrete under cyclic axial compression. *Eng. Struct.* **2009**, *31*, 308–321. [CrossRef]
36. Raut, N.K.; Kodur, V.K.R. Response of High-Strength Concrete Columns under Design Fire Exposure. *J. Struct. Eng.* **2011**, *137*, 69–79. [CrossRef]
37. Handoo, S.K.; Agarwal, S.; Agarwal, S.K. Physicochemical, mineralogical, and morphological characteristics of concrete exposed to elevated temperatures. *Cem. Concr. Res.* **2002**, *32*, 1009–1018. [CrossRef]
38. Du, H.; Jiang, Y.; Liu, G.; Yan, R. CT image-based analysis on the defect of polypropylene fiber reinforced high-strength concrete at high temperatures. *J. Wuhan Univ. Technol. Mater. Sci. Ed.* **2017**, *32*, 898–903. [CrossRef]
39. *EN1994-1-2: 2005*; Eurocode 4-Design of Composite Steel and Concrete Structures Part 1-2: General Rules—Structural Fire Design. European Committee for Standardization: Brussels, Belgium, 2005.
40. Lie, T.T.; Irwin, R.J. Fire Resistance of Rectangular Steel Columns Filled with Bar-Reinforced Concrete. *J. Struct. Eng.* **1995**, *121*, 797–805. [CrossRef]
41. *EN1992-1-2: 2004*; Eurocode 2: Design of Concrete Structures. European Committee for Standardization: Brussels, Belgium, 2004.
42. Lee, J.; Fenves, G.L. Plastic-damage model for cyclic loading of concrete structures. *J. Eng. Mech.* **1998**, *124*, 892–900. [CrossRef]
43. Lubliner, J.; Oliver, J.; Oñate, S.O. A plastic-damage model for concrete. *Int. J. Solids Struct.* **1989**, *25*, 299–326. [CrossRef]
44. *GB50010-2010*; Code for Design of Concrete Structures. China Architecture & Building Press: Beijing, China, 2011.
45. *ANSI/AISC 360-10*; Specification for Structural Steel Buildings. American Institute of Steel Construction: Chicago, IL, USA, 2010.
46. Fu, B.; Chen, S.; Liu, X.; Feng, D. A probabilistic bond strength model for corroded reinforced concrete based on weighted averaging of non-fine-tuned machine learning models. *Constr. Build. Mater.* **2022**, *318*, 125767. [CrossRef]

47. Özkal, F.M.; Polat, M.; Yağan, M.; Öztürk, M.O. Mechanical properties and bond strength degradation of GFRP and steel rebars at elevated temperatures. *Constr. Build. Mater.* **2018**, *184*, 45–57. [CrossRef]
48. Bergman, T.L.; Lavine, A.S.; Incropera, F.P.; Dewitt, D.P. *Introduction to Heat Transfer*; John Wiley & Sons: Hoboken, NJ, USA, 2011.
49. Kim, S.; Kim, H. A new metric of absolute percentage error for intermittent demand forecasts. *Int. J. Forecast.* **2016**, *32*, 669–679. [CrossRef]
50. GB50011-2010; Code for Seismic Design of Buildings. China Architecture & Building Press: Beijing, China, 2010.
51. Gernay, T. Fire resistance and burnout resistance of reinforced concrete columns. *Fire Saf. J.* **2019**, *104*, 67–78. [CrossRef]

buildings

MDPI

Article

Seismic Fragility Assessment of Cable-Stayed Bridges Crossing Fault Rupture Zones

Junjun Guo [1], Yitong Gu [1], Weihong Wu [2], Shihyu Chu [3] and Xinzhi Dang [1,*]

1 State Key Laboratory of Disaster Reduction in Civil Engineering, Tongji University, 1239 Siping Road, Shanghai 200092, China; guojj@tongji.edu.cn (J.G.); guyitong@tongji.edu.cn (Y.G.)

2 Gansu Transportation Planning Survey and Design Institute Co., Ltd., Lanzhou 730010, China; wwh_3040@163.com

3 Department of Civil Engineering, National Cheng Kung University, No. 1 University Road, Tainan 701, Taiwan; sychu@mail.ncku.edu.tw

* Correspondence: 021tjdangxz@tongji.edu.cn

Abstract: Current studies lack probabilistic evaluations on the performance of fault-crossing bridges. This paper conducts seismic fragility analyses to evaluate the fragility of cable-stayed bridges with the effects of fault ruptures. Synthetic across-fault ground motions are generated using existing simulation methods for the low-frequency pulses and high-frequency residuals. Incremental dynamic analysis is utilized to generate the seismic responses of the bridge. The optimal intensity measure (IM) for a cable-stayed bridge that crosses a fault is identified based on the coefficient of determination (R^2). Root-mean-square velocity (V_{rms}) is found to be the best IM for cable-stayed bridges traversed by fault ruptures, instead of the commonly used ones such as peak ground acceleration or velocity (PGA or PGV). Fragility curves for the critical components of fault-crossing cable-stayed bridges, including pylons, cables, and bearings, are developed using the IM of V_{rms}, and are subsequently compared with those for the cable-stayed bridge near faults. Results show that the bearings on transition piers are the most vulnerable component for fault-crossing cable-stayed bridges because of the rotation of their girder. Compared to cable-stayed bridges near faults, pylons and bearings are more vulnerable in the transverse direction for cable-stayed bridges crossing faults, whereas the vulnerability of cables is comparable.

Keywords: cable-stayed bridge; crossing faults; synthetic ground motions; intensity measure; fragility analysis

Citation: Guo, J.; Gu, Y.; Wu, W.; Chu, S.; Dang, X. Seismic Fragility Assessment of Cable-Stayed Bridges Crossing Fault Rupture Zones. *Buildings* **2022**, *12*, 1045. https://doi.org/10.3390/buildings12071045

Academic Editor: Antonio Formisano

Received: 24 June 2022
Accepted: 14 July 2022
Published: 19 July 2022

Copyright: © 2022 by the authors. Licensee MDPI, Basel, Switzerland. This article is an open access article distributed under the terms and conditions of the Creative Commons Attribution (CC BY) license (https://creativecommons.org/licenses/by/4.0/).

1. Introduction

The large velocity pulse and large ground displacement in near-fault regions usually lead to significant damage to bridges traversed by fault ruptures. These two devastating effects, known as "forward directivity" and "fling-step", have been shown in previous earthquakes [1,2]. In order to avoid bridges suffering from undesirable calamities, many seismic design codes recommend preventing the construction of bridges that traverse crossing faults. However, it is sometimes inevitable to construct fault-crossing bridges [2,3]. As a result, it is crucial to study the seismic behavior of bridges crossing fault ruptures.

Although considerable research has been conducted on the seismic performance of fault-crossing girder bridges [3–11], there is limited research on evaluating the responses of fault-crossing long-span bridges [12,13]. Zeng et al. [14] studied the seismic responses of a cable-stayed bridge, and they found that the magnitude of the permanent ground displacement can significantly affect the bridge's responses. Gu et al. [15] evaluated the seismic responses of cable-stayed bridges that cross faults. However, most of the previous studies on fault-crossing bridges utilize a deterministic method with a few ground motions as the input, and thus cannot account for the uncertainties of both bridges and earthquakes.

A major challenge that limits scholars from probabilistic analyses on fault-crossing bridges is the lack of actual motions recorded on both sides of the fault rupture. Researchers have made great efforts to address this issue. Generally, the approaches to obtaining ground motions with distinct characteristics of pulse-like earthquakes can be divided into two categories: (1) performing baseline corrections to raw records and (2) generating ground motions using numerical models.

For the first approach, the main objective is to preserve the permanent ground static offset from actual records. Wu et al. [16] developed an improved baseline correction method by modifying the models proposed by Iwan [17] and Boore [18]. The improved method is applied to the Chichi and Chengkung earthquake data to compute the coseismic displacements, and the results show favorable agreement with GPS measurements. Lin et al. [19] proposed a new baseline correction scheme to obtain ground motions with target final displacements. By considering six simulated fling-step ground motions, Zhang et al. [11] studied the seismic behavior of girder bridges crossing faults.

For the second approach, the main objective is to build an artificial across-fault ground motion with directivity or fling pulse. Various researchers have proposed low-frequency pulse models. Menun [20] developed an analytical model defined by five parameters for ground motions in the fault-normal (FN) direction. Mavroeidis et al. [21] introduced a mathematical model and calibrated that the model can capture the characteristics of near-fault motions. Makris et al. [22] demonstrated that a one-cosine acceleration pulse causes the velocity pulse. Hoseini Vaez et al. [23] presented a new model to simulate the velocity pulses. Kamai et al. [24] simulated the velocity pulses with a half-period sine wave. Burks et al. [25] defined the fault-parallel (FP) fling by a ramp function and derived a predictive model for the parameters. Yadav et al. [26] modeled the fling-step velocity pulse using a function related to the pulse's amplitude, duration, and location. Hamidid et al. [27] used the Green function to simulate the ground motions. In particular, the model proposed by Mavroeidis et al. [21] successfully simulates the FN and FP pulses of the ground motion across a strike-slip fault [28]. In addition to the low-frequency pulse models, the specific barrier model [29] and stochastic model [30–34] are proposed to represent the high-frequency content of the motions.

Since various numerical models have been developed so far, it is possible to investigate fault-crossing bridges with probabilistic seismic-risk assessment approaches. Developing structure fragility curves is one practical and effective tool for probabilistic seismic-risk assessments. In the past two decades, much research has been carried out to evaluate the vulnerability of bridges. However, most previous studies focused on girder bridges. Pang et al. [35] performed a vulnerability analysis of cable-stayed bridges considering various uncertainties. Zhong et al. [13,36–38] studied cable-stayed bridges' fragility in selecting optimal intensity measures and the effect of spatially distributed motions. Wu et al. [39] conducted the fragility of a concrete cable-stayed bridge subjected to far-field motions. Wang et al. [40] utilized the vulnerability to estimate the effect of an innovative bearing on the performance of a cable-stayed bridge subjected to ground motions with velocity pulses. The fragility curves are established using the incremental dynamic analyses (IDA) method. Li et al. [41] assessed the fragility of a cable-stayed bridge adopting synthetic offshore multi-support ground motions. Wei et al. [42] conducted seismic fragility analysis of a multipylon cable-stayed bridge with super-high piers. Nevertheless, fragility assessments for fault-crossing cable-stayed bridges are still insufficient.

This study aims to conduct the fragility assessment of cable-stayed bridge crossing faults. First, using the existing mathematical models, fling-step motions are simulated by superimposing the long-period pulses and their high-frequency residuals. Then, a numerical model is constructed with the OpenSeesPy platform [43]. The synthetic ground motions are adopted as inputs to perform nonlinear analysis. Subsequently, the appropriate intensity measure is selected for cable-stayed bridges crossing faults, and the fragility curves are developed. Moreover, the vulnerability of cable-stayed bridges subjected to fling-step and velocity pulse motions are compared. Finally, several conclusions of this study are presented.

2. Seismic Fragility Methodology

Seismic fragility is a conditional probability that gives a bridge's likelihood to meet or exceed a certain level of damage for a given intensity measure (IM) [44]. Assuming both the demand (*D*) and capacity (*C*) follow lognormal distributions, the fragility function can be represented as follows:

$$P = P[D \geq C|IM] = \Phi\left[\frac{\ln(D) - \ln(C)}{\sqrt{\beta_D^2 + \beta_C^2}}\right] \quad (1)$$

where *D* and *C* are the mean values of demand and capacity, respectively; β_D and β_C are the logarithmic standard deviations, and $\Phi[\cdot]$ is the standard normal cumulative distribution function.

To develop the fragility curves, it is necessary to obtain the demand (*D*) beforehand. *D* is estimated using a probabilistic seismic-demand model (PSDM) in fragility analysis. PSDM expresses the relationship between the demand *D* and the IM. Conventionally, *D* and IM are assumed to exhibit a linear regression relationship in the logarithmic space as:

$$\ln(D) = a\ln(IM) + b \quad (2)$$

where *a* and *b* are regression coefficients. Alternatively, Pan et al. [45] apply a quadratic regression to fit the data. Zhong et al. [13] used quadratic regression to obtain the PSDM for a cable-stayed bridge. The quadratic PSDM takes the form:

$$\ln(D) = a\ln^2(IM) + b\ln(IM) + c \quad (3)$$

where *a*, *b*, and *c* are the regression coefficients. Thus, once the damage state and component capacity S_C are determined, the fragility curve can be derived by using Equation (1).

3. Simulation of the Ground Motions

This study generates synthetic across-fault ground motions by combining the simulated coherent (long-period) and incoherent (high-frequency) components. Specifically, the high-frequency components in horizontal directions (FN and FP directions) are simulated with a stochastic model proposed by Dabaghi et al. [32–34], whereas the long-period components are simulated according to the pulse model proposed by Mavroeidis et al. [21]

3.1. Ground-Motion Models

Dabaghi et al. [32] pointed out that the residual of a velocity pulse motion after removal of the pulse is generally a broadband time series. A model proposed by Rezaeian et al. [30,31] and adopted by Dabaghi et al. [32–34] can be used to describe these broadband motions. The expression of this MFW model is:

$$a_{MFW} = q(t)\left\{\frac{1}{\sigma_h(t)}\int_{-\infty}^{t} h[t - \tau, \lambda(\tau)]w(\tau)d\tau\right\} \quad (4)$$

where $w(t)$ is a white-noise process, $\sigma_h(t)$ is the standard deviation of the process defined by the integral, *q*(*t*) is a time-modulating function that characterizes the root-mean-square of the acceleration process, and $h[t - \tau, \lambda(\tau)]$ is the unit-impulse response function (IRF) of a linear filter with the time-varying parameter $\lambda(\tau) = [w_f(\tau), \xi_f(\tau)]$, given as follows:

$$h[t - \tau, \lambda(\tau)] = \begin{cases} \frac{\omega_f(\tau)}{\sqrt{1-\xi_f^2(\tau)}} e^{-\xi_f(\varnothing)\omega_f(\tau)(t-\tau)} \sin\left[\omega_f(\tau)\sqrt{1-\xi_f^2(\tau)}(t-\tau)\right], \tau \leq t \\ 0, \text{ elsewhere} \end{cases} \quad (5)$$

where τ is the time of the pulse, $\xi_f(\tau)$ is the filter damping that represents the bandwidth of the acceleration process (regarded as a constant), and $\omega_f(\tau)$ is the filter frequency:

$$\omega_f(\tau)= 2\pi\left[f_{\text{mid}}+f'(\tau-t_{\text{mid}})\right] \tag{6}$$

where f_{mid} is the filter frequency at the middle of the motion t_{mid}, and f' is the frequency change rate with time.

For the modulating function $q(t)$, a four-parameter piecewise function is presented as:

$$q(t) = \begin{cases} 0,\ t \le t_{0,q} \\ c\left(\frac{t-t_{0,q}}{t_{\max,q}-t_{0,q}}\right)^{ff},\ t_{0,q} < t \le t_{\max,q} \\ ce^{[-\beta(t-t_{\max,q})]},\ t_{\max,q} < t \end{cases} \tag{7}$$

where $t_{0,q}$ is the starting time, $t_{\max,q}$ is the time that the modulating function arrives at its peak, α is the order of the polynomial at the start of the function, β is the decaying rate of an exponential function in the end phase of the function, and c controls the amplitude of the modulating function. These four parameters ($t_{\max,q}$, α, β, c) are related to the Arias intensity of the motion.

The model put forward by Mavroeidis et al. [21] (M03 model) has been widely employed for generating the long-period pulse of near-fault motions. Currently, when simulating the fling of the motion, the M03 model is adopted with its original form as follows:

$$v_{\text{pul}}(t) = \begin{cases} \frac{V_p}{2}\left[1+\cos\left(\frac{2\pi}{\gamma T_p}(t-t_0)\right)\right]\cos\left[\frac{2\pi}{T_p}(t-t_0)+v\right],\ t_0-\frac{\gamma}{2}T_p< t \le t_0+\frac{\gamma}{2}T_p \\ 0,\ \text{otherwise} \end{cases} \tag{8}$$

where V_p is the pulse magnitude, T_p is the period, γ is a variable controlling the oscillation number of a pulse, ν is the phase angle, and t_0 is the epoch of the peak of the envelope.

For the simulation of the velocity pulse, the form of the MP model proposed by Dabaghi et al. [32] is adopted:

$$v_{\text{pul}}(t) = \begin{cases} \left\{\frac{V_p}{2}\cos\left[2\pi\left(\frac{t-t_{\max,p}}{T_p}\right)+v\right]-\frac{D_r}{\gamma T_p}\right\}\left\{1+\cos\left[\frac{2\pi}{\gamma}\left(\frac{t-t_{\max,p}}{T_p}\right)\right]\right\},\ t_{\max,p}-\frac{\gamma}{2}T_p < t < t_{\max,p}+\frac{\gamma}{2}T_p \\ 0,\ \text{elsewhere} \end{cases} \tag{9}$$

where t_0 in Equation (8) is replaced by $t_{\max,p}$, and D_r is a permanent displacement represented with $D_r= V_pT_p\frac{\sin(\nu+\gamma\pi)-\sin(\nu-\gamma\pi)}{4\pi(1-\gamma^2)}$.

Additionally, the FN component of the simulated motions is checked, adopting the criteria put forward by Baker [46] to ensure that it contains a velocity pulse. The three criteria can be found in [46] and are given here for convenience:

(1) The equation of the pulse indicator (larger than 0.85) is given as:

$$\text{Pulse indicator} = \frac{1}{1+e^{-23.3+14.6(\text{PGV ratio})+20.5(\text{energy ratio})}}$$

where the PGV ratio is defined as the PGV of the residual record divided by the original record's PGV and the energy ratio with a similar definition.

(2) The pulse occurs at the early stage of the motion, as indicated by the time when the original record reaches 20% of its total cumulative squared velocity (CSV) and is greater than the time at which the pulse reaches 10% of its CSV.

(3) The PGV of the motion is larger than 30 cm/s.

3.2. Determination of the Input Parameters of the Models

Input parameters of the models for high-frequency components and the forward-directivity pulse are determined through linear predictive equations developed by

Dabaghi et al. [32–34], according to the source and site characteristics. Specifically, seven parameters are selected for the predictive equations, including the type of faulting (F), the moment magnitude (M), the depth to the top of the rupture plane (Z_{TOR}), the closest distance from the site to the fault rupture (R_{RUP}), the shear-wave velocity of the top 30 m of soil at the site (V_{s30}), and directivity parameters s (or d) and θ (or φ). The schematic of the directivity parameters is plotted in Figure 1. Detailed procedures for the regression of these model parameters can be found in [32–34].

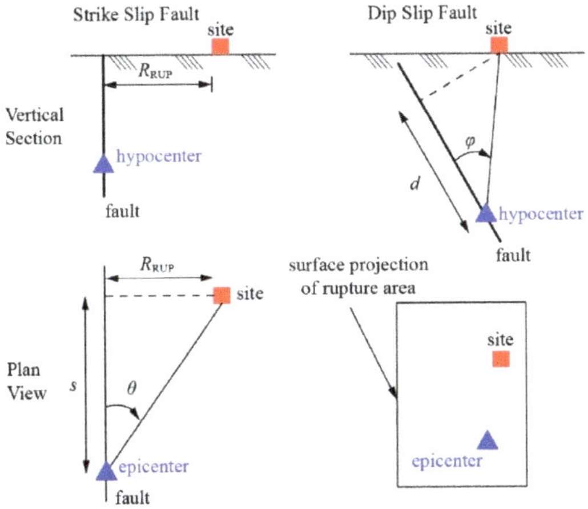

Figure 1. Directivity parameters [47].

Input parameters of the fling model are determined following the guidelines recommended by Yang et al. [28]. It is noteworthy that choosing equations for estimating the parameters for the fling model is not unique. In this study, the predictive equations proposed by Abrahamson [48] are selected.

Table 1 lists the predictive equations for estimating the input parameters.

Table 1. Predictive equations for the input parameters.

Model	Predictive Equation	Reference
MFW model	$\alpha, \beta, c, t_{max,q}, f_{mid}, f' = \mathbf{F}(F, M, Z_{TOR}, R_{RUP}, V_{s30}, s, \theta)$ $t_{0,q} = 0$	[32–34]
Directivity pulse model	$V_p, T_p, \gamma, \nu, t_{max,p} = \mathbf{F}(F, M, Z_{TOR}, R_{RUP}, V_{s30}, s, \theta)$	[32–34]
Fling model	$V_p = 2D_{site}/[T_p/(2+\varepsilon)], \varepsilon \in (0,0.1]$ $\log(T_p) = \log(2+\varepsilon) - 3.00 + 0.50M, \varepsilon \in (0,0.1]$ $\log(D_{site})_{avg} = -1.70 + 0.50M,$ $(D_{site})_{max} \approx (D_{site})_{avg}/\lambda, \lambda \in [0.2,0.8]$ $\gamma = 1 + \varepsilon, \varepsilon \in (0,0.1]$ $\nu \approx 0$ or π $t_0 \geq \gamma T_p/2$	[28]

$(D_{site})_{max}$ is used to estimate V_p.

4. Vulnerability Analysis of a Fault-Crossing Bridge

4.1. Depiction of the Analysis Model

A typical cable-stayed bridge is selected as the case study bridge. The bridge has a length of 174 + 352 + 174 m. There are 168 stay cables, whose tensile strength is 1670 MPa, with a fan-typed configuration. The cable force is optimized based on the method proposed by Guo et al. [49].

Wind-resistance bearings are arranged between the girder and pylons in the transverse direction. Note that the dampers, restrainers, and transverse retainers are not considered in this study.

The numerical model of this case bridge is constructed with the OpenSeesPy platform [43], as illustrated in Figure 2. The dimension of the critical components is given in Figure 3. The elastic beam-column element is adopted for the main girder, cap beams, and cross girders, as they are assumed to remain elastic. The plasticity fiber model is used for the sections of pylons and piers. Fiber sections of the tower are presented in Figure 4. The P-delta effect of the pylons is considered. The truss element is used to model the cables. The pile foundation is simulated with three translational and three rotational springs. For the spherical bearing, the no-tension uniaxial material is adopted to simulate the vertical behavior of the bearings. The elastic no-tension material is used to model its axial stiffness for the wind-resistance bearing. The damping ratio of the example bridge is 3%. The first five natural vibration periods of the model are 10.999, 4.537, 3.092, 2.940, and 1.913 s. The first three modes are longitudinal floating vibration, vertical vibration, and transverse floating vibration, respectively.

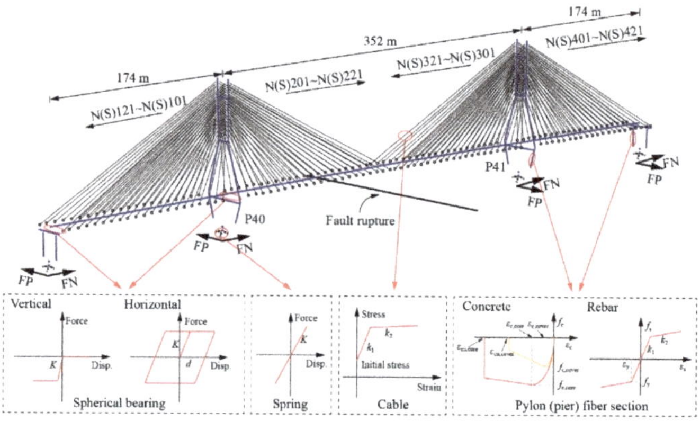

Figure 2. Configuration of the case study bridge.

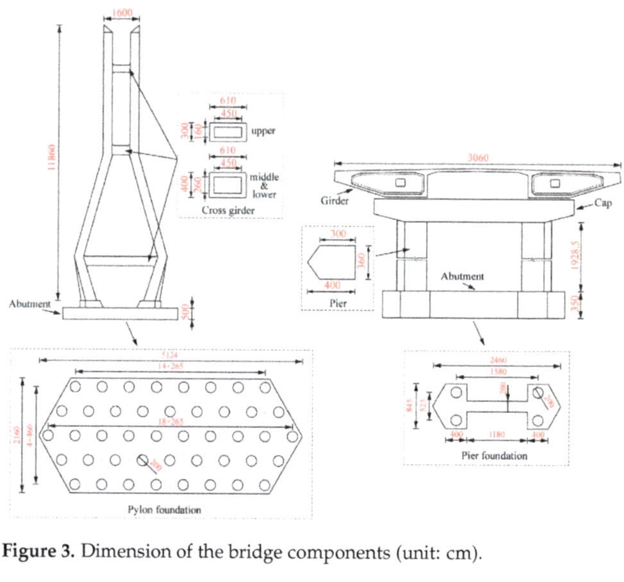

Figure 3. Dimension of the bridge components (unit: cm).

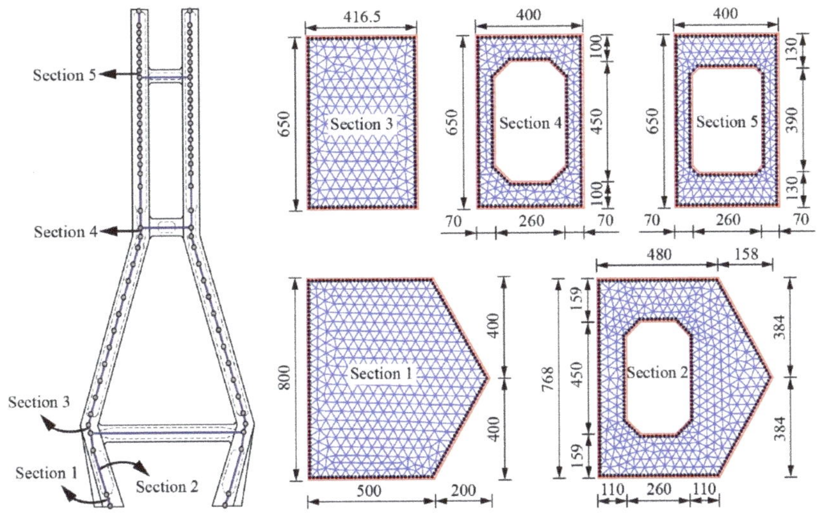

Figure 4. Fiber sections of the pylon (unit: cm).

Material uncertainties are considered for the numerical model of the bridge. In addition, the uncertainty of the actual initial cable forces due to the construction error is also considered. In this study, the compressive strength of the concrete (f_c), yield strength of the rebar (f_y), initial stiffness of the rebar (E), the ratio between the post-yield and initial stiffness, and the ratio between actual and designed initial cable forces (R) are modeled as random variables. Table 2 summarizes the statistical information of these uncertainty parameters.

Table 2. Uncertainty parameters of the bridge and their probability distribution.

Variable	Distribution *	Reference
$f_{c,pylon}$	Normal (C50, μ = 32.35 MPa, cov = 0.18)	[50]
$f_{c,pier}$	Normal (C40, μ = 26.75 MPa, cov = 0.18)	[50]
f_y	Lognormal (HRB400, μ = 400 MPa, cov = 0.08)	[51]
E	Lognormal ($\mu = 2 \times 10^8$ MPa, cov = 0.033)	[52]
b	Lognormal (μ = 0.005, cov = 0.2)	[35,52]
R	Uniform ($\mu = 1$, $\sigma = 0.1$)	/

* μ denotes the mean; cov denotes the coefficient of variation; σ denotes the standard deviation.

4.2. Ground Motions

It is pointed out that the model developed by Dabaghi et al. should be used within the boundaries of the database, which are $6 \leq M \leq 7.5$, $5 < R_{RUP} \leq 25$ km, and $400 < V_{s30} < 1000$ m/s [32]. Therefore, 25 ground-motion pairs (in FN and FP directions) are simulated with their input parameters randomly generated within the corresponding ranges. It is worth noting that for strike-slip faults, R_{RUP}, s, and θ are correlated with $R_{RUP} = s\tan\theta$ if a straight line can represent the rupture and if the site is located alongside the rupture, as shown in Figure 1. In the present study, only the case of vertical strike-slip faults with surface rupture is considered, and the site is assumed to be perpendicular to the epicenter. As a result, F, Z_{TOR}, s, and θ are constrained as 0, 0, 0, and 90°, respectively, in the simulation. Models of the low-frequency pulses and high-frequency residuals are programmed in Python language. Figure 5 presents the time histories of one of the synthetic motions. The "forward-directivity" pulse in the FN direction and the "fling-step" pulse in the FP direction can be successfully simulated. The random parameters of the 25 synthetic ground-motion pairs are listed in Table 3.

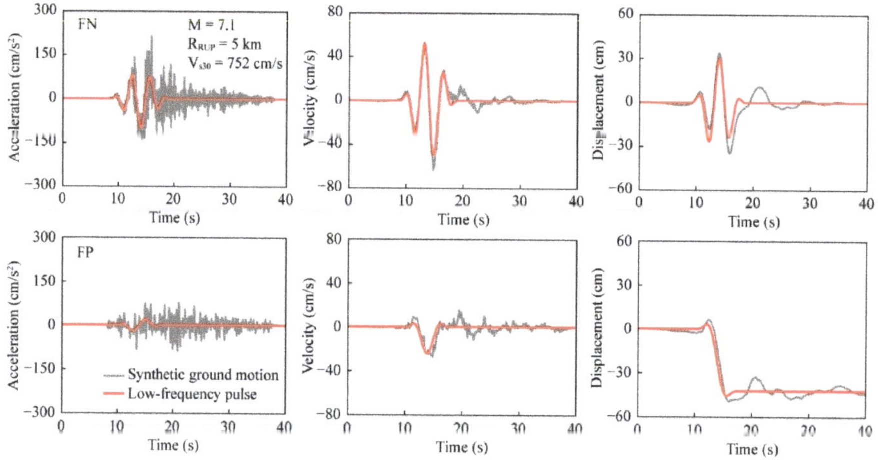

Figure 5. Time histories of a synthetic ground-motion pair.

Table 3. Input parameters of the synthetic ground motions.

GM	*M*	R_{RUP} (km)	V_{s30} (m/s)	GM	*M*	R_{RUP} (km)	V_{s30} (m/s)
1	6	10	702	14	7.3	5.1	873
2	7	18.9	446	15	7.2	6.5	508
3	7.1	5	752	16	7.2	8.2	524
4	6.7	15.1	642	17	7	7.9	691
5	7.3	7.8	993	18	6.7	9	651
6	7.1	5.8	954	19	7.5	23.4	445
7	6.3	14.2	691	20	6	15.9	588
8	6.9	8.2	846	21	6.7	10.8	630
9	7.1	20.3	882	22	7.1	8.2	675
10	6.1	7	542	23	6.3	14.3	947
11	7.3	7.2	622	24	6.9	5.8	455
12	6.2	12.4	905	25	6.9	15.1	587
13	6.8	5.9	903				

$F = 0$ for strike-slip faults; $Z_{TOR} = 0$; $s = 0$; $\theta = 90°$.

The synthetic motion pairs are then scaled using a factor ranging from 0.5 to 3 with an increment of 0.5. Thus, 150 ground-motion pairs are generated for the nonlinear time-history analysis. Figure 6 presents the spectra displacement of the input ground motions in FN and FP directions. It shows that the component in the FP direction has larger displacements than that in the FN direction in the long-period range.

According to previous studies, cable-stayed bridges are suggested to cross the fault perpendicularly [15]. Therefore, in this study, the fault rupture should cross the bridge in its middle span with an angle of 90°. The time-history series applies to the bridge supports in the FN and FP directions. Because this study focuses on the vertical strike-slip scenario, the ground dislocation is assumed to distribute equally at the two sides of the fault. As a result, the FN ground motions are the same on each fault side, whereas the FP ground motions have equal amplitudes but with reversed polarity [53] (as shown in Figure 7). To exclude the interference of the effect of vertical ground motions and mainly consider the effect of the FN and FP components, the vertical motions are not considered in this study. The wave passage effect is not considered as well.

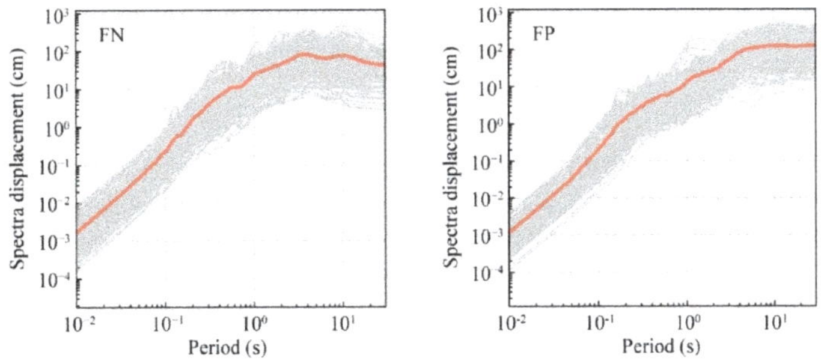

Figure 6. Response spectra of the 150 ground motions.

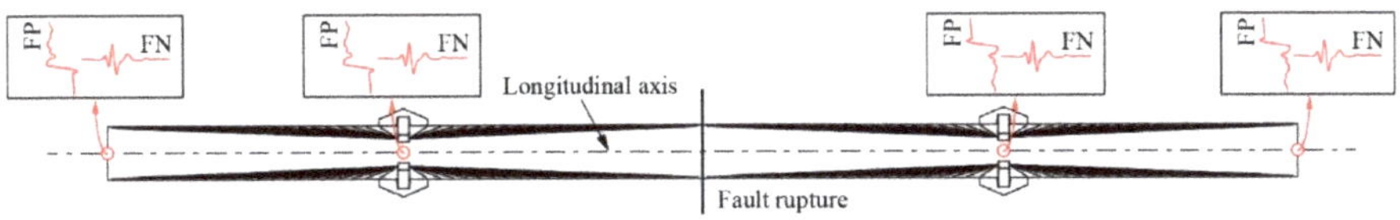

Figure 7. Input ground motions for cable-stayed bridges crossing faults.

4.3. Optimal IM Determination and PSDMs Establishment

The selection of optimal IMs is essential for establishing reliable PSDMs. In this study, 12 IM candidates are examined to identify the optimal one for the fault-crossing cable-stayed bridge. Table 4 gives the definitions of the studied IMs. The IMs are divided into three groups: acceleration-related, velocity-related, and displacement-related. Note that only structure-independent IMs are used in the current study.

Table 4. Description of the considered IMs.

Type	IM	Definition	Reference
Acceleration-related	PGA	Peak ground acceleration Max$\|a(t)\|$, $a(t)$ is the acceleration time history	/
	SMA	Sustained maximum acceleration third largest peak in $a(t)$	[54]
	CAV	Cumulative absolute velocity $\int_0^{t_{tot}} \|a(t)\| dt$, t_{tot} is the total duration	[55]
	I_a	Arias intensity $\frac{\pi}{2g} \int_0^{t_{tot}} a^2(t) dt$,	[56]
	A_{rms}	Root-mean-square of acceleration $\sqrt{\frac{1}{t_{tot}} \int_0^{t_{tot}} a^2(t) dt}$	/
Velocity-related	PGV	Peak ground velocity Max$\|v(t)\|$, $v(t)$ is the velocity–time history	/
	SMV	Sustained maximum acceleration third largest peak in $v(t)$	[54]
	CAD	Cumulative absolute displacement $\int_0^{t_{tot}} \|v(t)\| dt$, t_{tot} is the total duration	[55]
	V_{rms}	Root-mean-square of velocity $\sqrt{\frac{1}{t_{tot}} \int_0^{t_{tot}} v^2(t) dt}$	/
	I_v	Velocity intensity $\frac{\pi}{2g} \int_0^{t_{tot}} v^2(t) dt$	[57]
Displacement-related	PGD	Peak ground displacement Max$\|d(t)\|$, $d(t)$ is the displacement–time history	/
	D_{rms}	Root-mean-square of displacement $\sqrt{\frac{1}{t_{tot}} \int_0^{t_{tot}} d^2(t) dt}$	/

The geometric mean of the IMs for the FN and FP input ground motions is adopted here to represent the IM of a motion pair:

$$IM = \sqrt{IM_{FN} \times IM_{FP}} \tag{10}$$

During the analyses, responses of the following components (engineering demand parameters, EDPs) are recorded: (1) curvature of Section 1, φ_{S1}; (2) curvature of Section 3, φ_{S3}; (3) bearing displacement between girder and pier $\sigma_{b,pier}$; (4) bearing displacement between girder and pylon $\sigma_{b,pylon}$; (5) force of the longest backstay cable (N121) F_{bc}; and (6) force of the longest forestay cable (N221) F_{fc}. The upper sides of the pylon are less vulnerable than the lower sides, and the pier has a small effect on the seismic responses of the bridge compared to the pylon [38], so their responses are not considered in this study. Quadratic polynomials are adopted to construct the PSDMs.

Because the regression model is quadratic, the criteria of an optimal IM for linear regressions (practically, efficiency, proficiency, and sufficiency [57]) are not appropriate here. In this study, the determination coefficient (R^2) of the regression is used to judge whether an IM is optimal or not. The definition of R^2 is given as follows:

$$R^2 = \frac{\Sigma(\hat{y} - \bar{y})^2}{\Sigma(y - \bar{y})^2} = 1 - \frac{\Sigma(y - \hat{y})^2}{\Sigma(y - \bar{y})^2} \tag{11}$$

where y is the sample value, $\hat{y}$ is the regression value, and $\bar{y}$ is the mean of the sample. The larger R^2, the better the regression is.

Figure 8 plots the determination coefficients for the regressions in the case of each IM candidate. As shown in Figure 8, the largest R^2 varies for different EDPs. Additionally, R^2 can be significantly different for an EDP in different directions. For example, PGV performs well in predicting the transverse bearing displacements, pylon responses, and the backstay cable's force but may not provide reliable predictions on bearing displacements in the longitudinal direction and the force of forestay cables. Therefore, it is difficult to identify the optimal IM from R2 directly. The multicriteria decision-making (MCDM) method is adopted to tackle this issue. In each MCD, there are several alternatives and criteria. The alternative with the highest score is selected as the best one and is placed in the first rank [58]. In this research, different IMs are considered alternatives, and the corresponding determination coefficients R2 for different components are chosen as criteria.

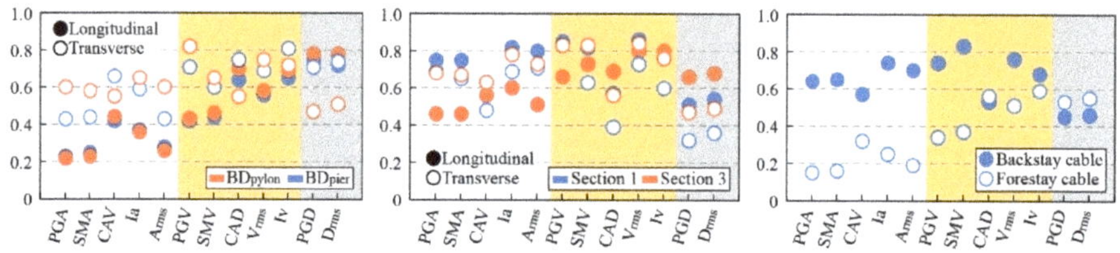

Figure 8. Determination coefficients for the regressions of various EDPs. The background colors denote the type of the IMs (white: acceleration-related; yellow: velocity-related; gray: displacement-related).

The weights of different components are assigned based on 10-point scale qualitative evaluations, as shown in Table 5. Moreover, the weights of a component in different directions are considered to be the same for simplicity. Compared with the bearings, cables and pylons are considered more critical because their failure would cause bridge collapse. As a result, the weights of pylons and cables are both assigned as 10, whereas the weights of bearings are assigned 3, 5, or 7.

Table 5. Assignment of values for a 10-point scale [58].

Attribute Evaluation	Value
Extremely unimportant	0
Very unimportant	1
Unimportant	3
Average	5
Important	7
Very important	9
Extremely important	10

After assigning the weights, a weighted sum method calculates the scores and ranks the IMs. Results of the MCDM method are tabulated in Table 6.

Table 6. Results of the MCDM method.

Weight of Criteria			Rank of the Optimal IMs		
Pylons	Cables	Bearings	1st	2nd	3rd
10	10	7	V_{rms}	I_v	PGV
10	10	5	V_{rms}	I_v	PGV
10	10	3	V_{rms}	I_v	PGV

According to the rank of the IMs for different weight assignments, it is found that Vrms is the optimal one among the 12 candidates. Results also demonstrate that the velocity-dependent IMs perform better than the others for all the weight assignments. Moreover, it is worth mentioning that PGA, which has been used as the optimal IM for seismic fragility analyses in previous studies [41,59], is inappropriate for cable-stayed bridges that cross fault ruptures. The PSDMs established for each EDP using V_{rms} are illustrated in Figure 9, and the corresponding regression coefficients are listed in Table 7. It is worth noting that the hazard curve for Vrms is unavailable. However, it is possible to establish the correlation between Vrms and an IM with available hazard curves, with the development of ground-motion prediction equations, using similar methods introduced in [60].

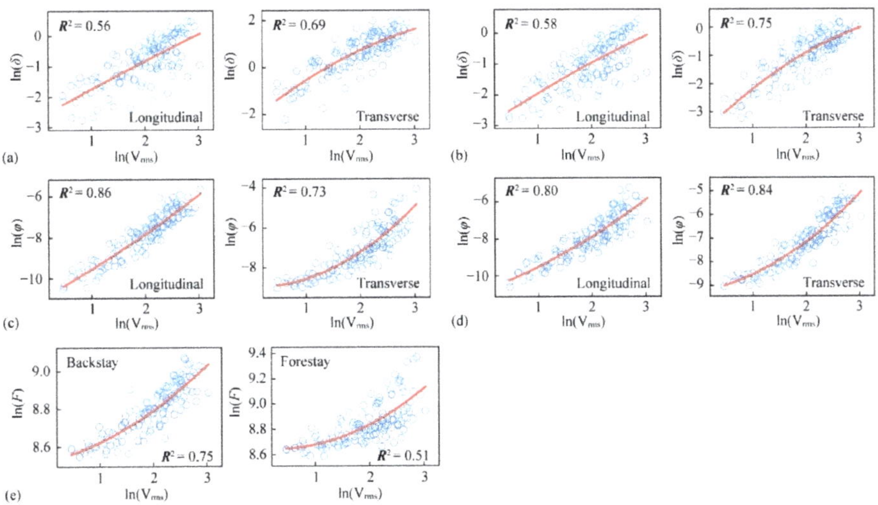

Figure 9. PSDMs for different components: (**a**) bearing at girder–pier location; (**b**) bearing at girder–pylon location; (**c**) Section 1 of the pylon; (**d**) Section 3 of the pylon; and (**e**) cables.

Table 7. Regression coefficients for the PSDMs.

EDP	Longitudinal			EDP	Transverse			EDP	/		
	a	*b*	*c*		*a*	*b*	*c*		*a*	*b*	*c*
$\sigma_{b,pier}$	−0.022	0.979	−2.696	$\sigma_{b,pier}$	−0.196	1.864	−2.202	F_{bc}	0.035	0.065	0.522
$\sigma_{b,pylon}$	−0.059	1.165	−3.054	$\sigma_{b,pylon}$	−0.203	1.883	−3.856	F_{fc}	0.065	−0.038	8.652
φ_{S1}	0.099	1.408	−11.050	φ_{S1}	0.471	−0.058	−8.963				
φ_{S3}	0.183	1.091	−10.791	φ_{S3}	0.326	0.408	−9.252				

Formulation for the quadratic regression: $\ln(S_D) = a\ln^2(IM) + b\ln(IM) + c$.

4.4. Definition of the Damage Index

Four limit states are considered, namely slight, moderate, extensive, and complete. For the pylon sections, the value of each damage state proposed by Feng [61] is adopted in this study. Figure 10 illustrates the moment-curvature curve for a pylon section. The slight damage state is characterized by the equivalent yield curvature (φ_{ey}), whereas the complete one is characterized by the ultimate curvature (φ_u). Taking the difference between φ_{ey} and φ_u as $\Delta\varphi$, the curvatures that represent moderate and extensive damage states are defined as ($\varphi_{ey} + \Delta\varphi/3$) and ($\varphi_{ey} + 2\Delta\varphi/3$), respectively. For the spherical bearings, the slight limit state is defined to be the maximum allowable bearing displacement under normal service conditions; the distance between the center of the bearing and the edge of the masonry plate is defined as the threshold of the extensive damage state, and the threshold of the moderate damage state is defined as the median of slight and extensive damage states; when the center of the bearing moves over the edge of the substructure (i.e., the cap or cross girder), the bearing is considered to be completely damaged. Figure 11 gives the definition of the damage states for the bearing. For the cable forces, the threshold of the complete damage state is defined as the cable's breaking force. The difference between the initial and cable-breaking force is quartered, and the quartering points are defined as slight, moderate, and extensive damage thresholds.

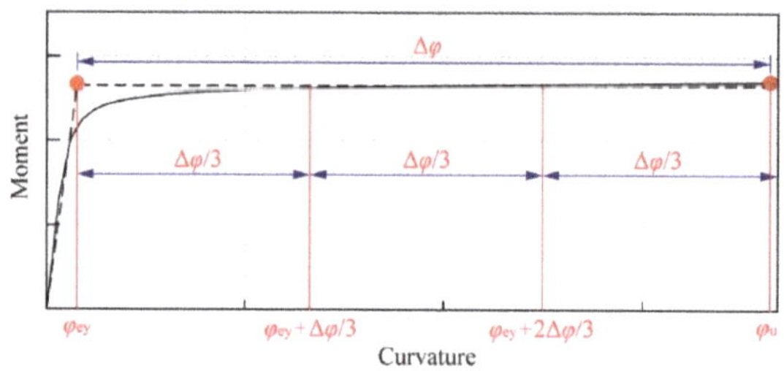

Figure 10. Limit states for the pylons.

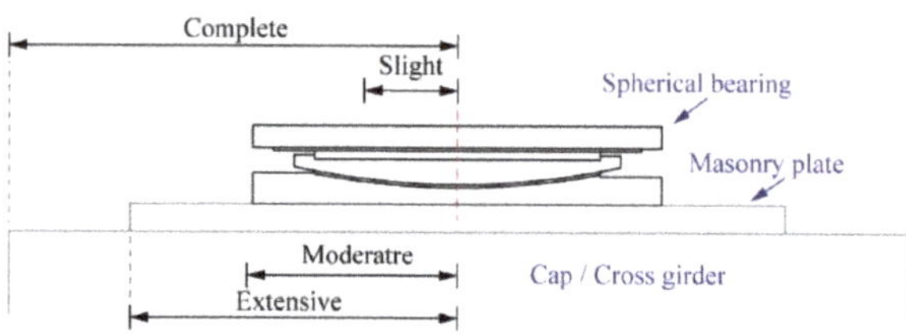

Figure 11. Definition of the damage states for bearings.

Table 8 summarizes the damage index of different damage states for the EDPs, where S_C and β_C mean the median and logarithmic standard deviation, respectively. Note that the thresholds of slight, moderate, and extensive damage states for the bearings are the same in longitudinal and transverse directions. However, the bearing's complete damage state threshold is larger in the transverse direction because of the adequate distance between the bearing's center line and the cap's transverse edge.

Table 8. Damage index for different damage states.

EDP	Slight		Moderate		Extensive		Complete	
	S_C	β_C	S_C	β_C	S_C	β_C	S_C	β_C
φ_{S1}, longitudinal	1	0.04	6.1	0.05	11.2	0.05	16.3	0.06
φ_{S1}, transverse	1	0.04	7.6	0.04	14.3	0.04	20.9	0.04
φ_{S3}, longitudinal	1	0.15	6.7	0.06	12.5	0.06	18.2	0.06
φ_{S3}, transverse	1	0.13	7.1	0.04	13.2	0.04	19.3	0.04
σ_b, longitudinal (mm)	300	0.35	450	0.35	600	0.35	800	0.35
σ_b, transverse (mm)	300	0.35	450	0.35	600	0.35	2500	0.35
F_{bc} (kN)	8586	0.10	11,785	0.10	14,987	0.10	18,188	0.10
F_{fc} (kN)	8279	0.10	11,196	0.10	14,114	0.10	17,031	0.10

β_C for bearing displacements and cable forces refers to [13].

4.5. Component Fragility Curves

Utilizing the aforementioned PSDMs, the component fragility curves for the fault-crossing bridge under different limit states are developed, as presented in Figure 12. As shown in Figure 12, the bearings at girder–pier locations are the most fragile components for all damage states, especially in the transverse direction. Comparing the bearings at different locations, their vulnerabilities are similar in the longitudinal direction but different in the transverse direction. This can be attributed to the rotation of the deck caused by the permanent displacement.

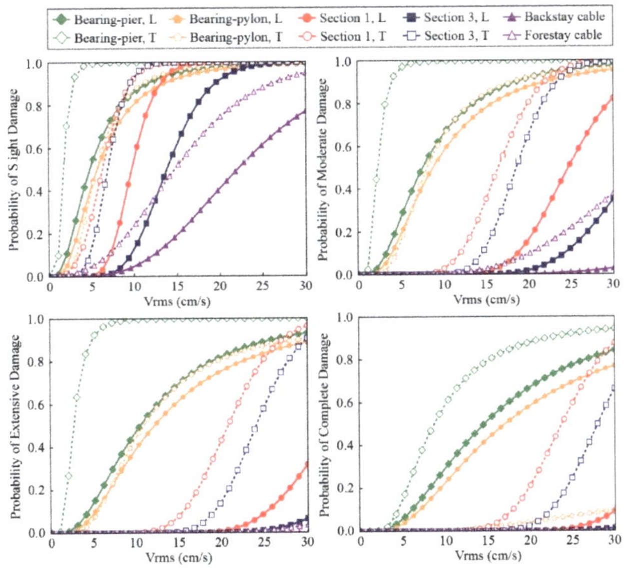

Figure 12. Component fragility curves for the fault-crossing bridge.

Figure 12 also shows that the pylon sections are more vulnerable in the transverse direction, and the pylon bottom section (Section 1) is more vulnerable than the section of the connection zone between the pylon and lower cross-girder (Section 3). Although the capacity of Section 1 is stronger than that of Section 3 (as shown in Figure 4), the pylon bottom can suffer larger seismic forces under earthquakes, which results in its higher vulnerability.

As for the cables, they are likely to experience only slight and moderate damage. Furthermore, the forestay one appears more vulnerable than the backstay one. This is because the girder's inertia force is undertaken alone by the forestay cables in the middle span and the backstay cables and transition piers in the side span.

5. Comparison of the Fragility of the Bridge Subjected to Fling-Step and Near-Faults Motions

To investigate the influence of traversing a fault rupture on the vulnerability of cable-stayed bridges, fragility curves of the same case bridge near the fault rupture are developed as a comparison. The finite element model, input ground motions, selected IM, and EDPs are the same as those described in previous sections. The only difference is that the transverse excitations are applied to the bridge with equal amplitudes and the same polarity for the near-fault scenario, as shown in Figure 13.

Figure 14 compares the component fragility curves of the bridge subjected to near-fault and fling-step motions. Fragility curves of the cables under extensive and complete damage states are not plotted because of their low damage-exceedance probabilities. Table 9 presents the median value (the V_{rms} associated with the 50% exceedance probability) of the fragility curves across the four limit states. Note that a larger median V_{rms} means the component is less vulnerable.

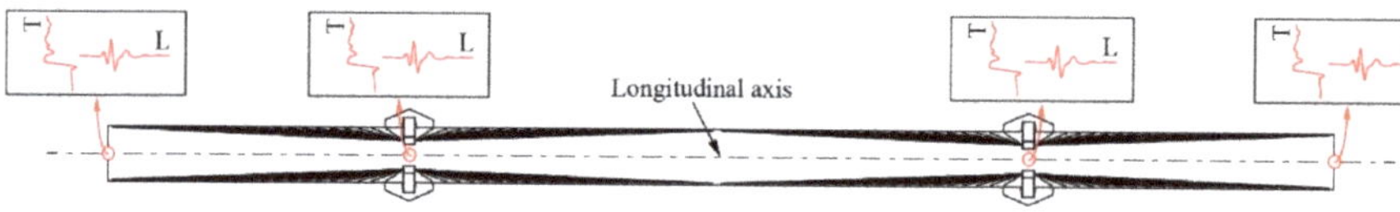

Figure 13. Input ground motions for near-fault cable-stayed bridges.

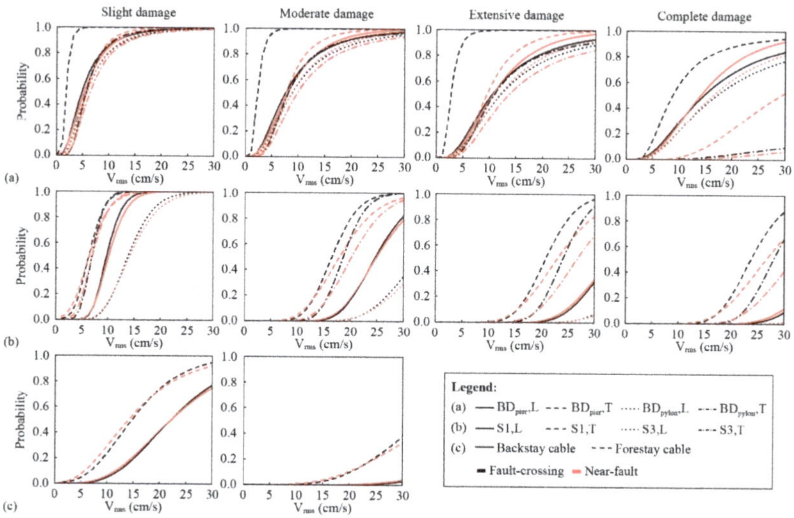

Figure 14. Comparison between the component fragility curves of fault-crossing and near-fault cable-stayed bridges for (**a**) bearings, (**b**) pylons, and (**c**) cables.

Table 9. Median V_{rms} across different damage states for fault-crossing and near-fault bridges (unit: cm/s).

EDP	Slight		Moderate		Extensive		Complete	
	FC	NF	FC	NF	FC	NF	FC	NF
$\sigma_{b,pier}$, longitudial	4.86	5.24	7.62	8.00	10.56	10.63	14.71	13.95
$\sigma_{b,pier}$, transverse	1.77	5.70	2.28	7.63	2.76	9.43	8.73	29.02
$\sigma_{b,pylon}$, longitudial	5.71	5.75	8.81	8.76	12.17	11.90	17.08	16.27
$\sigma_{b,pylon}$, transverse	5.65	6.37	8.15	9.76	10.97	13.62	/	/
φ_{S1}, longitudial	9.77	10.12	24.67	24.88	33.08	32.93	39.49	38.97
φ_{S1}, transverse	6.34	6.33	16.45	17.80	20.76	22.91	23.72	26.49
φ_{S3}, longitudial	13.85	14.12	32.82	34.64	42.35	45.30	49.31	53.18
φ_{S3}, transverse	6.96	6.96	18.61	20.13	24.10	26.75	28.00	31.57
F_{bc}	22.04	21.73	59.95	59.08	/	/	/	/
F_{fc}	14.80	14.02	33.76	36.57	/	/	/	/

Figure 14a depicts that the transverse vulnerability of the bearings at the girder–pier location is significantly different for the two scenarios. Bearings at that location are much more vulnerable to fault-crossing cable-stayed bridges due to the deck's rotation, which is, as mentioned earlier, caused by permanent ground dislocations. As presented in Table 9, the median root-mean-square velocities of the bearings' longitudinal vulnerabilities are similar for the two case bridges. However, Figure 14a shows that when the intensity of ground motions is high, bearings of the bridge subjected to near-fault motions are a little more vulnerable in the longitudinal direction.

Figure 14b compares the vulnerabilities of the pylons for fault-crossing and near-fault bridges. It illustrates that the pylon's vulnerabilities for the two scenarios are comparable under the slight damage state. However, the pylon's transverse vulnerabilities of fault-crossing bridges become higher under the other three damage states, whereas its longitudinal vulnerabilities remain comparable. According to Table 9, the median root-mean-square velocities of the transverse moderate, extensive, and complete damage states of Section 1 for near-fault bridges are 8%, 10%, and 12% larger than those for fault-crossing bridges. Similarly, differences in median values of these damage states of Section 3 between the two cases are 8%, 11%, and 13%, respectively. This is because the transverse seismic responses of the pylon for the fling-step bridge contain not only dynamic ones but also static ones exerted by the ground dislocation. Thus, the transverse seismic demand of the pylon for fault-crossing bridges can be more significant than that for near-fault bridges. In other words, there is a high risk of suffering severe damages in the transverse direction for the pylons if the bridge crosses a fault.

Figure 14c and Table 9 show that the vulnerabilities of the stay cables are almost the same under slight and moderate damage states, whether the bridge is across a fault rupture or not. Such phenomena can be attributed to the fact that the cable force is mainly affected by the vertical and longitudinal deformations of the girder and pylons, and these deformations of the near-fault bridges are supposed to be similar to those of the fault-crossing bridges.

6. Conclusions

This study aims at the seismic fragility assessment for cable-stayed bridges crossing faults. Synthetic fling-step motions are generated and applied to the numerical models. The optimal ground-motion intensity measure for the bridge is identified among 12 candidates. Fragility analysis is conducted, and the results are compared. The conclusions are summarized as follows:

(1) According to the coefficient of determination R^2 and the multicriteria decision-making (MCDM) method, the root-mean-square velocity (V_{rms}) is identified as the optimal IM for the cable-stayed bridge crossing faults.

(2) Bearings are the most fragile components for the fault-crossing bridge, especially in the transverse direction of those on transition piers. The pylon bottom is more

vulnerable than the connection zone between the pylon and cross-girder. In contrast, the cables are not likely to suffer severe damage.

(3) Compared with the bridge subjected to near-fault motions, the vulnerability of pylons and bearings of the fault-crossing bridge becomes higher in the transverse direction. However, the vulnerability of the cables is comparable.

(4) This study limits the fragility analysis of cable-stayed bridges subjected to vertical strike-slip faults with a fault crossing angle of 90°. The vulnerability of cable-stayed bridges crossing dip-slip and oblique-slip faults or existing bridges with other fault crossing angles needs further investigation.

Author Contributions: Conceptualization, J.G., Y.G., W.W., S.C. and X.D.; methodology, J.G., Y.G.; investigation, W.W. and X.D.; writing—original draft preparation, J.G. and Y.G.; writing—review and editing, S.C. and X.D.; supervision, X.D.; formal analysis, J.G. and Y.G.; validation, W.W. and X.D. All authors have read and agreed to the published version of the manuscript.

Funding: This research was funded by the National Key Research and Development Program of China (2019YFE0112300); The Shanghai Post-doctoral Excellent Program (2021333); The Post-doctoral Innovation Practice Base Program of Shanghai Yangpu District; and the National Natural Science Foundation of China (51978512).

Conflicts of Interest: The authors declare no conflict of interest.

References

1. Hui, Y.; Wang, K.; Wu, G.; Li, C. Seismic responses of bridges crossing faults and their best crossing angles. *J. Vib. Shock* **2015**, *34*, 6–11. (In Chinese)
2. Yang, S.; Mavroeidis, G.P. Bridges crossing fault rupture zones: A review. *Soil Dyn. Earthq. Eng.* **2018**, *113*, 545–571. [CrossRef]
3. Yang, S.; Mavroeidis, G.P.; Tsopelas, P. Seismic response study of ordinary and isolated bridges crossing strike-slip fault rupture zones. *Earthq. Eng. Struct. Dyn.* **2021**, *50*, 2841–2862. [CrossRef]
4. Park, S.W.; Ghasemi, H.; Shen, J.; Somerville, P.G.; Yen, W.P.; Yashinsky, M. Simulation of the seismic performance of the Bolu Viaduct subjected to near-fault ground motions. *Earthq. Eng. Struct. Dyn.* **2004**, *33*, 1249–1270. [CrossRef]
5. Goel, R.K.; Asce, F.; Chopra, A.K.; Asce, M. Role of Shear Keys in Seismic Behavior of Bridges Crossing Fault-Rupture Zones. *J. Bridg. Eng.* **2008**, *13*, 398–408. [CrossRef]
6. Goel, R.K.; Chopra, A.K. Nonlinear analysis of ordinary bridges crossing fault-rupture zones. *J. Bridg. Eng.* **2009**, *14*, 216–224. [CrossRef]
7. Goel, R.K.; Chopra, A.K. Linear analysis of ordinary bridges crossing fault-rupture zones. *J. Bridg. Eng.* **2009**, *14*, 203–215. [CrossRef]
8. Ucak, A.; Mavroeidis, G.P.; Tsopelas, P. Assessment of Fault Crossing Effects on a Seismically Isolated Multi-Span Bridge. In Proceedings of the 15th World Conference on Earthquake Engineering (15WCEE), Lisbon, Portugal, 24–28 September 2012.
9. Ucak, A.; Mavroeidis, G.P.; Tsopelas, P. Behavior of a seismically isolated bridge crossing a fault rupture zone. *Soil Dyn. Earthq. Eng.* **2014**, *57*, 164–178. [CrossRef]
10. Yang, H.; Li, J. Response analysis of seismic isolated bridge under influence of fault-crossing ground motions. *J. Tongji Univ. (Nat. Sci.)* **2015**, *43*, 1144–1152. (In Chinese) [CrossRef]
11. Zhang, F.; Li, S.; Wang, J.; Zhang, J. Effects of fault rupture on seismic responses of fault-crossing simply-supported highway bridges. *Eng. Struct.* **2020**, *206*, 110104. [CrossRef]
12. Sharabash, A.M.; Andrawes, B.O. Application of shape memory alloy dampers in the seismic control of cable-stayed bridges. *Eng. Struct.* **2009**, *31*, 607–616. [CrossRef]
13. Zhong, J.; Jeon, J.-S.; Yuan, W.; DesRoches, R. Impact of Spatial Variability Parameters on Seismic Fragilities of a Cable-Stayed Bridge Subjected to Differential Support Motions. *J. Bridg. Eng.* **2017**, *22*, 04017013. [CrossRef]
14. Zeng, C.; Song, G.; Jiang, H.; Huang, L.; Guo, H.; Ma, X. Nonliear Seismic Response Characteristics of Fault-crossing Single-tower Cable-stayed Bridge. *China J. Highw. Transp.* **2021**, *34*, 230–245. (In Chinese)
15. Gu, Y.; Guo, J.; Dang, X.; Yuan, W. Seismic performance of a cable-stayed bridge crossing strike-slip faults. *Structures* **2022**, *35*, 289–302. [CrossRef]
16. Wu, Y.M.; Wu, C.F. Approximate recovery of coseismic deformation from Taiwan strong-motion records. *J. Seismol.* **2007**, *11*, 159–170. [CrossRef]
17. Iwan, W.D.; Moser, M.A.; Peng, C.-Y. Some observations on strong-motion earthquake measurement using a digital accelerograph. *Bull. Seismol. Soc. Am.* **1985**, *75*, 1225–1246. [CrossRef]
18. Boore, D.M. Effect of baseline corrections on displacements and response spectra for several recordings of the 1999 Chi-Chi, Taiwan, earthquake. *Bull. Seismol. Soc. Am.* **2001**, *91*, 1199–1211. [CrossRef]
19. Lin, Y.; Zong, Z.; Tian, S.; Lin, J. A new baseline correction method for near-fault strong-motion records based on the target final displacement. *Soil Dyn. Earthq. Eng.* **2018**, *114*, 27–37. [CrossRef]

20. Charles, M. An Analytical Model for Near-Fault Ground Motions and the Response of SDOF Systems. In Proceedings of the 7th U.S. National Conference on Earthquake Engineering, Boston, MA, USA, 21–25 July 2002.
21. Mavroeidis, G.P.; Papageorgiou, A.S. A mathematical representation of near-fault ground motions. *Bull. Seismol. Soc. Am.* **2003**, *93*, 1099–1131. [CrossRef]
22. Makris, N.; Black, C.J. Dimensional Analysis of Rigid-Plastic and Elastoplastic Structures under Pulse-Type Excitations. *J. Eng. Mech.* **2004**, *130*, 1006–1018. [CrossRef]
23. Hoseini Vaez, S.R.; Sharbatdar, M.K.; Ghodrati Amiri, G.; Naderpour, H.; Kheyroddin, A. Dominant pulse simulation of near fault ground motions. *Earthq. Eng. Eng. Vib.* **2013**, *12*, 267–278. [CrossRef]
24. Kamai, R.; Abrahamson, N.; Graves, R. Adding fling effects to processed ground-motion time histories. *Bull. Seismol. Soc. Am.* **2014**, *104*, 1914–1929. [CrossRef]
25. Burks, L.S.; Baker, J.W. A predictive model for fling-step in near-fault ground motions based on recordings and simulations. *Soil Dyn. Earthq. Eng.* **2016**, *80*, 119–126. [CrossRef]
26. Yadav, K.K.; Gupta, V.K. Near-fault fling-step ground motions: Characteristics and simulation. *Soil Dyn. Earthq. Eng.* **2017**, *101*, 90–104. [CrossRef]
27. Hamidi, H.; Khosravi, H.; Soleimani, R. Fling-step ground motions simulation using theoretical-based Green's function technique for structural analysis. *Soil Dyn. Earthq. Eng.* **2018**, *115*, 232–245. [CrossRef]
28. Yang, S.; Mavroeidis, G.P.; Ucak, A. Analysis of bridge structures crossing strike-slip fault rupture zones: A simple method for generating across-fault seismic ground motions. *Earthq. Eng. Struct. Dyn.* **2020**, *49*, 1281–1307. [CrossRef]
29. Papageorgiou, A.S.; Aki, K. A specific barrier model for the quantitative description of inhomogenous faulting and the prediction of strong ground motion. *Bull. Seismol. Soc. Am.* **1983**, *73*, 693–722. [CrossRef]
30. Rezaeian, S.; Kiureghian, A. Der A stochastic ground motion model with separable temporal and spectral nonstationarities. *Earthq. Eng. Struct. Dyn.* **2008**, *37*, 1565–1584. [CrossRef]
31. Rezaeian, S.; Kiureghian, A. Der Simulation of synthetic ground motions for specified earthquake and site characteristics. *Earthq. Eng. Struct. Dyn.* **2010**, *39*, 1155–1180. [CrossRef]
32. Dabaghi, M.; Der Kiureghian, A. *Stochastic Modeling and Simulation of Near-Fault Ground Motions for Performance-Based Earthquake Engineering*; University of California: Berkeley, CA, USA, 2014.
33. Dabaghi, M.; Der Kiureghian, A. Stochastic model for simulation of near-fault ground motions. *Earthq. Eng. Struct. Dyn.* **2016**, *46*, 963–984. [CrossRef]
34. Dabaghi, M.; Der Kiureghian, A. Simulation of orthogonal horizontal components of near-fault ground motion for specified earthquake source and site characteristics. *Earthq. Eng. Struct. Dyn.* **2018**, *47*, 1369–1393. [CrossRef]
35. Pang, Y.; Wu, X.; Shen, G.; Yuan, W. Seismic Fragility Analysis of Cable-Stayed Bridges Considering Different Sources of Uncertainties. *J. Bridg. Eng.* **2014**, *19*, 04013015. [CrossRef]
36. Zhong, J.; Pang, Y.; Jeon, J.S.; Desroches, R.; Yuan, W. Seismic fragility assessment of long-span cable-stayed bridges in China. *Adv. Struct. Eng.* **2016**, *19*, 1797–1812. [CrossRef]
37. Zhong, J.; Jeon, J.S.; Ren, W.X. Risk assessment for a long-span cable-stayed bridge subjected to multiple support excitations. *Eng. Struct.* **2018**, *176*, 220–230. [CrossRef]
38. Zhong, J.; Jeon, J.-S.; Shao, Y.-H.; Chen, L. Optimal Intensity Measures in Probabilistic Seismic Demand Models of Cable-Stayed Bridges Subjected to Pulse-Like Ground Motions. *J. Bridg. Eng.* **2019**, *24*, 04018118. [CrossRef]
39. Wu, W.; Li, L.; Shao, X. Seismic Assessment of Medium-Span Concrete Cable-Stayed Bridges Using the Component and System Fragility Functions. *J. Bridg. Eng.* **2016**, *21*, 04016027. [CrossRef]
40. Wang, J.; Li, S.; Zhang, F. Seismic fragility analyses of long-span cable-stayed bridge isolated by SMA Wire-based Smart Rubber Bearing in near-fault regions. *China J. Highw. Transp.* **2017**, *30*, 30–39. (In Chinese) [CrossRef]
41. Li, C.; Li, H.N.; Hao, H.; Bi, K.; Chen, B. Seismic fragility analyses of sea-crossing cable-stayed bridges subjected to multi-support ground motions on offshore sites. *Eng. Struct.* **2018**, *165*, 441–456. [CrossRef]
42. Wei, B.; Hu, Z.; He, X.; Jiang, L. Evaluation of optimal ground motion intensity measures and seismic fragility analysis of a multi-pylon cable-stayed bridge with super-high piers in Mountainous Areas. *Soil Dyn. Earthq. Eng.* **2020**, *129*, 105945. [CrossRef]
43. Zhu, M.; McKenna, F.; Scott, M.H. OpenSeesPy: Python library for the OpenSees finite element framework. *SoftwareX* **2018**, *7*, 6–11. [CrossRef]
44. Mangalathu, S.; Jeon, J.-S.; Padgett, J.E.; DesRoches, R. Performance-based grouping methods of bridge classes for regional seismic risk assessment: Application of ANOVA, ANCOVA, and non-parametric approaches. *Earthq. Eng. Struct. Dyn.* **2017**, *46*, 2587–2602. [CrossRef]
45. Pan, Y.; Agrawal, A.K.; Ghosn, M. Seismic Fragility of Continuous Steel Highway Bridges in New York State. *J. Bridg. Eng.* **2007**, *12*, 689–699. [CrossRef]
46. Baker, J.W. Quantitative classification of near-fault ground motions using wavelet analysis. *Bull. Seismol. Soc. Am.* **2007**, *97*, 1486–1501. [CrossRef]
47. Somerville, P.G.; Smith, N.F.; Graves, R.W.; Abrahamson, N.A. Modification of empirical strong ground motion attenuation relations to include the amplitude and duration effects of rupture directivity. *Seismol. Res. Lett.* **1997**, *68*, 199–222. [CrossRef]
48. Abrahamson, N. Velocity pulses in near-fault ground motions. In Proceedings of the UC Berkeley—CUREE Symposium in Honor of Ray Clough and Joseph Penzien, Berkeley, CA, USA, 9–11 May 2002; pp. 40–41.

49. Guo, J.; Yuan, W.; Dang, X.; Alam, M.S. Cable force optimization of a curved cable-stayed bridge with combined simulated annealing method and cubic B-Spline interpolation curves. *Eng. Struct.* **2019**, *201*, 109813. [CrossRef]
50. Stefanidou, S.P.; Kappos, A.J. Methodology for the development of bridge-specific fragility curves. *Earthq. Eng. Struct. Dyn.* **2017**, *46*, 73–93. [CrossRef]
51. Nielson, B.G.; DesRoches, R. Analytical seismic fragility curves for typical bridges in the central and southeastern United States. *Earthq. Spectra* **2007**, *23*, 615–633. [CrossRef]
52. Wang, X.; Shafieezadeh, A.; Ye, A. Optimal intensity measures for probabilistic seismic demand modeling of extended pile-shaft-supported bridges in liquefied and laterally spreading ground. *Bull. Earthq. Eng.* **2018**, *16*, 229–257. [CrossRef]
53. Dreger, D.; Hurtado, G.; Chopra, A.; Larsen, S. *Near-Fault Seismic Ground Motions*; Report No. UCB/EERC-2007/03; Earthquake Engineering Research Center, University of California: Berkeley, CA, USA, 2007.
54. Nuttli, O.W. *The Relation of Sustained Maximum Acceleration and Velocity to Earthquake Intensity and Magnitude*; Report No. S-73-1; U.S. Army Engineer Waterways Experiment Station: Vicksburg, MI, USA, 1979.
55. Reed, J.W.; Kassawara, R.P. A criterion for determining exceedance of the operating basis earthquake. *Nucl. Eng. Des.* **1990**, *123*, 387–396. [CrossRef]
56. Arias, A. *A Measure of Earthquake Intensity*; MIT Press: Cambridge, MA, USA, 1969.
57. Padgett, J.E.; Nielson, B.G.; Desroches, R. Selection of optimal intensity measures in probabilistic seismic demand models of highway bridge portfolios. *Earthq. Eng. Struct. Dyn.* **2008**, *37*, 711–725. [CrossRef]
58. Hedayati Dezfuli, F.; Alam, M.S. Multi-criteria optimization and seismic performance assessment of carbon FRP-based elastomeric isolator. *Eng. Struct.* **2013**, *49*, 525–540. [CrossRef]
59. Zhang, C.; Lu, J.; Zhou, Z.; Yan, X.; Xu, L.; Lin, J. Lateral seismic fragility assessment of cable-stayed bridge with diamond-shaped concrete pylons. *Shock Vib.* **2021**, *2021*, 2847603. [CrossRef]
60. Bradley, B.A. Empirical correlations between cumulative absolute velocity and amplitude-based ground motion intensity measures. *Earthq. Spectra* **2012**, *28*, 37–54. [CrossRef]
61. Feng, R.; Yuan, W.; Sextos, A. Probabilistic loss assessment of curved bridges considering the effect of ground motion directionality. *Earthq. Eng. Struct. Dyn.* **2021**, *50*, 3623–3645. [CrossRef]

buildings

MDPI

Article

Effect of an Innovative Friction Damper on Seismic Responses of a Continuous Girder Bridge under Near-Fault Excitations

Junjun Guo [1], Huaifeng Li [2], Changyong Zhang [2], Shihyu Chu [3] and Xinzhi Dang [1,*]

1 State Key Laboratory of Disaster Reduction in Civil Engineering, Tongji University, 1239, Siping Road, Shanghai 200092, China; guojj@tongji.edu.cn
2 Shandong Provincial Communications Planning and Design Institute, Jinan 250031, China; bgs@sdcpd.com (H.L.); kjxx@sdcpd.com (C.Z.)
3 Department of Civil Engineering, National ChengKung University, No. 1 University Road, Tainan 701, Taiwan; sychu@mail.ncku.edu.tw
* Correspondence: 021tjdangxz@tongji.edu.cn

Abstract: Continuous girder bridges have been extensively constructed in China over the past 30 years, and these bridges tend to experience severe damage under ground motions with velocity pulses. In the current research, an innovative linear friction damper (LFD) is proposed to mitigate the seismic damages of continuous girder bridges subjected to near-fault ground motions. The OpenSees platform is adopted to establish the numerical model of a continuous girder bridge in the near-fault region. Sixteen ground motions with velocity pulses are selected from the PEER ground motion database. The wavelet method is used to extract the maximum velocity pulse from the two orthogonal components of a ground motion. The effects of the initial gap, the coefficient of friction, and the spring stiffness of LFD on the seismic responses of the bridge are investigated by the response surface method (RSM). The seismic responses of the bridge for the original system (Non-isolated), LFD system (Isolated-LFD), and lead rubber bearing (LRB) system (Isolated-LRB), such as force–displacement relationship, bearing displacement, and pier curvature, are obtained after conducting a series of nonlinear time history analyses. The numerical results reveal that this innovative device (LFD) can effectively control the relative displacements between the superstructure and substructure of the bridge. Meanwhile, the seismic responses of the piers can be significantly decreased compared with the non-isolated system.

Keywords: near-fault isolation; velocity pulses; continuous girder bridges; linear friction damper; lead rubber bearing

Citation: Guo, J.; Li, H.; Zhang, C.; Chu, S.; Dang, X. Effect of an Innovative Friction Damper on Seismic Responses of a Continuous Girder Bridge under Near-Fault Excitations. *Buildings* **2022**, *12*, 1019. https://doi.org/10.3390/buildings12071019

Academic Editors: Bo Wang and Rita Bento

Received: 7 June 2022
Accepted: 5 July 2022
Published: 15 July 2022

Copyright: © 2022 by the authors. Licensee MDPI, Basel, Switzerland. This article is an open access article distributed under the terms and conditions of the Creative Commons Attribution (CC BY) license (https://creativecommons.org/licenses/by/4.0/).

1. Introduction

Continuous girder bridges have been widely constructed in China over the past 30 years. These bridges are characterized by expansion joints between adjacent segments to conform to displacement induced by temperature and fixed bearings at one of the piers to resist braking force. Unseating of the superstructure and pier collapse of girder bridges have been frequently observed in the past with strong earthquakes [1]. In the 2008 Wenchuan earthquake, the Huilan bridge was severely damaged, and it was found that the bearing on the top of the shortest pier was damaged first, and then, the pier with the largest flexural stiffness failed [2,3]. The girder may experience unseating at the expansion joint for a continuous girder bridge. Thus, it is essential to maintain the integrity of a continuous girder bridge. The inertial force of the superstructure should not be transmitted mainly through the pier with a fixed bearing.

Near-fault ground motions possessing strong velocity pulses impose high seismic demands for structures. The directivity effects occur when the fault ruptures toward the direction of the site at a speed approximately equal to the shear wave velocity [4]. Velocity pulses are characterized by a long period and short duration, leading to significantly larger seismic responses than far-field ground motions [5]. For example, girder bridges will experience excessive relative displacement at expansion joints and severe damage at piers with fixed bearings. Passive control devices have been used to isolate the bridge deck from the substructure and to provide additional damping to dissipate input energy. These isolation systems, including bearings [6–11] and dampers [12–14], are widely adopted to control the seismic responses of bridges. However, these isolation devices might perform poorly under near-fault ground motions because large displacement caused by long-period velocity pulses can be difficult to accommodate by these systems [5]. Providakis [15,16] investigated the effect of supplemental damping on LRB seismic isolators under near-fault ground motions. The results revealed that the supplemental damping must be carefully controlled to avoid some adverse effects. Panchal and Jangid [17] studied the application of a variable friction pendulum system on structures under near-fault ground motions. It was found that this device can effectively control the seismic response of buildings subjected to near-fault ground motions. Dicleli et al. [18] conducted a parametric analysis to select the optimal isolator properties for bridges under near-fault ground motions. It was observed that energy dissipation devices are required to decrease the isolator displacements. Li et al. [19] found that shape memory alloy wire-based lead rubber bearings effectively control the deck displacement and reduce the internal forces of the pylon of cable-stayed bridges. Based on the research above, the isolation systems used to reduce the seismic responses of structures under near-fault ground motions should be characterized by large displacement capacity. Meanwhile, they should be able to control the relative displacement between the superstructure and substructure; thus, pounding and unseating can be avoided.

This paper aims to propose a linear friction damper (LFD) and validate the effectiveness of this device installed at continuous girder bridges in the near-fault region. First, this device's working mechanism, numerical validation, and design process are introduced in detail. Second, the numerical model of a continuous girder bridge is established using the OpenSees program. Third, 16 seismic records with strong velocity pulses are selected from the PEER website as input ground motions. Then, a parametric analysis of LFD is conducted to determine the appropriate parametric values for this device. Finally, the seismic responses of the bridge with the original system (Non-isolated), lead rubber bearing system (Isolated-LRB), and LFD system (Isolated-LFD), such as bearing displacement and pier curvature, are obtained after performing a series of nonlinear time history analyses. Comparisons are made between the original system and the isolation systems to evaluate the effectiveness of LFD in mitigating the overall seismic responses of bridges subjected to ground motions with velocity pulses.

2. Linear Friction Damper (LFD)

2.1. Working Mechanism of LFD

Figure 1 depicts a self-centering linear friction damper (LFD) composed of a guiding system and sliding block. The sliding block is rigidly connected with the superstructure and can move along the axial direction of the guiding rod. The guiding system comprises a slot, guiding rods, hinge plates, baffles, friction blocks, and springs. The bottom of the slot is fixed on the substructure, and both ends of the guiding rod are rigidly connected to the slot. The springs providing restoring force are rigidly connected with the slot and baffles. The hinge rods hinge plates, and the friction blocks are hinged. Thus, relative displacement will take place when they are subjected to force action. The working mechanism of LFD can be summarized as below: After the initial gap between the sliding block and slot is filled, the sliding block starts to push the hinge plate, and the friction blocks are forced to move outside. This movement is strongly restricted by the side plate of the slot. Thus, normal pressure between the friction block and slot takes place. Moreover, this force linearly increases with the movement of the sliding block. Meanwhile, the corresponding spring is also compressed to resist the force transmitted from the sliding block. In the unloading process, the direction of normal pressure will change to the opposite direction. The force–displacement relationship of LFD is illustrated in Figure 2. In the current study, the angle between the hingle plate and the hingle rod is assumed to be 135 degrees. In the loading process (Figure 2a), the external force is resisted by the spring force and the friction force in the direction of displacement. In the preliminary research of the device, for brevity, it is assumed that the friction coefficient complies with the Coulomb friction law. Thus, the force–displacement relationship of LFD in the loading process can be obtained as below:

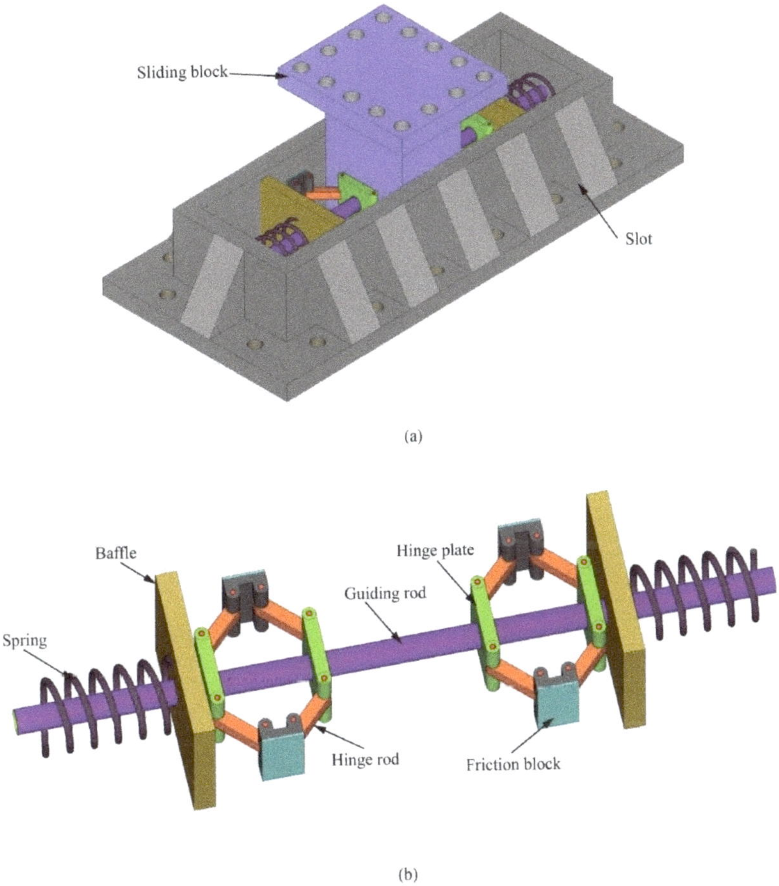

Figure 1. Schematic illustration of the linear friction damper (LFD): (**a**) general view of LFD; (**b**) internal details of LFD.

$$F = kd + 2f \tag{1}$$

$$f = \frac{kd + F}{2}\mu \tag{2}$$

$$F = \frac{1+\mu}{1-\mu}kd \tag{3}$$

where μ is the friction coefficient, k is the spring stiffness, and d is the displacement. It can be found that the stiffness of LFD in the loading process is $k(1+\mu)/(1-\mu)$, which is larger than that of spring due to the contribution of the friction effect. In the unloading process (Figure 2b), the direction of the friction force f is reversed, and the spring force kd

is balanced by the external force F and friction force f. As a result, the force–displacement relationship of LFD in the unloading process can be derived as below:

$$F = \frac{1-\mu}{1+\mu}kd \tag{4}$$

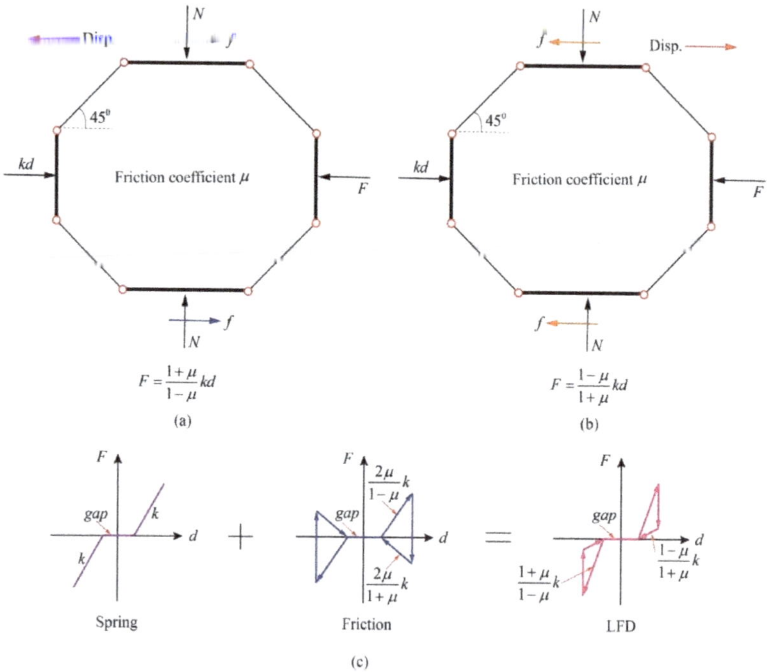

Figure 2. The working process of the LFD: (**a**) loading process; (**b**) unloading process; (**c**) constitutive model.

Based on the aforementioned analysis, the force–displacement relationship of LFD comprises two parts: spring and linear friction (Figure 2c). It is found that an initial gap can be set in LFD, and the device does not play its role until the initial gap is exhausted. The spring system can provide strong restoring force. Therefore, LFD can accommodate large displacements caused by velocity pulses and can effectively control the corresponding displacements.

2.2. Numerical Validation of LFD

A numerical experiment was conducted to validate the theoretical force–displacement relationship of LFD. Important features of the FE model are shown in Figure 3a. Only half of the model is considered for the numerical experiment. The slot, hinge rods, and friction blocks are modeled using linear eight-node solid elements. For brevity, the hinge parts are not precisely modeled, and constraint is adopted to simulate the hinge behavior. In addition, the spring in LFD is simulated by a linear elastic material with a stiffness of 2500 N/mm. The contact and friction behaviors between the slot and friction block are elaborately simulated. The Coulomb law is adopted to simulate the contact property of the surfaces in the tangential direction, and the friction coefficient is set to 0.2. Based on the previous research [20], the friction coefficient depends on contact pressure, sliding velocity, and air temperature. In the preliminary research of this new device, for brevity, the Coulomb friction law is used. The displacements of 50, 100, and 150 mm are applied to the loading point to obtain the force–displacement relationship of the device. Figure 3b illustrates the Mises stress of the device in the loading process. Figure 4 compares the theoretical constitutive model and that of the numerical experiment. It is found that the theoretical model is consistent with that of the numerical experiment. Moreover, this theoretical institutive model will be adopted in the following sections.

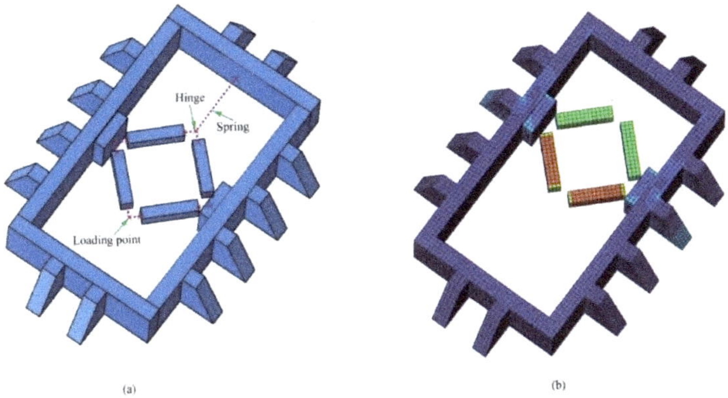

Figure 3. The numerical experiment of LFD: (**a**) arrangement; (**b**) Mises stress in the loading process.

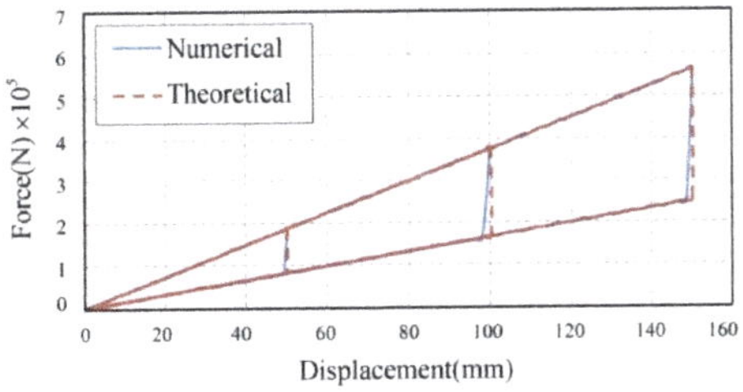

Figure 4. The theoretical constitutive model compared to that of the numerical experiment of LFD.

2.3. Design Process of LFD

LFDs can be installed either in the longitudinal or transverse direction of bridges, such as girder and cable-stayed bridges. The design flow chart of LFD is demonstrated in Figure 5. According to the force–displacement relationship, it is found that the critical parameters of LFD are the initial gap d_0, friction coefficient μ, and spring stiffness k. The effects of spring stiffness on the seismic responses will be discussed in the following section in detail. For a specific case, first, the initial parametric values of LFD are assumed (for example, the friction coefficient can be determined by the material property of the contact surface, the initial gap can be set based on the displacement induced by temperature, and the initial stiffness can be set to 10,000 kN/m). LFDs and bidirectional sliding bearings replace the original bearing arrangement of the bridge. Second, a nonlinear time history analysis is performed to obtain the seismic responses of the critical components, such as superstructure displacement and pier curvature. Then, whether the seismic responses exceed their corresponding limits is examined. If the seismic responses are over their limit values, the parametric values of LFD should be reassumed. Otherwise, the current parametric values will be selected. Finally, the dimensional details of LFD are determined based on the parametric values and seismic responses of LFD.

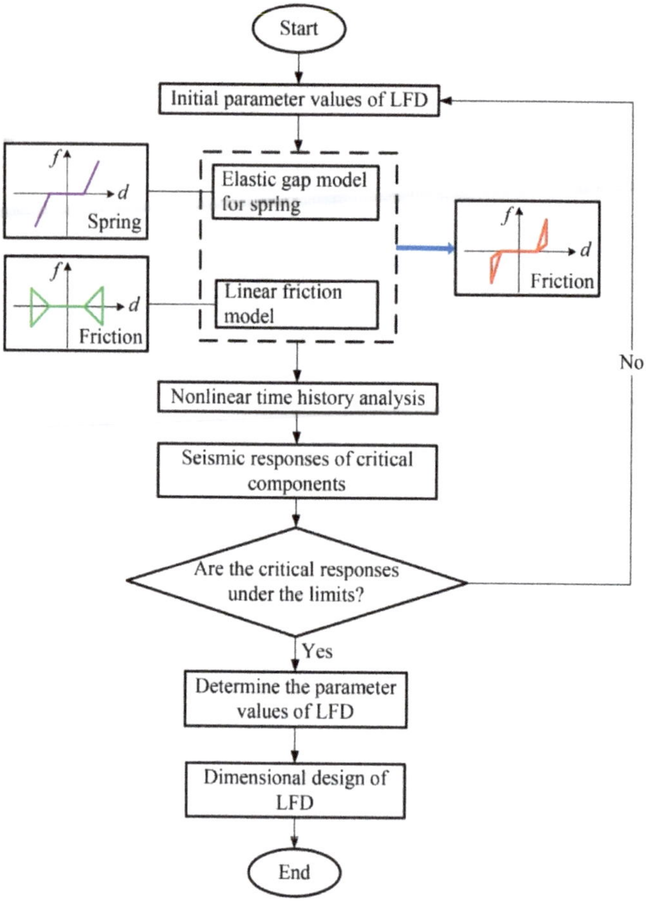

Figure 5. Design flow chart of the bridge installed with the LFD.

3. Continuous Girder Bridge

3.1. Geometry Description

In order to evaluate the isolation effect of continuous girder bridges equipped with LFDs when subjected to near-fault ground motions, the current study adopts a three-span continuous box-girder bridge with typical configurations of girder bridges in China. Figure 6a illustrates the bridge's elevation with a main and side span length of 100 and 60 m, respectively. The concrete strength of the bridge is 29.2 MPa, and the yield strength of the reinforcement is 345 MPa. The height of the bridge girder varies from 3 to 6.5 m, and expansion joints are installed to accommodate displacement induced by temperature. The original bearing arrangement (Non-isolated) is presented in Figure 6b, including fixed bearing, unidirectional and bidirectional sliding bearings. As a result, the superstructure can freely deform in the longitudinal and transverse directions. Lead rubber bearings (LRBs), due to their high damping properties, are used worldwide [21]. Therefore, in the current study, the comparisons are made between the LRBs system (Isolated-LRB) and the LFDs system (Isolated-LFD) to evaluate the effectiveness of LFDs in mitigating the seismic responses of the bridge under near-fault ground motions. Figure 6c demonstrates the cross-sections of the box-girder and piers.

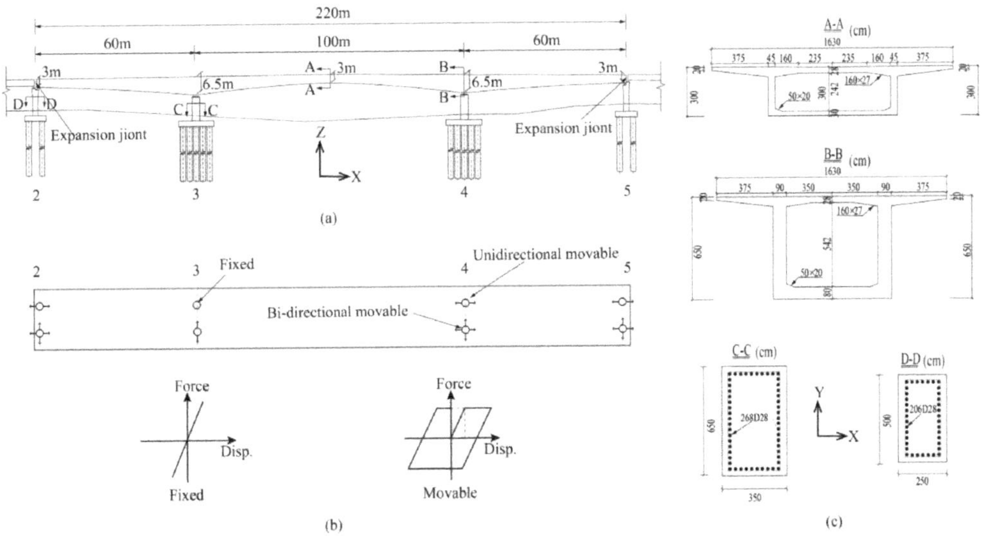

Figure 6. Description of the girder bridge: (**a**) elevation and (**b**) plan of the prototype bridge and (**c**) section properties.

3.2. Numerical Modeling

The OpenSees (the Open System for Earthquake Engineering Simulation) [22] platform is adopted to establish the numerical model of the bridge, as can be seen in Figure 7. In the current study, the superstructure is modeled as a spine, and the elastic beam–column element is used to simulate it because the deck is expected to maintain elasticity during seismic excitation. In order to account for pier nonlinearity, fiber sections composed of confined concrete, unconfined concrete, and longitudinal reinforcement are employed to establish piers. The model of Mander et al. [23] is adopted to simulate the confined concrete. Lumped linear springs, including three translational and rotational springs, model the pile foundations. In addition, the pile caps mass is lumped at their centers. Pot rubber bearings are used for the original bearing system (Non-isolated), and the bearing arrangement can be seen in Figure 6b. For the lead rubber bearing system (Isolated-LRB), the bearings in piers 3 and 4 are replaced with LRBs, as shown in Figure 7. Based on the vertical load capacity of LRB, the elastic stiffness (K1), post-yield stiffness (K2), and characteristic strength (Qd) of LRBs are set to 30,000 kN/m, 4500 kN/m, and 771 kN, respectively. For the linear friction damper system (Isolated-LFD), the bearings in pier 3 and 4 are replaced with bidirectional movable bearings and LFDs, as seen in Figure 7. In the current research, the ground motions excite the bridge only in the longitudinal direction, and the other two directions are not considered.

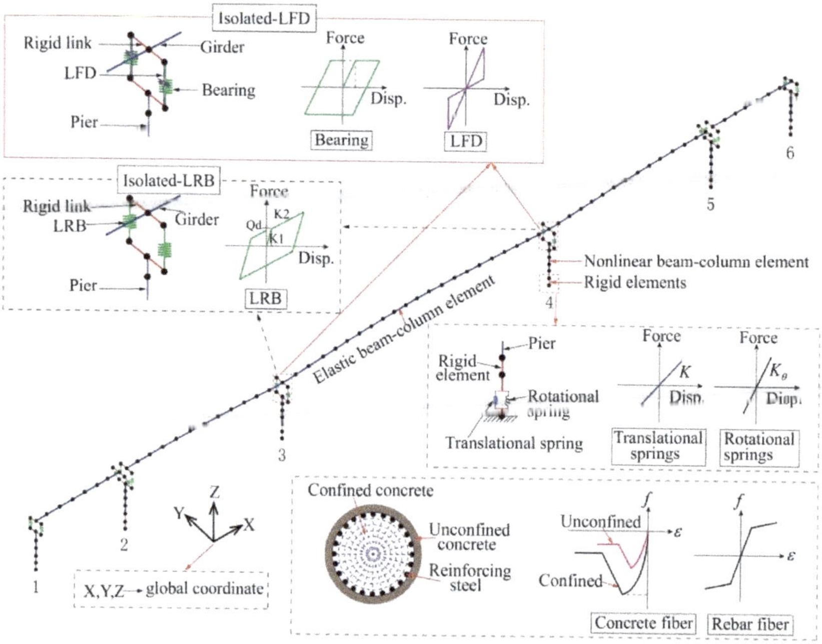

Figure 7. Numerical model of the prototype bridge.

4. Near-Fault Ground Motions

To evaluate the directivity effects on the seismic responses of the prototype bridge, sixteen ground motions with velocity pulses were selected from the Pacific Earthquake Engineering Research Center database in the current study, as seen in Table 1. The magnitudes and pulse periods in Table 1 are varied from 5.8 to 7.62 and from 0.57 to 9.33 s, respectively. The original ground motion refers to a linear combination of two orthogonal components of a ground motion in the strongest velocity pulse direction. In the present study, the procedure developed by Baker [24,25] was adopted to extract the largest velocity pulse from an original ground motion. This procedure mainly contains two steps: First, Daubechies wavelet of order four is adopted to calculate the wavelet transform coefficients of the two orthogonal components of a ground motion, and the sum of the squares of these two coefficients sequences is compared to determine the largest velocity pulse direction. A linear combination of the two orthogonal components in the largest velocity pulse direction yields the original ground motion. Second, wavelet analysis is performed for the original ground motion to extract the largest velocity pulse, and the details are referred to in [24]. The peak ground accelerations (PGA), peak ground velocities (PGV), and pulse periods (Tp) in Table 1 correspond to their original ground motions, as discussed previously. Figure 8 presents the largest velocity pulses extracted from the ground motions in Table 1, and these pulses are characterized by different shapes, peak ground velocities, and pulse periods. Thus, Daubechies wavelet of order four can effectively extract velocity pulses from ground motions. Figure 9 illustrates the velocity spectra of the original ground motion, extracted pulse, and residual ground motion for ground motions in Table 1. It can be found that velocity pulses possess long periods. In addition, the pulse periods are not always consistent with the maximum spectral periods of the original ground motions. For instance, the pulse period of the 13th ground motion in Table 1 is larger than its corresponding maximum spectral period of original ground motion. Therefore, it is more reasonable to determine pulse period from the aspect of time–frequency than from the view of frequency. In the current research, the original ground motions in Table 1 are applied to the longitudinal direction of the prototype in the following sections.

Table 1. Ground motions with forward-directivity effects.

No.	Record Number	Earthquake Name	Year	Magnitude	Rrup (km)	PGA (g)	PGV (cm/s)	Tp (s)
1	4126	Parkfield-02_CA	2004	6	3.79	0.85	43.2	0.57
2	1052	Northridge-01	1994	6.69	7.26	0.53	56.1	0.73
3	568	San Salvador	1986	5.8	6.3	0.70	68.3	0.81
4	1004	Northridge-01	1994	6.69	8.44	0.75	77.8	0.92
5	1106	Kobe_Japan	1995	6.9	0.96	0.86	105.1	1.09
6	77	San Fernando	1971	6.61	1.81	1.38	121.7	1.64
7	723	Superstition Hills-02	1987	6.54	0.95	0.46	143.8	2.39
8	828	Cape Mendocino	1992	7.01	8.18	0.71	96.6	3.00
9	982	Northridge-01	1994	6.69	5.43	0.38	101.5	3.14
10	1085	Northridge-01	1994	6.69	5.19	0.84	113.9	3.61
11	1176	Kocaeli_Turkey	1999	7.51	4.83	0.28	90.6	4.95
12	1244	Chi-Chi_Taiwan	1999	7.62	9.94	0.39	108.8	5.34
13	1501	Chi-Chi_Taiwan	1999	7.62	9.78	0.18	78.9	6.55
14	879	Landers	1992	7.28	2.19	0.72	132.1	5.12
15	1502	Chi-Chi_Taiwan	1999	7.62	16.59	0.13	52.3	8.46
16	1482	Chi-Chi_Taiwan	1999	7.62	19.89	0.20	57.9	9.33

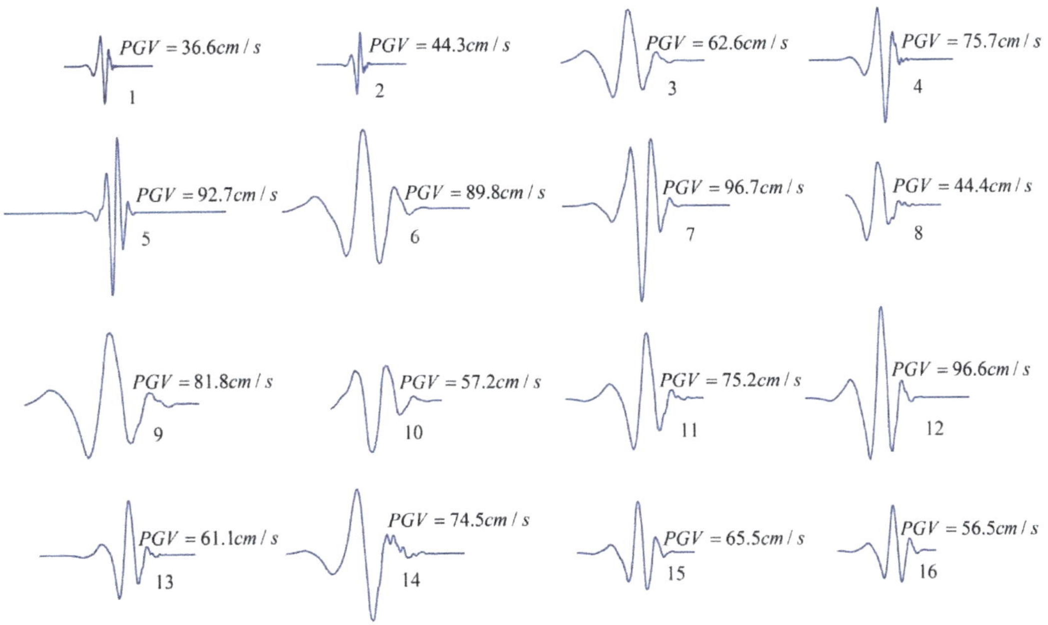

Figure 8. The strongest velocity pulses extracted from ground motions in Table 1.

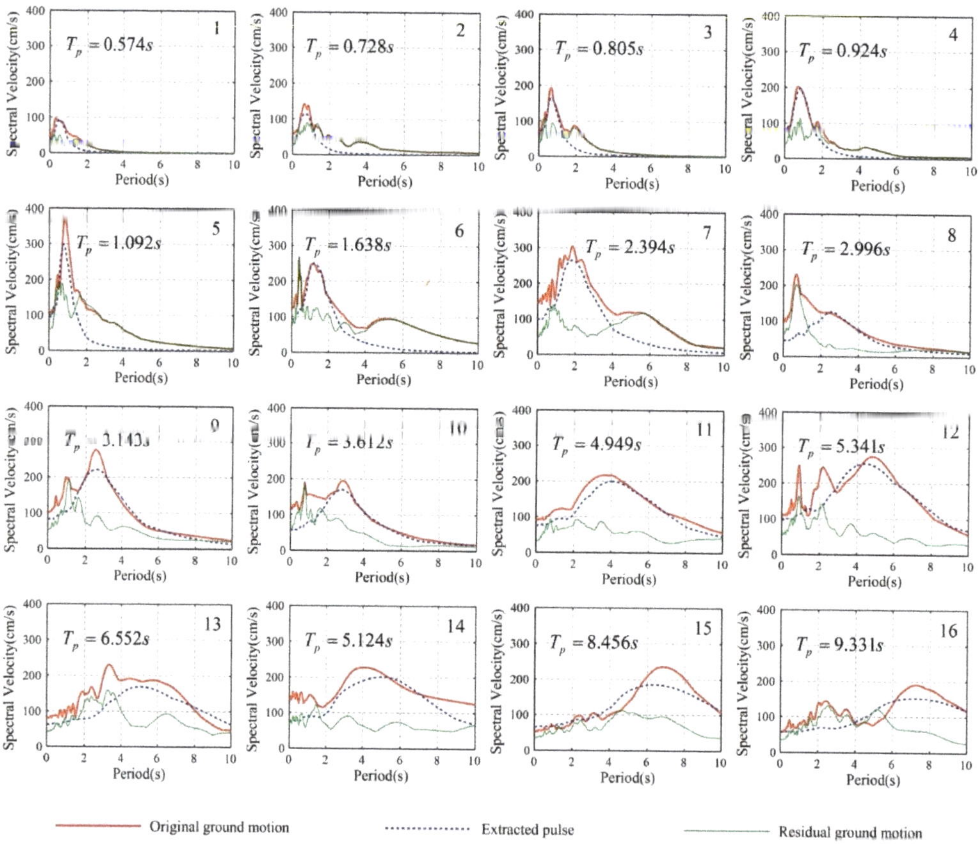

Figure 9. Velocity spectra of the original ground motion, extracted pulse, and residual ground motion for ground motions in Table 1.

5. Numerical Results and Discussion

5.1. Parametric Analysis of LFD

In this section, the Response Surface Method (RSM) [26] is used to obtain the optimal parametric values of LFD. The fundamental idea of RSM is to adopt a polynomial with multiple variables to simulate the real response surface [27]. More detailed information on RSM can be referred to in the work of Towashiraporn [28].

In the current study, the initial gap (x_1), the coefficient of friction (x_2), and the spring stiffness (x_3) of LFD are regarded as design parameters. For each ground motion, 15 experimental design points are generated based on RSM, and nonlinear time histories are conducted to obtain the maximum pier curvatures and bearing displacements of the bridge at the design points. Figure 10 illustrates the contour maps of the maximum pier curvatures and bearing displacements with respect to the initial gap (x_1), the coefficient of friction (x_2), and the spring stiffness (x_3) of LFD ground motions 1 and 7 in Table 1. It is found that the maximum pier curvatures increase with the increase in spring stiffness. Moreover, the initial gap (x_1) significantly influences the bearing displacement. It also indicates that the coefficient of friction (x_2) has little effect on the seismic responses of the bridge compared to the other two parameters. The large displacement accommodation should be considered for ground motions with velocity pulses.

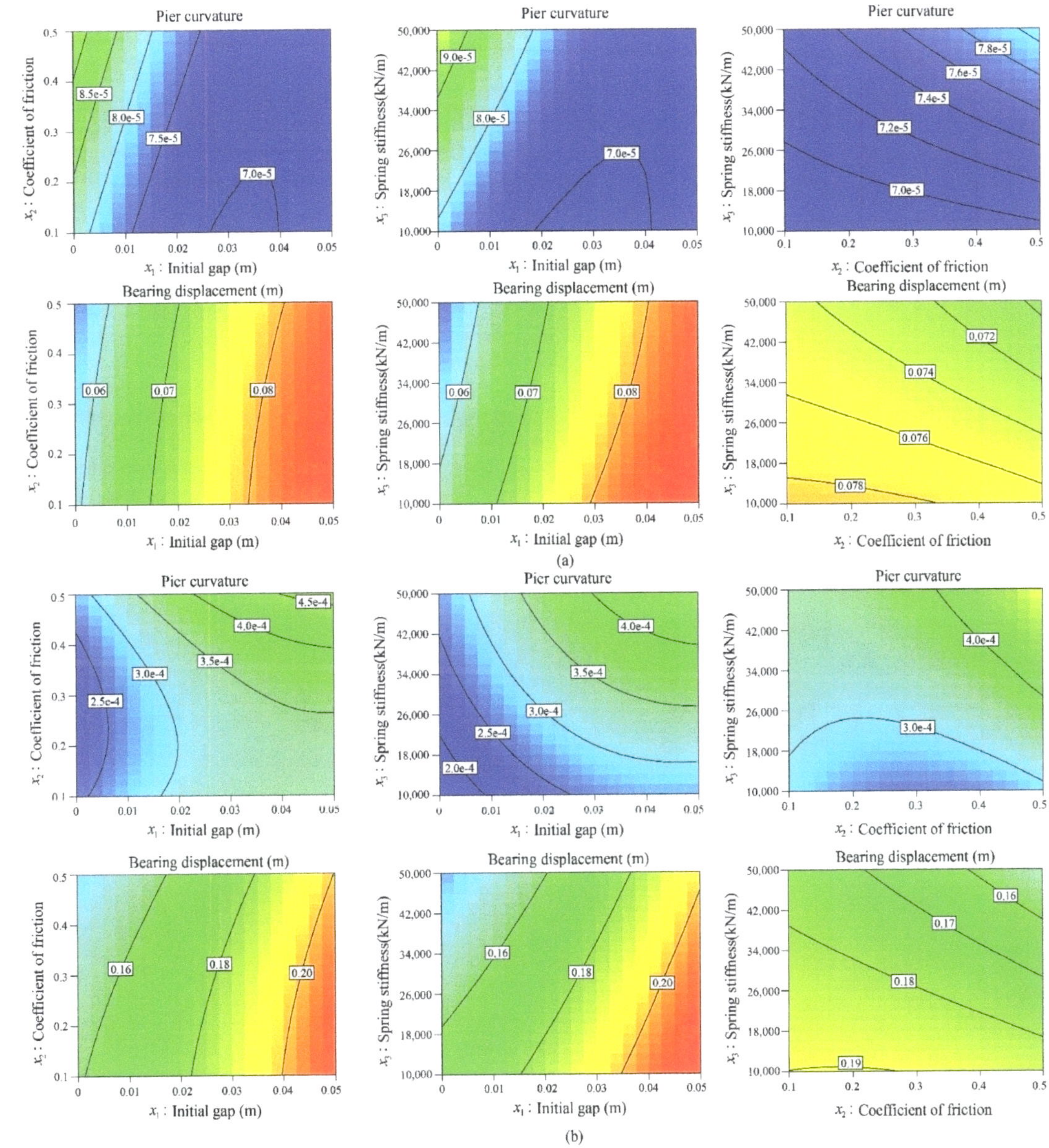

Figure 10. Contour maps of maximum pier curvature and bearing displacement with respect to the variables of initial gap (x_{21}), coefficient of friction (x_2), and spring stiffness (x_3) of LFD: (**a**) wave 1 and (**b**) wave 7 in Table 1.

Furthermore, the maximum displacement between the superstructure and the substructure of the bridge should be limited to avoid pounding and unseating of the girders. However, this benefit is at the expense of increasing the pier damage. Therefore, proper device parameters should control the maximum relative displacement between the superstructure and the substructure to a reasonable range and should not significantly increase the seismic demand of the pier. Based on the above discussions, the spring stiffness of LFD is set to 30,000 kN/m. In addition, the friction coefficient and initial gap of LFD are set to 0.2 and 0, respectively.

5.2. Force–Displacement Relationship of Components in the Longitudinal Direction

Figure 11 illustrates the force–displacement relationships of the bearing and pier of the prototype bridge with the original system (Non-isolated), LFD system (Isolated-LFD), and LRB system (Isolated-LRB) for waves 4, 9, and 14 in Table 1. For the original system (Non-isolated), the elastic perfectly plastic material is adopted to simulate the force–displacement relationship of the sliding bearing. The bearing starts to move when its friction force reaches a critical value. Meanwhile, for the LFD system (Isolated-LFD), the bearing system's constitutive model comprises two parts: sliding bearing and LFD. In addition, for the LRB system (Isolated-LRB), the bearing yields when the shear force reaches the characteristic strength (Qd) of the bearing. It can be found that the LRB system (Isolated-LRB) usually experiences a larger bearing displacement than the other two systems. This phenomenon can explain that for the LRB system (Isolated-LRB), the second stiffness (K2) of the LRB bearing is much smaller than its initial stiffness. When a ground motion with a strong velocity pulse excites the bridge in the longitudinal direction, the small post-yielding stiffness of the LRB bearing cannot control the relative displacement induced by a large velocity pulse. Meanwhile, the LFD system (Isolated-LFD) can provide large constant stiffness in the loading process. Thus, the relative displacement between the superstructure and the substructure of the bridge can be effectively controlled.

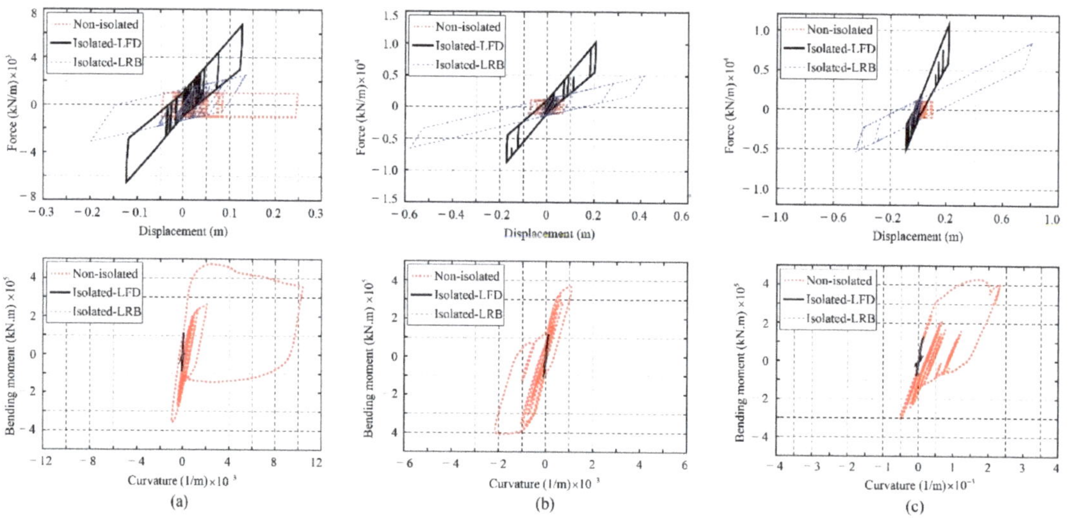

Figure 11. Force–displacement relationship of the continuous girder bridge for the original system (Non-isolated), LFD system (Isolated-LFD), and LRB system (Isolated-LRB) with (**a**) wave 4, (**b**) wave 9, and (**c**) wave 14 in Table 1.

It also can be seen from Figure 11 that the pier force–displacement relationships of the original system (Non-isolated) and isolation system (Isolated-LFD, Isolated-LRB) are significantly different. The pier with the fixed bearing experiences a huge hysteretic loop, which severely damages the pier. The piers maintain elasticity for the isolation systems (Isolated-LFD, Isolated-LRB). The seismic responses of the piers for the isolation systems are much smaller than that for the non-isolated system. Therefore, the substructure of the isolation systems (Isolated-LFD, Isolated-LRB) can effectively be protected from damage. This benefit can be explained below: For the original system (Non-isolated), the superstructure and the substructure are rigidly connected by the fixed bearing, and the inertial force of the superstructure is mainly transmitted by the fixed bearing. However, for the isolation system (Isolated-LFD, Isolated-LRB), the inertial force of the superstructure is mainly transmitted both from piers 3 and 4. Conversely, the effects of velocity pulses are mitigated by the large relative displacement between the superstructure and substructure provided by the isolation system.

5.3. Bearing Displacement in the Longitudinal Direction

Figure 12 presents the bearing displacement time histories at pier 4 for the original system (Non-isolated), LFD system (Isolated-LFD), and LRB system (Isolated-LRB). It can be observed that large displacement pulses occur for all systems. This is because the large displacement caused by ground motions with velocity pulses must be undertaken either by the superstructure or the substructure. For the original system (Non-isolated), this displacement is taken by the pier with a fixed bearing. In contrast, for the isolation system (Isolated-LFD, Isolated-LRB), this displacement is accommodated by LFD and LRB. The relative displacement between the superstructure and the substructure of the LRB system (Isolated-LRB) is usually larger than that of the LFD system (Isolated-LFD). Moreover, the explanation of this phenomenon can be found in Section 5.2. Figure 13 shows the maximum bearing displacement at pier 4 for the original system (Non-isolated), LFD system (Isolated-LFD), and LRB system (Isolated-LRB) with 16 ground motions in Table 1. For the LRB system (Isolated-LRB), the large relative displacement between the superstructure and the substructure frequently occurs, and this large displacement far exceeds the allowable displacement of LRB (0.3 m). As a result, pounding and unseating of the superstructure of the bridge will happen under ground motion with strong velocity pulses. Meanwhile, LRBs are expected to experience severe damage. However, the maximum bearing displacements for the LFD system (Isolated-LFD) can be effectively controlled.

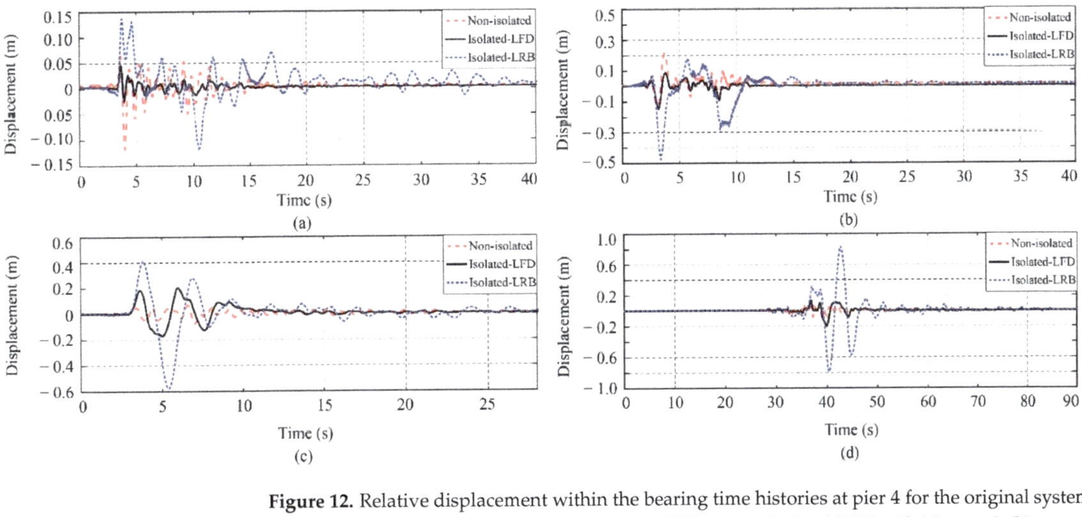

Figure 12. Relative displacement within the bearing time histories at pier 4 for the original system (Non-isolated), LFD system (Isolated-LFD), and LRB system (Isolated-LRB) with (**a**) wave 2, (**b**) wave 6, (**c**) wave 9, and (**d**) wave 12 in Table 1.

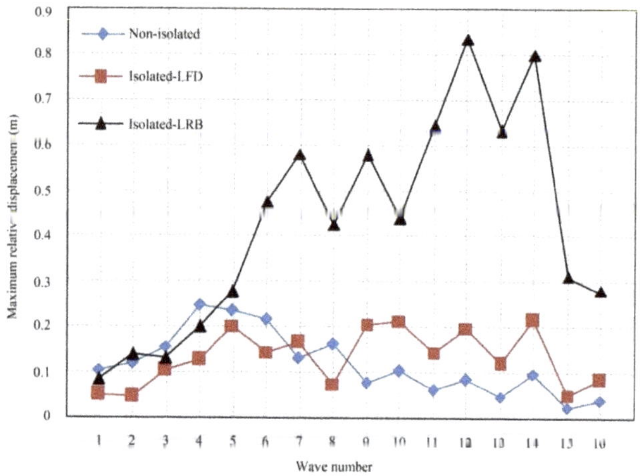

Figure 13. Maximum bearing displacement at pier 4 for the original system (Non-isolated), LFD system (Isolated-LFD), and LRB system (Isolated-LRB) with ground motions in Table 1.

5.4. Pier Curvature in the Longitudinal Direction

Piers are expected to enter strong nonlinearity for ground motions with velocity pulses. Thus, curvature is adopted to evaluate the seismic responses of piers. The ratios of maximum pier curvature to its yielding curvature for ground motions in Table 1 are presented in Table 2. It shows that the maximum curvatures of pier 3 for the isolation systems are much smaller than those for the original system. Moreover, most of the maximum curvature in pier 3 for the isolation systems decreases by 90%. The first yield curvature of the section at the bottom of pier 3 is 0.00082 1/m, and the maximum pier curvatures for the isolation systems are smaller than the first yield curvature of the pier section. The maximum curvature in pier 4 for the original system (Non-isolated) is relatively small because the sliding bearing is applied at the top of pier 4. Thus, pier 4 is isolated from the superstructure, and only friction force can be transmitted. Based on the aforementioned discussion, it can be concluded that the proposed device (LFD) can effectively reduce the seismic responses of the substructure due to its capability of accommodating large relative displacement induced by velocity pulses. For structures subjected to ground motions with velocity pulses, accommodating large displacement is more important than dissipating energy.

Table 2. The ratios of maximum pier curvature to yielding curvature for the three systems.

Earthquake Number	Pier 3			Pier 4		
	Non-Isolated	Isolated-LRB	Isolated-LFD	Non-Isolated	Isolated-LRB	Isolated-LFD
1	0.09	0.03	0.10	0.07	0.03	0.09
2	**5.67**	0.02	0.57	0.07	0.02	0.55
3	**6.99**	0.03	0.11	0.07	0.03	0.11
4	**12.62**	0.03	0.14	0.07	0.03	0.15
5	**11.11**	0.03	0.93	0.07	0.03	0.76
6	**11.36**	0.07	0.42	0.07	0.07	0.43
7	**4.10**	0.05	0.28	0.07	0.06	0.28
8	**7.72**	0.05	0.68	0.07	0.06	0.69
9	**2.64**	0.05	0.14	0.07	0.05	0.16
10	**3.95**	0.06	0.16	0.07	0.06	0.18
11	**1.24**	0.06	0.10	0.07	0.06	0.08
12	**2.91**	0.07	0.10	0.07	0.08	0.10
13	**1.29**	0.05	0.10	0.07	0.06	0.08
14	**2.82**	0.10	0.14	0.07	0.10	0.16
15	0.22	0.03	0.10	0.07	0.03	0.08
16	0.88	0.03	0.10	0.07	0.03	0.08

6. Conclusions

The current research investigates the effects of the proposed linear friction damper (LFD) on the seismic responses of a continuous girder bridge subjected to ground motions with velocity pulses. First, the working mechanism and the force–displacement relationship of LFD are described in detail. Then, a numerical experiment is designed to validate the proposed institutive model of LFD. In addition, the design process of LFD for a specific case is also illustrated. Second, three-span continuous girder bridges for the original system (Non-isolated), LFD system (Isolated-LFD), and LRB system (Isolated-LRB) are established by OpenSees. Third, 16 ground motions with velocity pulses are selected from the PEER ground motion database. Daubechies wavelet of order 4 is adopted to extract the maximum velocity pulse from the two orthogonal components of a ground motion. Then, the effects of the initial gap, the coefficient of friction, and the spring stiffness of LFD on the seismic responses of the bridge are analyzed. Finally, the seismic responses in the longitudinal direction of the bridge, including the force–displacement relationship, relative displacement within the bearing, and curvature at the bottom of the pier, are calculated and intensively discussed. Based on the numerical results, the main conclusions are summarized below:

The bearings of the LRB system usually experience large relative displacement due to its small post-yielding stiffness for ground motions with strong velocity pulses. Piers with fixed bearings are severely damaged, and this disadvantage is eliminated for the isolation systems (Isolated-LFD, Isolated-LRB). This phenomenon can be explained because for the original system (Non-isolated), the superstructure and substructure are rigidly connected by the fixed bearing. The inertial force is mainly transmitted through the pier with a fixed bearing. However, for the isolation systems (Isolated-LFD, Isolated-LRB), the velocity pulses are accommodated. Thus, piers are isolated from the inertial force of the superstructure.

According to the bearing displacement analysis, the relative displacement within the bearing can be effectively controlled for the LFD system (Isolated-LFD). Thus, pounding and unseating of the superstructure can be mitigated. The LFD system (Isolated-LFD) can balance between controlling the maximum bearing displacement and mitigating the damages of the pier compared with the other two systems (Non-isolated, Isolated-LRB).

The current research focuses on the basic concept of LFD, and the Coulomb friction law is adopted in the numerical experiment. In future research, a friction coefficient depending on the sliding velocity and contact pressure will be used in the constitutive model of LFD. In addition, more accurate hysteresis models available in the literature [29,30] will be adopted to improve the accuracy of the results.

Author Contributions: Conceptualization, H.L., C.Z., S.C. and X.D.; methodology, C.Z.; investigation, H.L.; writing—original draft preparation, J.G.; writing—review and editing, S.C.; supervision, X.D.; formal analysis, J.G.; validation, X.D. All authors have read and agreed to the published version of the manuscript.

Funding: This research was funded by the National Key Research and Development Program of China (2019YFE0112300); the National Natural Science Foundation of China (51978512); The Shanghai Post-doctoral Excellent Program (2021333); The Post-doctoral Innovation Practice Base Program of Shanghai Yangpu District; and the Transportation science and technology plan of Shandong province (2017B75).

Institutional Review Board Statement: Not applicable.

Informed Consent Statement: Not applicable.

Data Availability Statement: Not applicable.

Conflicts of Interest: The authors declare no conflict of interest.

References

1. Yen, W.P.; Chen, G.; Yashinski, M.; Hashash, Y.; Holub, C.; Wang, K.; Guo, X. Lessons in bridge damage learned from the Wenchuan earthquake. *Earthq. Eng. Eng. Vib.* **2009**, *8*, 275–285. [CrossRef]
2. Kawashima, K.; Takahashi, Y.; Ge, H.; Wu, Z.; Zhang, J. Reconnaissance report on damage of bridges in 2008 Wenchuan, China, earthquake. *J. Earthq. Eng.* **2009**, *13*, 965–996. [CrossRef]
3. Sun, Z.; Wang, D.; Guo, X.; Si, B.; Huo, Y. Lessons learned from the damaged Huilan interchange in the 2008 Wenchuan earthquake. *J. Bridge Eng.* **2011**, *17*, 15–24. [CrossRef]
4. Sehhati, R.; Rodriguez-Marek, A.; ElGawady, M.; Cofer, W.F. Effects of near-fault ground motions and equivalent pulses on multi-story structures. *Eng. Struct.* **2011**, *33*, 767–779. [CrossRef]
5. Ismail, M.; Casas, J.R.; Rodellar, J. Near-fault isolation of cable-stayed bridges using RNC isolator. *Eng. Struct.* **2013**, *56*, 327–342. [CrossRef]
6. Ates, S.; Constantinou, M.C. Example of application of response spectrum analysis for seismically isolated curved bridges including soil-foundation effects. *Soil Dyn. Earthq. Eng.* **2011**, *31*, 648–661. [CrossRef]
7. Bhuiyan, A.R.; Alam, M.S. Seismic performance assessment of highway bridges equipped with superelastic shape memory alloy-based laminated rubber isolation bearing. *Eng. Struct.* **2013**, *49*, 396–407. [CrossRef]
8. Galindo, C.M.; Belda, J.G.; Hayashikawa, T. Non-linear seismic dynamic response of curved steel bridges equipped with LRB supports. *Steel Constr.* **2010**, *3*, 34–41. [CrossRef]
9. Monzon, E.V.; Buckle, I.G.; Itani, A.M. Seismic Performance and Response of Seismically Isolated Curved Steel I-Girder Bridge. *J. Struct. Eng.* **2016**, *142*, 04016121. [CrossRef]
10. Tanaka, R.; Galindo, C.M.; Hayashikawa, T. Nonlinear seismic dynamic response of continuous curved highway viaducts with different bearing supports. *World Acad. Sci. Eng. Technol.* **2009**, *5*, 327–333.
11. Losanno, D.; Palumbo, F.; Calabrese, A.; Barrasso, T.; Vaiana, N. Preliminary investigation of aging effects on recycled rubber fiber reinforced bearings (RR-FRBs). *J. Earthq. Eng.* **2021**, 1–18. [CrossRef]
12. Kataria, N.P.; Jangid, R.S. Seismic protection of the horizontally curved bridge with semi-active variable stiffness damper and isolation system. *Adv. Struct. Eng.* **2016**, *19*, 1103–1117. [CrossRef]
13. Dicleli, M.; Milani, A.S. *Steel Hysteretic Damper Featuring Displacement Dependent Hardening for Seismic Protection of Structures*; Springer: New Delhi, India, 2015.
14. Chen, Z.Y.; Chen, W.; Bian, G.Q. Seismic Performance Upgrading for Underground Structures by Introducing Shear Panel Dampers. *Adv. Struct. Eng.* **2014**, *17*, 1343–1358. [CrossRef]
15. Providakis, C.P. Effect of LRB isolators and supplemental viscous dampers on seismic isolated buildings under near-fault excitations. *Eng. Struct.* **2008**, *30*, 1187–1198. [CrossRef]
16. Liao, W.-I.; Loh, C.-H.; Lee, B.-H. Comparison of dynamic response of isolated and non-isolated continuous girder bridges subjected to near-fault ground motions. *Eng. Struct.* **2004**, *26*, 2173–2183. [CrossRef]
17. Panchal, V.R.; Jangid, R.S. Variable friction pendulum system for near-fault ground motions. *Struct. Control Health Monit.* **2008**, *15*, 568–584. [CrossRef]
18. Karalar, M.; Padgett, J.E.; Dicleli, M. Parametric analysis of optimum isolator properties for bridges susceptible to near-fault ground motions. *Eng. Struct.* **2012**, *40*, 276–287. [CrossRef]
19. Li, S.; Dezfuli, F.H.; Wang, J.-Q.; Alam, M.S. Longitudinal seismic response control of long-span cable-stayed bridges using shape memory alloy wire-based lead rubber bearings under near-fault records. *J. Intell. Mater. Syst. Struct.* **2017**, *29*, 703–728. [CrossRef]
20. Dolce, M.; Cardone, D.; Croatto, F. Frictional Behavior of Steel-PTFE Interfaces for Seismic Isolation. *Bull. Earthq. Eng.* **2005**, *3*, 75–99. [CrossRef]
21. Alam, M.S.; Bhuiyan, M.A.R.; Billah, A.H.M.M. Erratum to: Seismic fragility assessment of SMA-bar restrained multi-span continuous highway bridge isolated by different laminated rubber bearings in medium to strong seismic risk zones. *Bull. Earthq. Eng.* **2012**, *10*, 1911–1913. [CrossRef]
22. Mckenna, F.; Scott, M.H.; Fenves, G.L. Nonlinear Finite-Element Analysis Software Architecture Using Object Composition. *J. Comput. Civ. Eng.* **2010**, *24*, 95–107. [CrossRef]
23. Mander, J.B.; Priestley, M.J.N.; Park, R. Theoretical Stress-Strain Model for Confined Concrete. *J. Struct. Eng.* **1988**, *114*, 1804–1826. [CrossRef]
24. Shahi, S.K.; Baker, J.W. An Efficient Algorithm to Identify Strong-Velocity Pulses in Multicomponent Ground Motions. *Bull. Seismol. Soc. Am.* **2014**, *104*, 2456–2466. [CrossRef]
25. Baker, J.W. Quantitative Classification of Near-Fault Ground Motions Using Wavelet Analysis. *Bull. Seismol. Soc. Am.* **2007**, *97*, 1486–1501. [CrossRef]
26. Box, G.E.P.; Wilson, K.B. *On the Experimental Attainment of Optimum Conditions*; Springer: New York, NY, USA, 1992; pp. 270–310.
27. Guo, J.J.; Zhong, J.; Dang, X.Z.; Yuan, W.C. Seismic performance assessment of a curved bridge equipped with a new type spring restrainer. *Eng. Struct.* **2017**, *151*, 105–114. [CrossRef]
28. Towashiraporn, P. Building seismic fragilities using response surface metamodels. *Ga. Inst. Technol.* **2004**, *104*, 2456–2466.

29. Vaiana, N.; Sessa, S.; Marmo, F.; Rosati, L. A class of uniaxial phenomenological models for simulating hysteretic phenomena in rate-independent mechanical systems and materials. *Nonlinear Dyn.* **2018**, *93*, 1647–1669. [CrossRef]
30. Vaiana, N.; Sessa, S.; Rosati, L. A generalized class of uniaxial rate-independent models for simulating asymmetric mechanical hysteresis phenomena. *Mech. Syst. Signal Process.* **2021**, *146*, 106984. [CrossRef]

buildings

MDPI

Article

Study on VIV Behavior of Two 5:1 Rectangular Cylinders in Tandem Based on Correlation Analysis

Kai Ma [1], Changyong Zhang [2], Xinzhi Dang [1,3] and Guoquan Zhang [1,*]

1 Department of Bridge Engineering, Tongji University, 1239 Siping Road, Shanghai 200092, China; 1510169@tongji.edu.cn (K.M.); 021tjdangxz@tongji.edu.cn (X.D.)
2 Shandong Provincial Communications Planning and Design Institute Group Co., Ltd., 2177 Tianchen Road, Jinan 250031, China; kjxx@sdcpd.com
3 Suzhou Tongyao Civil Technology Co., Ltd., 2 Hushan Road, Changshu 215533, China
* Correspondence: zgqtj@tongji.edu.cn

Abstract: To investigate the vortex-induced vibration (VIV) characteristics of two rectangular cylinders with a width-to-depth ratio of 5:1 in a tandem arrangement, sectional model wind tunnel tests that measure vibration responses and pressure distributions simultaneously were adopted. The ratio of the spacing between the cylinders to its width is 1.2. The analyses were performed considering VIV responses as well as the distribution characteristics of mean and rms pressure coefficients. Additionally, the time-frequency domain statistical parameters like correlation and contribution coefficients, phase lags between distributed and general vortex excited forces (VEFs), and amplitudes of VEF coefficients at predominant frequencies were calculated to analyze the physical VIV mechanism of two 5:1 rectangular cylinders in tandem. This study indicates that the influence of incidence angles on the dynamic responses is notable; the contribution of the distributed VEFs acting on the trailing surface of the upstream cylinder and the leading surface of the downstream one is significant to VIVs of the cylinders from wind pressure distribution characteristics and correlation analyses.

Keywords: 5:1 rectangular cylinders in tandem; sectional model testing; pressure measurement testing; VIV behavior; correlation analysis

Citation: Ma, K.; Zhang, C.; Dang, X.; Zhang, G. Study on VIV Behavior of Two 5:1 Rectangular Cylinders in Tandem Based on Correlation Analysis. *Buildings* **2024**, *14*, 85. https://doi.org/10.3390/buildings14010085

Academic Editors: Bo Wang, Bo Fu and Xinxin Wei

Received: 7 March 2023
Revised: 11 December 2023
Accepted: 25 December 2023
Published: 28 December 2023

Copyright: © 2023 by the authors. Licensee MDPI, Basel, Switzerland. This article is an open access article distributed under the terms and conditions of the Creative Commons Attribution (CC BY) license (https://creativecommons.org/licenses/by/4.0/).

1. Introduction

The aerodynamic interference of square and rectangular columns has attracted considerable scientific interest in the wind engineering field since the sections are very common in sharp-edged bluff bodies. Therefore, a lot of research was carried out to investigate the mean force coefficients, surface pressure coefficients, and Strouhal numbers, as well as the surface pressure coherence and cross-spectrum of square or rectangular columns [1–3]. In addition, Shimada and Ishihara (2002) analyzed the aerodynamic characteristics of rectangular cylinders with aspect ratios varying from 0.8 to 8.0 using a modified two-layer k-ε model in CFD simulations, and the results indicate that aspect ratios play an important role in the flow regime of rectangular cylinders [4]. Furthermore, to deeply investigate the aerodynamic performance of the bluff bodies, a benchmark on the aerodynamics of a rectangular 5:1 cylinder (BARC) was launched in 2008. The span-wise correlation of the pressure and aerodynamic forces on BARC and the impact of incidence angles, Reynolds-number sensitivity, as well as turbulence intensity of free-flow were analyzed using numerical simulations and wind tunnel tests [5–15]. Additionally, a very detailed overview of previous research was presented by Bruno et al. (2014) [16]. These studies can provide useful information on the aerodynamic behavior of rectangular cylinders in tandem with the present paper.

However, the above pieces of research were confined to single rectangular cylinders. As we know, the aerodynamic performance of twin rectangular cylinders plays an important guiding role in the analysis of wind-induced behavior of twin-deck sections in bridge aerodynamics. The studies of a group of rectangular (or square) cylinders in a tandem arrangement have been carried out in a lot of studies, that is, through wind tunnel experiments [17–21] and numerical simulations [22–29]. Moreover, the research on vortex dynamics of square cylinders placed in a staggered arrangement was carried out by Chatterjee and Biswas (2015) [30]. However, less research is focused on the vortex-induced vibration (VIV) behavior of twin rectangular cylinders, though this type of wind-induced phenomenon of twin-deck sections is of extensive concern in the wind engineering field, especially in bridge aerodynamics. In previous research [31], we analyzed the pressure and force distribution on a twin in-line arrangement of twin 5:1 rectangular sections at different gaps. A proper orthogonal decomposition (POD) technique was adopted to analyze the fluctuating wind pressure field around the cylinders under wind velocities corresponding to the maximum VIV amplitudes; the results indicated that the gap distance has a critically important impact on the VIV behavior of twin rectangular cylinders. In addition, POD analysis indicates that the first mode is dominant to the VIVs of the downstream cylinder, while the second mode is closely related to the VIVs of the upstream one.

In the present study, the VIV behavior of two 5:1 rectangular cylinders in tandem is further studied. The VIV responses, mean and rms pressure distributions, as well as the correlation between distributed and general vortex-excited forces (VEFs) are analyzed. This study can provide some useful information on the analysis of aerodynamic interference of parallel bridges.

2. Wind Tunnel Tests

The sectional model vibration and pressure measurement tests were conducted in uniform flow, and the turbulence intensity was less than 0.4% without turbulence-generating devices. Each cylinder was supported separately by eight coil springs that permit both heaving and torsional oscillations, and the stiffness of the springs as well as their spacings were designed to simulate the natural frequencies of vibration. The sketch of the testing model is depicted in Figure 1. The VIV responses were obtained through vibration measurement tests within a certain inflow velocity range. Pressures around the cylinders were measured using the DSM3000 Scanivalve system (Scanivalve Corporation, Washington, DC, USA). A sampling frequency of 300 Hz and a sampling interval of 40 s was adopted to collect data.

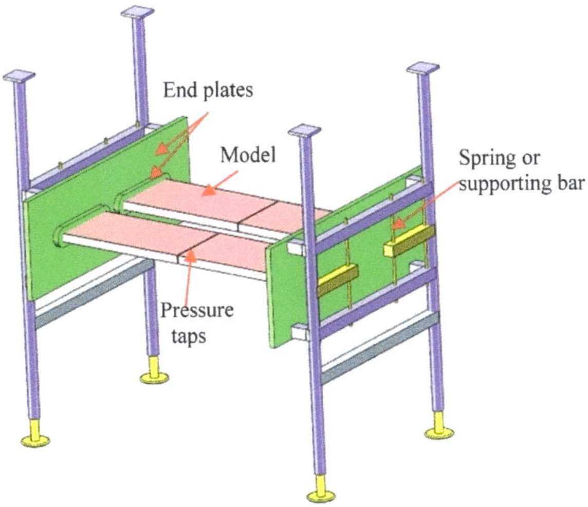

Figure 1. Sketch of testing apparatus.

In the tests, each cylinder is 1.7 m in length; the width-to-depth ratio (W/D) of each model is 5:1, that is, W equals 300 mm and D equals 60 mm. The Reynolds numbers are defined as Re = UB/ν, where U is the wind velocity and ν is the kinematic viscosity. The Reynolds numbers are 5.9×10^4 (U = 2.9 m/s), 7.3×10^4 (U = 3.6 m/s), and 1.0×10^5 (U = 5.0 m/s), respectively. Since the blockage ratio defined by the ratio between the height of the model to that of the wind tunnel test section is only 2.4%, no blockage correction was applied. The stiffness of the models is provided by longitudinal and transverse beams, which are covered by plate skins to simulate the appearance. The cylinders are characterized by sharp edges and smooth surfaces. For each model, 54 pressure taps were installed along the middle line to collect the pressure of its position. The fundamental bending and torsional natural frequencies are 3.997 and 8.221 Hz, and the corresponding damping ratios are 0.42% and 0.38%, respectively. The distributions of pressure taps and internal constitutions of each model are depicted in Figure 2.

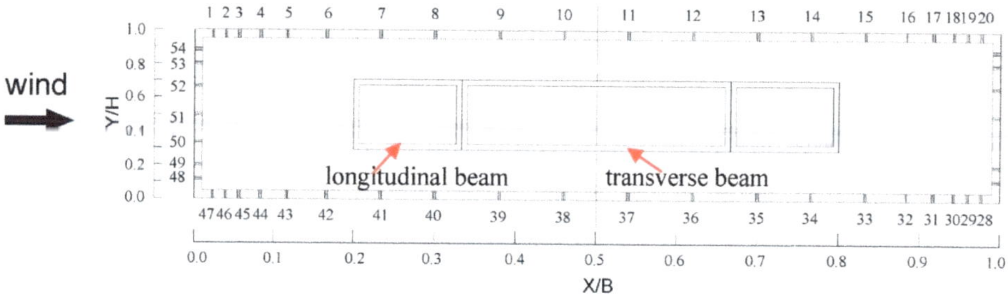

Figure 2. Model configuration and internal constitutions.

3. Results of Wind Tunnel Tests

3.1. VIV Responses under Different Angles of Incidence

The dynamic tests of twin rectangular cylinders were conducted at three angles of incidence, that is, 0°, +3°, and −3°. The positive angle of incidence is defined as the inflow pointing at the bottom surface of the cylinder and vice versa. The spacing between them is 1.2 times the section width, which is the most unfavorable case according to our previous wind tunnel tests with various spacings [31]. The results of bending and torsional VIV responses are taken as the square root two times the rms values and are presented in Figure 3. The reduced wind velocities are defined as U/f_vB and U/f_tB for heaving and torsional VIVs, where U is the wind velocity and f_v and f_t are the natural frequency for heaving and torsional vibrations, respectively.

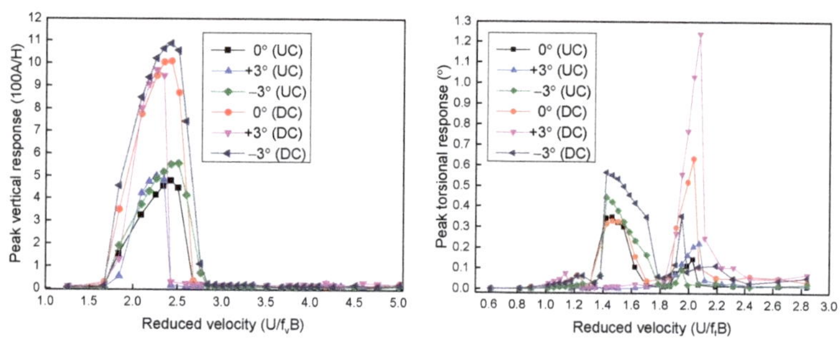

(**a**) (**b**)

Figure 3. VIV responses of the twin rectangular cylinders under different angles of incidence. (**a**) vertical VIV responses; (**b**) torsional VIV responses.

As shown from the heaving VIV responses of both cylinders, the starting wind velocities of both cylinders are identical for 0°, +3°, and −3° angles of incidence. The vibration amplitude is the largest at −3° for both cylinders. For the upstream cylinder (abbreviated as UC), the length of the lock-in region, the vibration amplitude, and the wind speed with the maximum response are the smallest at the null angle of incidence. The maximum amplitude decreases by 14% compared with that at −3°. For the downstream cylinder (abbreviated as DC), the lock-in region and the wind speed with the maximum response are identical for 0° and −3°. However, the velocities become lower for +3°, and the VIV amplitude reaches the minimum value, which decreases by 11% compared with that at −3°.

The lock-in regions for torsional VIVs are identical for both cylinders. There are two lock-in regions at 0° and −3°, that is, the range of U/f_tB = 1.30–1.70 (the first lock-in region) and U/f_tB = 1.86–2.07 (the second lock-in region). However, only a single lock-in region occurs at the +3° angle of incidence, which corresponds to the second one of the other two angles of incidence. For UC, the maximum VIV amplitudes reached in the first lock-in region are 0.35° and 0.44° at 0° and −3° angles of incidence, respectively, which are larger than that at +3°. The maximum VIV amplitudes reached in the second lock-in region are 0.14°, 0.1°, and 0.22° at 0°, −3°, and +3° angles of incidence, respectively. For DC, the peak amplitude in the first VIV region is 0.57° at the −3° incidence angle, which is approximately 1.7 times the value at the null angle of incidence (0.33°). However, for the second lock-in region, the maximum vibration amplitude of 0.64° is reached at 0° and is approximately 1.8 times the value of the −3° angle of incidence (0.36°), while it is only half the value of the +3° angle of incidence (1.24°).

As the results indicate, the effect of the incidence angle on VIV performance is complex. Generally, the bending VIV response is unfavorable at −3° for both UC and DC. The torsional VIV response for UC is larger at 0° and −3°, while a general trend can hardly be observed for DC.

3.2. Results of Pressure Measurement

The pressure data are measured at the steady-state flow, and the turbulence intensity is less than 0.4%. During the pressure measurement testing, both upstream and downstream cylinders are free to oscillate in the flow. The tests were performed at the null angle of incidence and under discrete wind velocities corresponding to different VIVs. Three typical wind velocities were set during pressure measurement, that is, the wind speed of 2.9 m/s (U/f_vB = 2.42) with the maximum bending VIVs and 3.6 m/s (U/f_tB = 1.46) and 5.0 m/s (U/f_tB = 2.03) with the peak torsional VIVs of the first and second lock-in regions. The mean and fluctuating pressure coefficients are defined as the time-average and rms values of the pressure data, respectively. The mean pressure coefficient distributions around the upstream and downstream cylinders are shown in Figure 4. According to our research, the pressure on the windward and leeward side faces has a minor effect on both bending and torsional VIVs. Therefore, the results only present the pressure distributions along the upper and lower surfaces of UC and DC; they are shown in Figures 4 and 5. The pressure measurements of the present study have been validated by comparing the aerostatic coefficients with the results from previous researchers [31,32].

As the results show, the mean pressure coefficient (C_p) distributions along the upper surface are similar to those on the lower surface for both cylinders, whereas the absolute C_p values on the lower surface are smaller than the upper surface. Since the pressure was measured when the cylinders were experiencing vortex-induced vibrations, the reason why the upper and lower surface pressures obtained at the null angle of incidence took different values is explained. For UC, the C_p values remain constant in the position $X/W \leq 0.5$, while the absolute values decrease steeply in the position $X/W > 0.5$ on both upper and lower surfaces. Additionally, the values in the region of $X/W < 0.4$ decreases as the wind speed increases; however, the trend is the opposite in the region of $X/W > 0.4$. Comparing the results under three wind velocities, the absolute value of C_p on the upper surface is smaller in the range of $X/W < 0.54$ for U = 5.0 m/s, while for $X/W > 0.62$, the value is at a minimum for U = 2.9 m/s. For DC, the difference between the C_p values is relatively minor and can be neglected compared with that of UC.

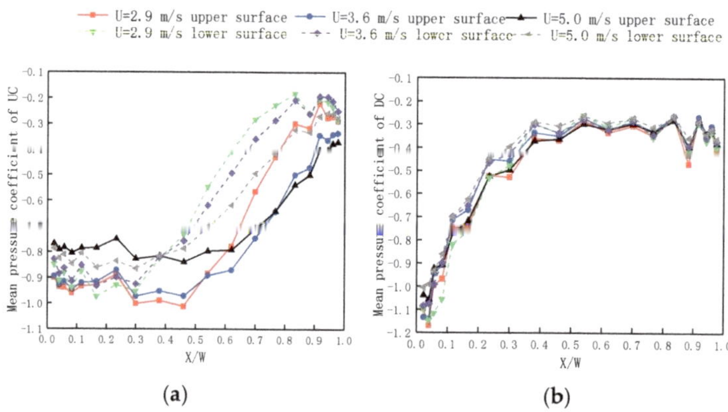

Figure 4. Mean pressure distributions of UC and DC. (**a**) the results of UC; (**b**) the results of DC.

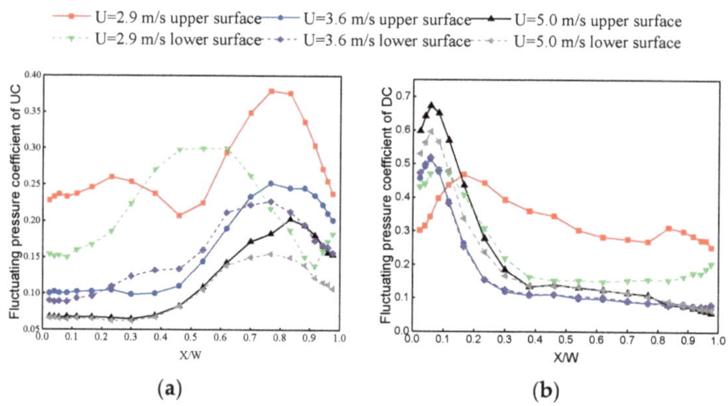

Figure 5. Fluctuating pressure distributions of UC and DC. (**a**) the results of UC; (**b**) the results of DC.

As for the fluctuating pressure coefficients (C_p') on the upper and lower surfaces of the cylinders, the distributions are similar for U = 3.6 and 5.0 m/s. However, the values are larger under U = 3.6 m/s for UC, while the results are larger under U = 5.0 m/s for DC. The C_p' distributions under U = 2.9 m/s are dramatically different from that of U = 3.6 and 5.0 m/s. For UC, the values along the upper surface and in the range of X/W $\leq$ 0.7 along the lower surface are larger for U = 2.9 m/s compared with the other two velocities. The C_p' distributions on the upper surface show symmetry about the centerline for U = 2.9 m/s. However, the values along the lower surface increase in the range of 0.46 $\leq$ X/W < 0.8 and decrease in 0.8 $\leq$ X/W $\leq$ 1.0, which is similar to the trend under U = 3.6 and 5.0 m/s. Nevertheless, there still exists some difference between them; for U = 2.9 m/s, the values first increase steadily and then experience a comparatively sharp decrease in the leading region, and the maximum value is observed at the location of X/W $\approx$ 0.46; meanwhile, the C_p' values remain steady in this region for U = 3.6 and 5.0 m/s.

For DC, the C_p' values decline on the lower surface for U = 2.9 m/s, and the maximum value moves in the flow direction as compared with that of U = 3.6 and 5.0 m/s. The C_p' values decrease steadily along the trailing surface for all three velocities. The distributions on the lower surfaces at U = 2.9 m/s show some similarity with the results under the other two velocities; meanwhile, the values for U = 2.9 m/s are relatively larger in most regions. Above all, the results indicate that the dynamic pressure measurements should be carried out under corresponding wind velocities in order to investigate the pressure field when rectangular cylinders in tandem undergo different types of VIVs.

4. Correlation Analysis

In the analysis of the VIV mechanism of a streamlined closed-box girder section, Hu et al. (2018) carried out synchronized measurements of dynamic responses, pressure, and force of the girder section and probed the correlation between the distributed aerodynamic forces and the general vortex-excited forces (VEFs) [33]. In that research, the distributed aerodynamic forces were directly obtained, without subtracting the mean components and projecting to the VEF direction (as depicted in the following flowchart in Figure 6). The same procedure is also employed in the pieces of research by Guo et al. (2012) and Guan et al. (2014) [34,35]. However, the correlation analyses ignored the fact that the aerodynamic forces are vectors in nature, and the results cannot reflect the contribution of the distributed pressure directly. For instance, the pressure on the side surfaces of rectangular cylinders has no contribution to the fluctuating lift acting on the cylinder when it is subjected to vertical VIVs, while the correlation values between them are not null obviously. To solve this problem, the distributed and general VEFs, which actually excite the VIVs of the cylinders, are directly obtained in the present study. The general and distributed VEFs are defined as follows: the pressure of each sampling point is multiplied by its tributary area to obtain the distributed aerodynamic forces. According to many previous pieces of research, the mean components contribute to the static wind forces acting on the cylinders. However, the VIVs are caused by the fluctuating wind pressure acting on the cylinders. Therefore, the mean components should not be involved in the analysis. Since the mean values are not related to the motion of the cylinders, the fluctuating aerodynamic forces are obtained after the mean components are subtracted from the original ones. After that, the modified aerodynamic forces are projected to the lift force direction to obtain the distributed vortex-excited lift time histories. Similarly, the procedure is applied to the fluctuating pressure data to obtain the distributed vortex-excited moment time histories. Furthermore, the general VEFs are defined by integrating the distributed VEFs over the whole surface. A flowchart of the calculation process of the distributed and general VEFs is shown in Figure 6.

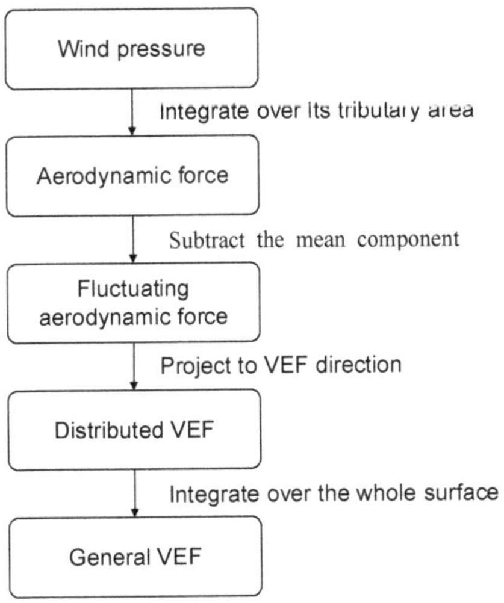

Figure 6. Flowchart of the calculation process of distributed and general VEFs.

The correlation between the distributed and general VEFs is analyzed using time-frequency domain statistical parameters as correlation and contribution coefficients and phase lags between the distributed and general VEFs, which are amplitudes of VEF coefficients at predominant frequencies. The correlation between the distributed VEFs acting on the ith pressure tap and the general VEFs on the whole cylinder is defined as follows:

$$Cor(F_i(t), F(t)) = \frac{Cov(F_i(t), F(t))}{\sigma(F_i(t))\sigma(F(t))} \tag{1}$$

where $\sigma(F_i(t))$ and $\sigma(F(t))$ denote the standard deviation of the distributed VEF at the ith pressure tap $(F_i(t))$ and the general VEF $(F(t))$, respectively. $Cov(F_i(t), F(t))$ is the covariance between them. The values of $|Cor(F_i(t), F(t))|$ satisfy $0 \leq |Cor(F_i(t), F(t))| \leq 1$; in particular, a value of unity represents perfect coherence, and the value of zero indicates two completely unrelated VEFs. For a non-null value, it means a certain level of coherence. The phase lags between distributed and general VEFs are obtained from the cross-spectrum analysis. The phase lag of zero degrees represents a perfect in-phase relation, and 90 degrees indicates no correlation, while 180 degrees means out-of-phase [36,37]. The results of the correlation coefficients for UC and DC are presented in Figures 7 and 8, respectively, and the results of the phase lags between the distributed and general VEFs are shown in Figures 9 and 10.

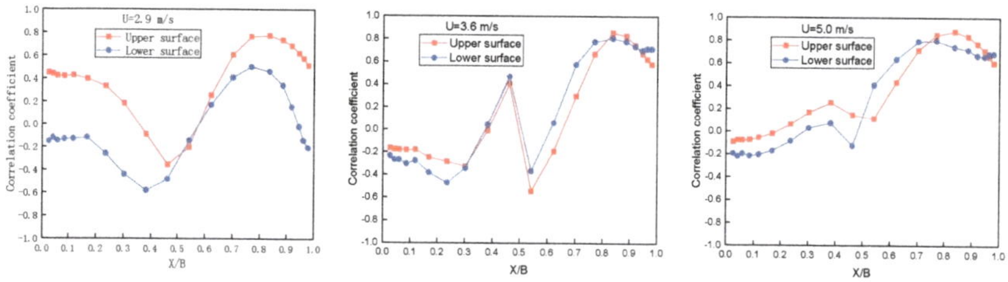

(a) (b) (c)

Figure 7. Correlation coefficients between distributed and general VEFs for UC. (**a**) U = 2.9 m/s; (**b**) U = 3.6 m/s; (**c**) U = 5.0 m/s.

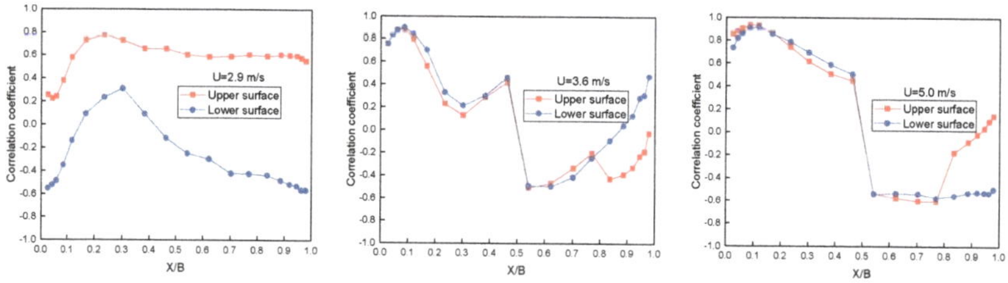

(a) (b) (c)

Figure 8. Correlation coefficients between distributed and general VEFs for DC. (**a**) U = 2.9 m/s; (**b**) U = 3.6 m/s; (**c**) U = 5.0 m/s.

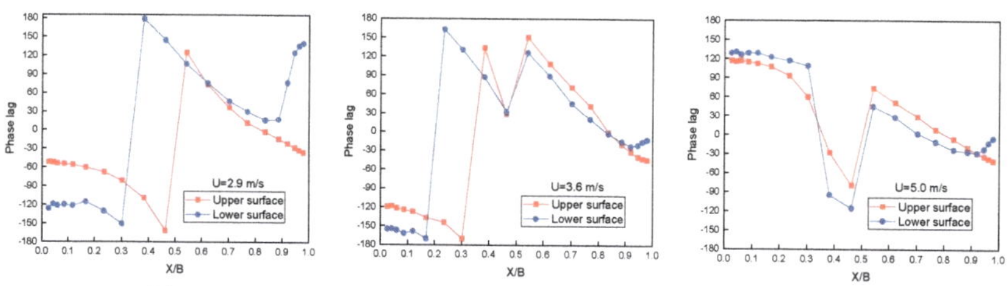

(a) (b) (c)

Figure 9. Phase lags between the distributed and general VEFs for UC. (**a**) U = 2.9 m/s; (**b**) U = 3.6 m/s; (**c**) U = 5.0 m/s.

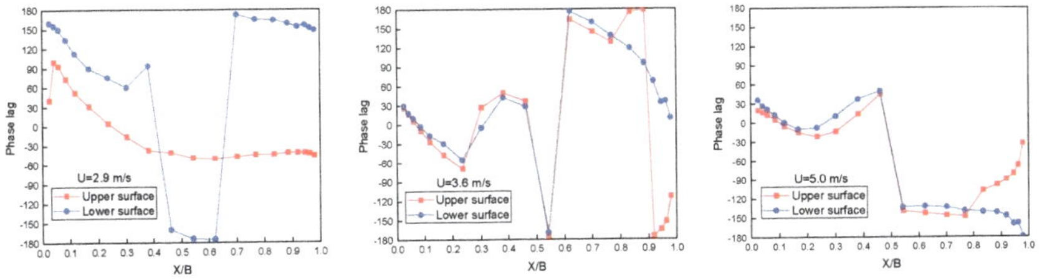

(**a**) (**b**) (**c**)

Figure 10. Phase lags between the distributed and general VEFs for DC. (**a**) U = 2.9 m/s; (**b**) U = 3.6 m/s; (**c**) U = 5.0 m/s.

As the results indicate, the correlation coefficients of UC are the largest in the trailing half region on both upper and lower surfaces, and the observation is more apparent for U = 3.6 and 5.0 m/s. For DC, the positive correlation between the distributed and general VEFs is more pronounced in the leading region for U = 3.6 and 5.0 m/s. However, for U = 2.9 m/s, the correlation coefficient of the upper surface is positive for DC, while the values are mainly negative in most regions of the lower surface.

For the phase lags between the distributed and general VEFs, the values are relatively smaller on the trailing surface of UC. The results at U = 2.9 and 3.6 m/s are similar in distributions, with the exception of some local regions. For DC, the values of phase lags on the leading surface are relatively smaller compared with other regions, and this observation is apparent, especially for U = 3.6 and 5.0 m/s. However, it needs to be pointed out that some exception exists in the phase lags on the upper surface for U = 2.9 m/s. Above all, the results further validate the conclusion that the distributed VEFs on the trailing surface of UC and the leading surface of DC are more closely related to the VIVs of the cylinders. In addition, this conclusion is also in line with our previous POD analysis [31].

The contribution of distributed VEFs on the general one depends not only on the amplitude of distributed VEFs but the correlation coefficient between them. The ratio between the rms values of the distributed VEFs and the general VEFs reflects the amplitude contribution of local VEF at each pressure tap, while the correlation coefficient reflects the correlation levels between them. The contribution coefficient is defined as the product of these parameters and expressed as follows:

$$C(F_i(t)) = Cor(F_i(t), F(t)) \cdot \frac{\sigma(F_i(t))}{\sigma(F(t))} \tag{2}$$

The results of VEF contribution coefficients for upstream and downstream cylinders are presented in Figures 11 and 12, respectively.

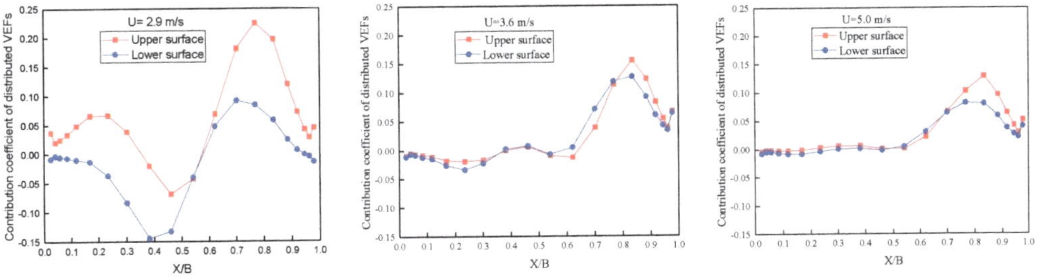

(**a**) (**b**) (**c**)

Figure 11. VEF contribution coefficient for testing point of UC. (**a**) U = 2.9 m/s; (**b**) U = 3.6 m/s; (**c**) U = 5.0 m/s.

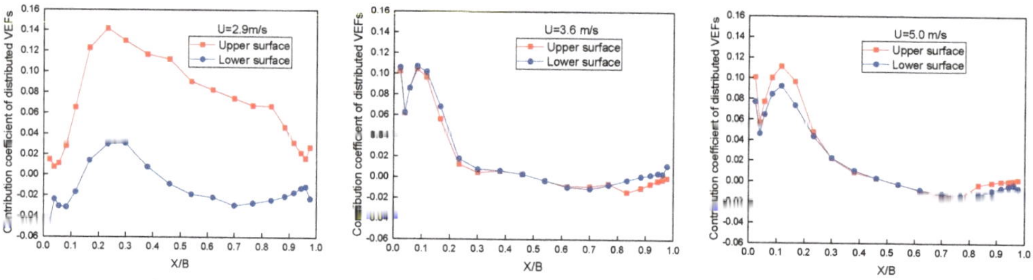

(a) (b) (c)

Figure 12. VEF contribution coefficient for testing point of DC. (**a**) U = 2.9 m/s; (**b**) U = 3.6 m/s; (**c**) U = 5.0 m/s.

As depicted in the figures, the contribution coefficient distributions of UC and DC are basically similar for the correlation coefficients, and the values are the maximum in the trailing half surface of UC and the leading half surface of DC. However, the distributions of the contribution coefficients are more smooth compared with Figures 8 and 9. Furthermore, the contributions of the distributed VEFs acting on the upper surface are more pronounced than those on the lower surface for most cases. This indicates that some difference in the flow structure on the upper and lower surfaces during VIVs of the cylinders exists. This conclusion is in line with the fluctuating pressure distributions (Figure 5) and our previous POD analysis [31].

The VEF coefficient amplitude at a predominant frequency (or natural vibration frequency) can be obtained from Fourier transformation, and it corresponds to the self-excited component in VEFs.

$$F_{iD} = \frac{2}{N}\mathcal{F}\{F_i(t)\} \tag{3}$$

where F_{iD} is the amplitude of the distributed VEFs at predominant frequency, N is the signal length of the distributed VEFs, and $\mathcal{F}\{\}$ is the operation of Fourier transformation.

The results of VEF coefficient amplitudes at predominant frequencies for upstream and downstream cylinders are depicted in Figures 13 and 14, respectively.

As shown in the figures, the distributions of VEF coefficient amplitudes at predominant frequencies (F_D) show similarities to Figures 12 and 13 for U = 3.6 and 5.0 m/s. Also, the values are relatively larger on the trailing half surface of UC and the leading half surface of DC. However, for U = 2.9 m/s, the difference between the distributions of F_D and Figures 12 and 13 is great. That is, for UC, the F_D distributions are symmetric around the centerline of the lower surface; there are two bumps on the upper surface, and the largest values are observed around the position of X/W = [0.54, 0.88]. For DC, the F_D values on the lower surface are smaller than that on the upper surface. The variation of F_D in the region of X/W = 0.23~0.83 on the upper surface is not that significant as compared with other regions, while the variation of F_D in the range of $0.38 \leq X/W \leq 1.0$ on the lower surface remains stable. The following observations can be reached from the comparison as follows: the vortex-excited forces in torsional VIVs of twin rectangular cylinders are mainly predominant by the self-excited components, while the contribution of the self-excited components is relatively smaller for the bending VIVs of the twin rectangular cylinders.

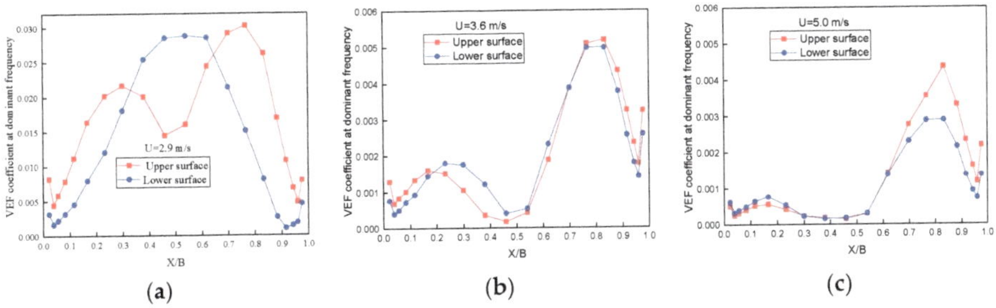

Figure 13. VEF coefficient amplitudes at a predominant frequency for UC. (**a**) U = 2.9 m/s; (**b**) U = 3.6 m/s; (**c**) U = 5.0 m/s.

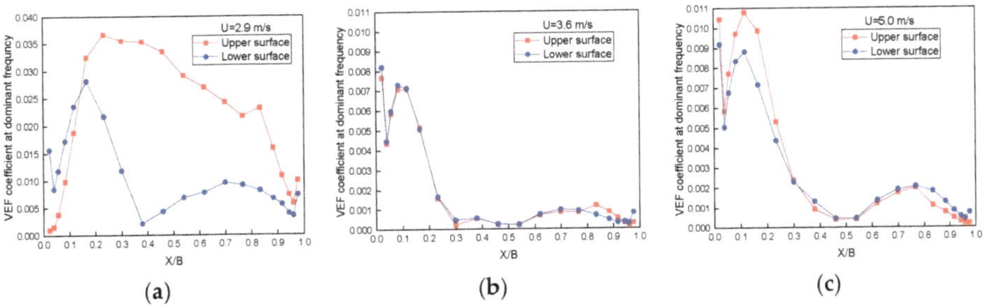

Figure 14. VEF coefficient amplitudes at a predominant frequency for DC. (**a**) U = 2.9 m/s; (**b**) U = 3.6 m/s; (**c**) U = 5.0 m/s.

5. Conclusions

The vortex-induced vibration performance, the pressure distributions, and the VIV mechanism for twin rectangular cylinders with a width-to-depth ratio of 5:1 in tandem arrangement are analyzed in this paper. The spacing between them is 1.2 times the width of one section. The main conclusions can be reached as follows:

(1) The results of dynamic tests indicate that the VIV performance of two 5:1 rectangular cylinders in tandem is different at three incidence angles. Generally, the vertical VIV response is unfavorable at the −3° angle of incidence. However, the torsional VIVs of the upstream body are more significant at 0° and −3°, while the general trend for the torsional VIV behavior of the downstream cylinder can hardly be obtained.

(2) The pressure measurements show that the mean and fluctuating pressure distributions of the cylinders under vertical VIVs are different from the results under torsional VIVs. Pressure measurements should be conducted under corresponding wind velocities to analyze the pressure field of twin rectangular cylinders when they undergo VIVs.

(3) The correlation analysis using time-frequency domain statistical parameters indicates that the contribution of distributed vortex-excited forces is significant in the region of the trailing surface of the upstream cylinder and the leading surface of the downstream cylinder.

Author Contributions: Conceptualization, K.M., C.Z., X.D. and G.Z.; methodology, K.M. and C.Z.; investigation, X.D. and G.Z.; writing-original draft preparation, K.M. and C.Z.; writing–review and editing, X.D. and G.Z.; supervision, G.Z. formal analysis, K.M. and C.Z.; validation, X.D. and G.Z. All authors have read and agreed to the published version of the manuscript.

Funding: This research was funded by the National Key Research and Development Program of China (2019YFE0112300); the National Natural Science Foundation of China (51978512); and the Transportation science and technology plan of Shandong province (2020B74); the Shanghai Science and Technology Plan Project (220H1179800).

Data Availability Statement: The data presented in this study are available on request from the corresponding author. The data are not publicly available due to privacy.

Conflicts of Interest: Changyong Zhang was employed by the company Shandong Provincial Communications Planning and Design Institute Group Co., Ltd. Author Xinzhi Dang was employed by the company Suzhou Tongyao Civil Technology Co., Ltd. The remaining authors declare that the research was conducted in the absence of any commercial or financial relationships that could be construed as a potential conflict of interest.

References

1. Taylor, I.; Vezza, M. Prediction of unsteady flow around square and rectangular section cylinders using a discrete vortex method. *J. Wind. Eng. Ind. Aerodyn.* **1999**, *82*, 247–269. [CrossRef]
2. Murakami, S.; Mochida, A. On turbulent vortex shedding flow past 2D square cylinder predicted by CFD. *J. Wind. Eng. Ind. Aerodyn.* **1995**, *54–55*, 191–211. [CrossRef]
3. Matsumoto, M.; Shirato, H.; Araki, K.; Haramura, T.; Hashimoto, T. Spanwise coherence characteristics of surface pressure field on 2-D bluff bodies. *J. Wind. Eng. Ind. Aerodyn.* **2003**, *91*, 155–163. [CrossRef]
4. Shimada, K.; Ishihara, T. Application of a modified k–ε model to the prediction of aerodynamic characteristics of rectangular cross-section cylinders. *J. Fluids Struct.* **2002**, *16*, 465–485. [CrossRef]
5. Bruno, L.; Coste, N.; Fransos, D. Simulated flow around a rectangular 5: 1 cylinder: Spanwise discretisation effects and emerging flow features. *J. Wind. Eng. Ind. Aerodyn.* **2012**, *104–106*, 203–215. [CrossRef]
6. Ricciardelli, F. Effects of the vibration regime on the spanwise correlation of the aerodynamic forces on a 5:1 rectangular cylinder. *J. Wind. Eng. Ind. Aerodyn.* **2010**, *98*, 215–225. [CrossRef]
7. Mariotti, A.; Salvetti, M.V.; Omrani, P.S.; Witteveen, J.A.S. Stochastic analysis of the impact of freestream conditions on the aerodynamics of a rectangular 5:1 cylinder. *Comput. Fluids* **2016**, *136*, 170–192. [CrossRef]
8. Mannini, C.; Weinman, K.; Soda, A.; Schewe, G. Three-dimensional numerical simulation of flow around a 1:5 rectangular cylinder. In Proceedings of the 5th European and African Conference on Wind Engineering, Florence, Italy, 19–23 July 2009.
9. Mannini, C.; Šoda, A.; Schewe, G. Numerical investigation on the three-dimensional unsteady flow past a 5:1 rectangular cylinder. *J. Wind. Eng. Ind. Aerodyn.* **2011**, *99*, 469–482. [CrossRef]
10. Mannini, C.; Šoda, A.; Schewe, G. Unsteady RANS modelling of flow past a rectangular cylinder: Investigation of Reynolds number effects. *Comput. Fluids* **2010**, *39*, 1609–1624. [CrossRef]
11. Bruno, L.; Fransos, D.; Coste, N.; Bosco, A. 3D flow around a rectangular cylinder: A computational study. *J. Wind. Eng. Ind. Aerodyn.* **2010**, *98*, 263–276. [CrossRef]
12. Berrone, S.; Garbero, V.; Marro, M. Numerical simulation of low-Reynolds number flows past rectangular cylinders based on adaptive finite element and finite volume methods. *Comput. Fluids* **2011**, *40*, 92–112. [CrossRef]
13. Mannini, C.; Marra, A.M.; Pigolotti, L.; Bartoli, G. The effects of free-stream turbulence and angle of attack on the aerodynamics of a cylinder with rectangular 5:1 cross section. *J. Wind. Eng. Ind. Aerodyn.* **2017**, *161*, 42–58. [CrossRef]
14. Schewe, G. Reynolds-number-effects in flow around a rectangular cylinder with aspect ratio 1:5. *J. Fluids Struct.* **2013**, *39*, 15–26. [CrossRef]
15. Schewe, G. Influence of the Reynolds-Number on Flow Induced Vibrations of Generic Bridge Sections. In Proceedings of the International Conference on Bridges, SECON HDGK, Dubrovnik, Croatia, 21–24 May 2006; pp. 351–358.
16. Bruno, L.; Salvetti, M.V.; Ricciardelli, F. Benchmark on the aerodynamics of a rectangular 5:1 cylinder: An overview after the first four years of activity. *J. Wind. Eng. Ind. Aerodyn.* **2014**, *126*, 87–106. [CrossRef]
17. Gnatowska, R. Aerodynamic characteristics of two-dimensional sharp-edged objects in tandem arrangement. *Arch. Mech.* **2008**, *60*, 475–490.
18. Havel, B.; Hangan, H.; Martinuzzi, R. Buffeting for 2D and 3D sharp-edged bluff bodies. *J. Wind. Eng. Ind. Aerodyn.* **2001**, *89*, 1369–1381. [CrossRef]
19. Liu, C.H.; Chen, J.M. Observations of hysteresis in flow around two square cylinders in a tandem arrangement. *J. Wind. Eng. Ind. Aerodyn.* **2002**, *90*, 1019–1050. [CrossRef]
20. Kolar, V.; Lyn, D.A.; Rodi, W. Ensemble-averaged measurements in the turbulent near wake of two side-by-side square cylinders. *J. Fluid Mech.* **1997**, *346*, 201–237. [CrossRef]
21. Chatterjee, D.; Biswas, G.; Amiroudine, S. Numerical simulation of flow past row of square cylinders for various separation ratios. *Comput. Fluids* **2010**, *39*, 49–59. [CrossRef]
22. Edamoto, K.; Kawahara, M. Finite element analysis of two- and three-dimensional flows around square columns in tandem arrangement. *Int. J. Numer. Methods Fluids* **1998**, *28*, 95–112. [CrossRef]
23. Bhattacharyya, S.; Dhinakaran, S. Vortex shedding in shear flow past tandem square cylinders in the vicinity of a plane wall. *J. Fluids Struct.* **2008**, *24*, 400–417. [CrossRef]
24. Bao, Y.; Wu, Q.; Zhou, D. Numerical investigation of flow around an inline square cylinder array with different spacing ratios. *Comput. Fluids* **2012**, *55*, 118–131. [CrossRef]

25. Kondo, N. Numerical Simulation of Aerodynamic Characteristics of Two Rectangular Cylinders in Side-by-side Arrangement. *Int. J. Comput. Fluid Dyn.* **2004**, *18*, 367–379. [CrossRef]
26. Agarwal, A.; Djenidi, L.; Antonia, R.A. Investigation of flow around a pair of side-by-side square cylinders using the lattice Boltzmann method. *Comput. Fluids* **2006**, *35*, 1093–1107. [CrossRef]
27. Kumar, S.R.; Sharma, A.; Agrawal, A. Simulation of flow around a row of square cylinders. *J. Fluid Mech.* **2008**, *606*, 369–397. [CrossRef]
28. Ul Islam, S.; Rahman, H.; Zhou, C.Y. Effect of gap spacings on flow past row of rectangular cylinders with aspect ratio 1.5. *Ocean. Eng.* **2016**, *119*, 1–15. [CrossRef]
29. Burattini, P.; Agrawal, A. Wake interaction between two side-by-side square cylinders in channel flow. *Comput. Fluids* **2013**, *77*, 134–142. [CrossRef]
30. Chatterjee, D.; Biswas, G. Dynamic behavior of flow around rows of square cylinders kept in staggered arrangement. *J. Wind. Eng. Ind. Aerodyn.* **2015**, *136*, 1–11. [CrossRef]
31. Ma, K.; Hu, C.X.; Zhou, Z.Y. Investigation of vortex-induced vibration of twin rectangular 5:1 cylinders through wind tunnel tests and POD analysis. *J. Wind. Eng. Ind. Aerodyn.* **2019**, *187*, 97–107. [CrossRef]
32. Liu, X.B. *Study on Aerodynamic Interference of Long-Span Bridges with Twin Separate Parallel Decks*; Hunan University: Changsha, China, 2010. (In Chinese)
33. Hu, C.X.; Zhao, L.; Ge, Y.J. Time-frequency Evolutionary Characteristics of Aerodynamic Forces around a Streamlined Closed-box Girder during Vortex-induced Vibration. *J. Wind. Eng. Ind. Aerodyn.* **2018**, *182*, 330–343. [CrossRef]
34. Guo, Z.W.; Zhao, L.; Ge, Y.J.; Chen, H.X. Mechanism analysis for vortex-induced vibration reduction of a flat streamlined steel box-shaped girder with airflow-suppressing board based on statistical property of surface pressure. *J. Vib. Shock.* **2012**, *31*, 89–94. (In Chinese)
35. Guan, Q.H.; Li, J.W.; Hu, Z.T.; Liu, J.X. Effects of railings on vortex induced vibration of a bridge deck section. *J. Vib. Shock.* **2014**, *33*, 150–156. (In Chinese)
36. Zhu, Z.W. LES prediction of aerodynamics and coherence analysis of fluctuating pressure on box girders of long-span bridges. *Comput. Fluids* **2015**, *110*, 169–180. [CrossRef]
37. Zhu, Z.W. Feasibility investigation on prediction of vortex shedding of flat box girder based on a 2D RANS model. *China J. Highw. Transp.* **2015**, *28*, 24–33. (In Chinese)

Disclaimer/Publisher's Note: The statements, opinions and data contained in all publications are solely those of the individual author(s) and contributor(s) and not of MDPI and/or the editor(s). MDPI and/or the editor(s) disclaim responsibility for any injury to people or property resulting from any ideas, methods, instructions or products referred to in the content.

MDPI
St. Alban-Anlage 66
4052 Basel
Switzerland
www.mdpi.com

Buildings Editorial Office
E-mail: buildings@mdpi.com
www.mdpi.com/journal/buildings

Disclaimer/Publisher's Note: The statements, opinions and data contained in all publications are solely those of the individual author(s) and contributor(s) and not of MDPI and/or the editor(s). MDPI and/or the editor(s) disclaim responsibility for any injury to people or property resulting from any ideas, methods, instructions or products referred to in the content.

www.ingramcontent.com/pod-product-compliance
Lightning Source LLC
Chambersburg PA
CBHW061833210926
51850CB00006B/880
9783725805594